Crossing Boundaries for Collaboration

Congratulations Rhonda on winning at "Taste of the Tropics"! May your future collaborations be similarly fruitful —

Crossing Boundaries for Collaboration

Conservation and Development Projects in the Amazon

Stephen G. Perz

LEXINGTON BOOKS
Lanham • Boulder • New York • London

Published by Lexington Books
An imprint of The Rowman & Littlefield Publishing Group, Inc.
4501 Forbes Boulevard, Suite 200, Lanham, Maryland 20706
www.rowman.com

Unit A, Whitacre Mews, 26-34 Stannary Street, London SE11 4AB

British Library Cataloguing in Publication Information Available

Library of Congress Cataloging-in-Publication Data Available

ISBN 978-1-4985-3566-3 (cloth : alk. paper)
ISBN 978-1-4985-3568-7 (pbk. : alk. paper)
ISBN 978-1-4985-3567-0 (electronic)

∞™ The paper used in this publication meets the minimum requirements of American National Standard for Information Sciences Permanence of Paper for Printed Library Materials, ANSI/NISO Z39.48-1992.

Printed in the United States of America

Contents

List of Figures

List of Tables

Acknowledgments

Naming one author for a book on collaboration across boundaries is little more than an ironic conceit. This volume resulted from many years of work with innumerable people who exert an enormous influence on everything that follows. I therefore gratefully acknowledge those who have trusted and taught me about collaboration across boundaries. The list is long and yet woefully incomplete. Here I recognize collaborators with whom I directly interacted over a period of time. I usually list people in alphabetical order. I have also simplified many compound names to the primary first and last names due to space constraints. Even so, the list is a small fraction of those who actually contributed to the complicated projects described hereinafter. I apologize repeatedly for the short lists; they fail to do justice to the true scale of the efforts involved.

Early influences on collaborative projects were many: Charles Wood introduced me to the Amazon. Robert Walker, Roberto Porro, Marcellus Caldas, Eugenio Arima, Alfredo Homma, and Adilson Serrão also opened crucial doors.

The UF Roadies group was an invaluable crucible for interdisciplinary debate. I have no complete list of names, but Grenville “Barney” Barnes, Michael Binford, Graeme Cumming, Robert Holt, Clyde Kiker, Marianne Schmink, Katie Sieving, Jane Southworth, Claudia Stickler, and Tracy Van Holt all made vital contributions.

Foster Brown facilitated my arrival in the MAP frontier. Valerio Gomes and Kennedy Souza provided important early introductions in Acre. My first MAP forum in 2002 was shared by folks named above as well as then-graduate students Angelica Almeyda, Andrea Chavez, Amy Duchelle, and Jeffrey Luzar. Influential people at the early MAP forums but not on specific projects include Christiane Ehringhaus, Miroslava Frias, Alfredo Garcia,

Bertha Ikeda, Santos Ikeda, Ernesto Raez, Guillermo Rioja, Julio Rojas, and Cora van Oosten. There were many, many others: I am pained that my poor memory cannot recall them all.

I thank the funders of the famously complicated projects described hereinafter, notably the National Science Foundation (NSF), the US Agency for International Development (USAID), and Higher Education for Development (HED) of the American Council on Education (ACE). On the NSF Roads project (NSF HSD Grant #0527511), many from Roadies contributed, along with Kelly Biedenweg, Rosa Cossio, Dave Elliott, Jeffrey Hoelle, Flavia Leite, Rafael O. Rojas, and Alexander Shenkin. In addition, I must recognize UFAC coordinators Veronica Passos (*minha salvadora*), Lucas Araújo Carvalho, Karla Rocha, and Marcos Silveira, along with Wendeson Castro, Adão Costa Silva, and Iracema de Souza Moll. At UNAMAD, coordinators included Jorge Castillo, Hugo Dueñas, and Americo Quevedo; in addition, Mercedes (Meche) Perales was a key contributor. At UAP, coordinators included Liliana Cabrera, Dean Kenji, Androncles Puerta, Daniel Rojas, and Guido Vasquez. Many more contributed: see Southworth, et al. (2011), Perz, et al. (2012), and Baraloto, et al. (2014).

The first Knowledge Exchange Train was an incubator for many collaborative relationships thereafter, particularly with Elsa Mendoza (*nossa estrela*). Beyond the participants already named, some key interactions involved Jacob Carballo, Jaime Chavez, Matthew Marsik (who also produced the maps in this book), Julio Mayna, and Armando Muñante. See also Mendoza, et al. (2007).

On the G-MAP Consortium (USAID CA RLA-A-00-06-00071-0), the project leaders taught me so much about collaboration across boundaries. The same goes for these players when we created M-P and MABE (USAID CA 512-A-00-08-00003-00). In addition to participants already named, I hereby acknowledge my enormous debts to Silvia Brilhante, Kelly Gouveia, Raul Pinedo, Vera Reis, Juan Fernando Reyes, and Monica de los Rios. Other influential players included Gabriel Alarcón, Moises Barbosa, Ruth Frisancho, Mishari Garcia, Galia Selaya, and Sonia Yufra. Bruno Sanguinetti stepped in at a crucial time in the transition to M-P and beyond. Under MABE, Vângela Nascimento, Paula Pinheiro, Miguel Scarcello, and Sonaira Souza were all important instructors. I am sure somebody will embarrass me by pointing out numerous others who I am foolishly overlooking.

The NSF Modeling project (NSF CNH Grant #1114924) introduced me to engineers, another breed: Rafael Muñoz-Carpena and Gregory Kiker are exemplary translators from stories into numbers and back again.

The MDD Consortium (USAID CA AID-OAA-A-11-00060) has been a roller coaster with many new players. Beyond people already listed, Antonio Bernales, Zenaida Chulla, Jhon Farfan, Oscar Guadalupe, Ana Hurtado, Luis Masias, Julia Quaedvlieg, Salvador Quispe, Norma Revoredo, Francisco Roman, Martha Rosero, and Lauren Samuels have all taught me lessons. The team and its collective network is, however, much more extensive.

The HED project (HED104-9748-LAC-12-04) had two editions, and both involved many inspiring people. In the first, I thank Nancy Acuña, Ludwing Arcienega, Marliz Arteaga, Nataly Ascarrunz, Jorge Azad, Napoleon Ferreira, Juan Carlos Gomes, Kaoru Kitajima, Ericka Llanos, Bonifacio Mostacedo, and Marisol Toledo. May we get another chance. The second edition involved a far-flung network, and I can only recognize a small subset of key players by name. Special thanks to Julio Araujo and Andrew Noss, who made it happen. Beyond participants already named, Elizabeth Anderson, Richard Bilsborrow, Ana Lemos, Bette Loiselle, Javier Maldonado, Carlos Mena, Lourdes Ruiz, and David Salisbury also made major contributions. In both editions, Ana Luz Porzecanski and Armando Valdes Velasquez were inspirations. Long live the scientific teaching revolution.

Contact personnel at funders have been a godsend. At NSF, Tom Baerwald, Sarah Ruth, and Alan Tessier were utter professionals, even when the funding decision was "decline." At USAID, I learned from numerous examples of grace under pressure, notably including Connie Campbell, Victor Merino, Beatriz Torres, and Nancy Diamond, among many others. At HED, Manuel Sanchez proved he is a master facilitator. Tully Cornick, Azra Nurkic, and Noopur Vyas, among others, went extra mile after mile for us.

Collaborating across boundaries also poses challenges at one's home institution. At UF, I want to recognize the following people for their patience in fighting so many bureaucratic fires I unwittingly set: Jessica Baker, Brandi Boniface, Jessica Caicedo, Peg Canaday, John Edmands, Cassandra Farley, Kat Carter Finn, Margarita Gandia, Alethea Geiger, Pat Greco, Ros Heath, John Henretta, Susanne Hill, Kelly Jacoby, Sheryl McIntosh, Andrew Noss, Ed Parker, Jocelyn Peskin, Brian Prindle, Lauren Samuels, Patricia Sampaio, Tiffany Schmidt, Constance Shehan, Brad Staats, Dawn Stewart, Dee Dee Teel, Leo Villalon, Barbara Zsembik, and probably most everyone else in the Division of Sponsored Programs, Contracts and Grants, Subawards, Travel, Purchasing, Tax Services, Human Resources, as well as the Center for Latin American Studies and the International Center.

Beyond all these, I have benefitted from collaborative relationships with other colleagues, administrators, and advisors at various partner institutions: Rusleyd Abreu, Omar Burgos, Margarida Carvalho, Francisco (Carlito) Cavalcanti, Manuel Delgado, Edgar Diaz, Arthur Leite, Rosélia Lopes, Claudio Maciel, Rene Mamani, Augusto Montes, Elder Morato, Milthon Muñoz, Vera Olinda, Carlos Peña, Miguel Peralta, Percy Pollito, Sabina Ribeiro, Fernando Schmidt, Jorge Vela, and Lisandro Vieira. One more time: my pathetic memory fails me in compiling an adequate list.

There is a far larger list of workshop participants, interviewees, public school faculty, labor unions, community associations, NGOs, governmental agencies, taxi drivers, local historians, landholders, store owners, and other people in the MAP frontier who have influenced my outlook on collaboration across boundaries. *Muchas gracias a tod@s de la frontera trinacional. Muito obrigado a tod@s da fronteira trinacional.*

Amidst all the other collaborators and partners, I must of course recognize my family. My wife Leslie has been unremittingly patient with my frequent travels and international phone calls. Our children, Samuel and Madelyn, are both junior authors who have provided inspiration for the writing here—copy-editing, too. My parents, Dennis and Marilyn Perz, and my grandfather, Austin Herr, have been very supportive in their curiosity about this book.

Let me conclude by thanking Brighid Stone and Lexington Books, an imprint of Rowman and Littlefield, for believing in this unusual book project. For the funders, lawyers, journalists, and the rest, I declare that the interpretations and opinions hereinafter are my own. I also accept responsibility for my miscomprehensions, simplifications, and other mistakes in what follows. I close by invoking a Brazilian Portuguese phrase favored by Charles Wood: *desculpe qualquer coisa.*

Preface

It was February 2006. "Will you do it?" they asked. "Will you lead the consortium?"

I said "Yes." I said yes because I wanted to see this consortium happen. The Brazilian partners asking the question proposed to address threats to biodiversity conservation via sustainable development initiatives in the Amazon basin, a big priority. It was an historic moment in their neighborhood, the southwestern Amazon: threats to conservation and development were arriving in many forms, ranging from large-scale infrastructure projects to extreme climatic events. However, the partners didn't have the capacity to manage the funds for an ambitious menu of activities. They needed collaborators, so they came to my university with the proposal to form a consortium.

I also said yes because nobody else stepped forward. Addressing threats to biodiversity is no simple trick. In fact, on reflection, it seemed like others who stood to be asked had taken a wary step backwards. Somehow, my name came up. Was I the right person in the right place at the right time, or perhaps the sacrificial lamb? In any event, at that point, none of us set to participate really knew what we were getting ourselves into. Had we known, we might all have taken a wary step backwards.

I now see that the call from the Brazilian partners that day was a defining moment, a fork in the road. That was the day we decided to follow the proverbial path less taken and collaborate across boundaries. The proposed consortium would bring together a group of organizations with distinct experiences and capacities. The lineup included experienced scholars and practitioners from multiple countries sharing the southwestern Amazon. The consortium's partner organizations would pursue a suite of activities to promote conservation and development by collaborating with local governments and rural communities. This was a big deal: the consortium in question constituted but a part of a major funding initiative for conservation and development across the Amazon basin. Being a part of this consortium was an adventure not to be missed. Indeed, the opportunity to support these partner organizations was for me a privilege.

The excitement came, however, with a certain trepidation. To be sure, we understood the basic mechanics of collaboration. We knew what our shared goals were, and we knew what governance structure we wanted. Collaboration in itself wasn't the source of the hope and the fear. Instead, the truly daunting concern stemmed from the fact that we were proposing to cross not one but several types of boundaries. Indeed, this was the master logic of the argument underlying our entire endeavor: conservation and development cannot be effective without spanning various kinds of divides.

At first glance, the great strength of our consortium was that it encompassed an array of scholars and practitioners who had already crossed boundaries. They knew each other and had worked together in the past. In that respect, the consortium merely formalized what already existed. But on second thought, there remained the open question of how well informal experience in crossing boundaries would map on to a more formal mode of collaborating. We knew this was an experiment.

The consortium included scholars with different disciplinary backgrounds, partner organizations ranging from universities to governmental agencies to non-governmental organizations, and personnel from different countries. We thus proposed to cross three types of boundaries: academic disciplines, organizational types, and sovereign nations. All of the standard challenges of collaboration—identifying effective practices for communication, jointly defining the problem, collectively articulating shared goals, and so on—suddenly looked much more formidable. This is because disciplines, organizational types, and countries have different histories, cultures, languages, institutional arrangements, and practices. These and other sources of difference are constituted and maintained by boundaries. In turn, those differences serve to undermine attempts to span divides. Bridging boundaries thus requires effort to address differences in the underlying mental models and daily practices at play. Understanding the mentalities and practices of "others" across divides permits navigation of their different definitions of what is important, how things get done, and so forth. Without that effort, collaboration across boundaries becomes an unproductive exercise in frustration.

We were aware of differences across boundaries, but in a positive sense: our point of departure resided in our design of the consortium to carefully leverage the complementary strengths of the different partners. I found out later that the management literature has a name for this: "collaborative advantage" (Huxham and Vangen 2005; Lank 2006), the notion that collaboration yields benefits to the extent that partners have complementary strengths which permit greater achievements than if they work separately.

That might be so, but collaborative advantage when crossing boundaries comes at the cost of managing the differences and tensions which exist at those boundaries. We learned this the hard way. It was not long before implicit differences in partner worldviews reared their fearsome heads in the form of bellicose disputes. When confronted with alien beliefs and threatening claims, partners became combatants who ardently sought to defend what they felt to be true and inviolable. This kicked off several years on a steep learning curve involving vociferous arguments, all manner of logistical headaches, and periodic revelations of hidden agendas. The exigencies of crossing boundaries ran well beyond the challenges of collaborating with others of the same discipline, organization, and country.

I was idealistic and naïve, but having gone through the experience of collaborating across boundaries, I'm still glad I said yes. For we did accomplish more together than would have been possible with everyone going it alone. Boundaries made for large complementarities in knowledge, experience, and capacities among partners, which served to multiply the advantages of collaboration. The greater administrative capacity permitted by the consortium, combined with the diverse know-how among local partners, allowed us to learn quickly, scale up activities, and respond to new opportunities. Just as important, we learned numerous lessons about collaboration across boundaries in its own right, and in the process we learned about ourselves, our institutions, and our countries, as well as about our partners, their institutions, and their countries. This revealed the larger strategic importance of collaboration across boundaries: the trials and tribulations served to build capacity to do it again in the future. Even if the technical outcomes had been poor, the capacity building process has value in its own right.

As a result, I found myself in a similar situation in 2011. "Will you do it again?" they asked. This time the goal was the same—biodiversity conservation in the Amazon—but the context was different. "The other partner doesn't want to lead this consortium. We know you didn't sign up for this, but will you lead, again?" It was another opportunity to form a consortium, this time with some familiar and also some new partners. The "Plan A" partner to lead the new consortium had backed out due to reasons tied to the complexities of crossing boundaries, in this case misunderstandings among types of organizations. This time it didn't seem so risky for me to say yes. Instead, the defining question went to the new partners: would they accept my home institution, an outside organization, as leader? The familiar partners spoke favorably to the new partners of the notion of my home institution leading the new consortium, based on the experience of the previous one—the one with all the heated squabbles among the partner-combatants. Now it was the turn of the new partners to make the key decision. On the word of the familiar partners, the new partners said yes. It was another experiment.

No two cases are alike. In 2012, we had another opportunity to create a consortium to help Amazonian universities build capacity to support biodiversity conservation. We first had to choose one (and only one) country in which to work, so interested faculty at my home university dutifully consulted with their partners in various countries. When it came time to decide who should coordinate, others not only took wary steps backwards, they then pointed their fingers at me, the guy who allegedly knew what to do. We thus pursued a familiar set of discussions with another array of partners. In yet another experiment, everybody signed on. Then the whole operation got shut down in a diplomatic dispute between two national governments. Nevertheless, the funders came back and I was asked another defining question: "Would you reprogram the project with partners in several other countries?" They seemed to think we knew what we were doing. Thus ensued another round of consultations with potential partners in various countries. By then, I was using lessons learned from previous consortia, hoping they still applied.

In the process of trying to collaborate across boundaries, I became an ersatz applied organizational sociologist. I didn't intend for this. The crucial driver was the shared recognition of common threats to conservation and development across boundaries. It became evident that somebody needed to adopt a suite of roles for the facilitation and management of collaborative networks: matchmaker, salesperson, midwife, diplomat, and occasionally scientist. Whether it was wise to try and adopt those roles remains an open question. Fortunately, others have also adopted those roles, and that makes us all better collaborators. There is also trust and patience among my network of collaborators that we learn from our mistakes. We don't step backwards, and as a result, we learn together.

Chapter One

Introduction

This book is about crossing boundaries for collaboration. In an increasingly integrated world, many challenges and problems require collaborative efforts to be addressed adequately. Examples include education, health care, community policing, and economic exchange. What is more, such collaboration increasingly requires joint action with "others" across any of numerous types of social divides, whether defined in terms of training background, type of organization, or nationality. In such cases, crossing boundaries is necessary for collaboration in order to address shared goals.

Among the various possibilities for a substantive focus to discuss collaboration across boundaries, I take up the case of environmental conservation and sustainable development. There has been no shortage of calls to collaborate across boundaries for these goals. One key element concerns the calls for larger investments in conservation and development. Sachs (2005) has argued that if rich countries spend 0.7 percent of their GDP on development assistance for poor countries, all countries will benefit economically, and extreme poverty can be eliminated by 2025. The UN (2014) has set its Sustainable Development Goals to build on previous environmental targets, such as the Millennium Development Goals, which highlight biodiversity conservation, sustainable water use, and related priorities. Sustainable development thus justifies international funding and UN actions. Much the same can be said about the UN's Framework Convention on Climate Change, which has a target of a 2°C increase in average global surface temperatures (UNFCC 2010). In parallel, the Belmont Forum white paper (2011) highlights key topics for scientific research on global environmental change to be prioritized by the national science funding agencies of governments around the world.

All such documents highlight that there remains a large gap between the need and the capacity for conservation and development work. All such documents also call for more collaboration across boundaries, construed in various ways, in order to fill the gap. The big problems of conservation and development in the twenty-first century are not local but transnational. Further, they are not merely problems for governments but all types of organizations and organized civil society. Finally, they are not simple problems with established disciplinary solutions, but instead require interdisciplinary approaches. The upshot is that the need to collaborate across national, organizational, and disciplinary boundaries for conservation far exceeds current capacity. The question is how best to respond.

There are many established approaches to conservation and development. These include state-based, government-based, market-based, movement-based, and community-based, among others. Each approach is defined by its focus on a particular social actor, which is held out as a key agent of change. Each approach has worked in specific circumstances for particular purposes, and has rightly won its advocates. But each approach has also revealed limitations and thus incurred specific criticisms (e.g., Torgerson 1990 on state-based approaches; Agrawal and Gibson 1999 on community-based approaches). Such criticisms have prompted questions about the validity of the approaches and engendered debate about their future prospects.

A characteristic common in many of the critiques is that the approaches put forward are too narrowly conceived as to the relevant social actors. Declarations of the "death of environmentalism" in the United States highlighted the problem that conservation organizations were insufficient to bring change by themselves (Nordhaus and Shellenberger 2007; Werbach 2004). Similarly, debate over development thought and practice has identified a set of dilemmas confronting governments and development policy (McMichael 2012; Pieterse 2001; Willis 2011). The underlying problem is that initiatives based on governments, markets, movements, and communities have the common shortcoming of invoking a single group or institution as the best means for pursuing conservation and development. Each is an institutionally "unitary" approach that neglects the assets of the others, with self-imposed limitations on what conservation and development action involves.

A key line of response to this sort of indictment has come in the form of sundry calls for more collaborative approaches to conservation and development. Such approaches highlight crossing of some sort of boundary, whether among disciplines, organizational types, or countries. In the conservation literature, an early manifestation came in the form of adaptive management, which highlighted multi-stakeholder collaboration in sequential management experiments (Holling 1978; Lee 1993). More recently, there has been a proliferation of proposals with names like "collaborative conservation" (Brick, et al. 2001), "co-management" (Carlsson and Berkes 2005; Armitage, et al.

2009), "collaborative management" (Buck, et al. 2001; Colfer 2010; Galat and Berkly 2014), "public ecology" (Robertson and Hull 2003), and "multi-stakeholder platforms" (Hemmati 2002; Warner 2007), along with other work on partnerships, alliances, and networks (Manring 2007; Poncelet 2004). In the development practice literature, "top-down" state-centric approaches have been countered by "bottom-up" approaches that variously highlight community participation (Chambers 1994; Hickey and Mohan 2004), social inclusion (Groves and Hinton 2004; Shortall 2008; Wood and Roberts 2005), dialogic democracy (Heath 2007), actor-oriented perspectives (Long 2001), and various forms of partnerships (Robinson, et al. 2000). Common to all such approaches is a commitment to collaborative action for regional development that features the role of local peoples, regardless of their education or other characteristics (Cernea 1991; Chambers 1997). Another indication of the zeitgeist surrounding collaborative approaches are proposals under the banner of various forms of "governance" (Batterbury and Fernando 2006; Bierman and Pattberg 2008; Lemos and Agrawal 2006; Newig and Fritsch 2009; Young 1997). In most renditions, the notion of governance goes beyond control by the state to explicitly incorporate the active participation of other stakeholders, with the typical list including businesses, NGOs, and organized civil society.

Despite the concurrence of opinion about collaboration for conservation and development, questions about the challenges of crossing boundaries have arisen. Some conservation scholar-practitioners have raised questions about operational models for collaboration (Knight, et al. 2006). Crossing boundaries presents its own challenges and can hardly be considered easy or automatic when the proposed partners come from very different backgrounds. Examples include the effectiveness of collaboration between scientists and local peoples (Shanley and Lopez 2009; Sheil and Lawrence 2004), the linkage of knowledge to action (Lauber, et al. 2011), and the relevance of conservation science for conservation practice (Laurance, et al. 2012). Issues thus arise about the viability of collaboration, precisely because crossing boundaries makes its own demands. Just as environmental conservation has become a typical example of a wicked problem (Balint, et al. 2011; Brown, et al. 2008), so too is collaboration itself (Weber and Khademian 2008). In the development literature, similar perils have appeared. There are well-known complexities of managing foreign aid among collaborating governmental partners (Gaillard 1994; Groves and Hinton 2004). Further, there are difficulties of managing social inclusion in on-the-ground development projects (e.g., Agarwal 2001; Cleaver 1999; Cooke and Kothari 2001; Shortall 2004).

So how is it that governments, communities, and other social actors invoked in unitary approaches to conservation and development are supposed to operate more effectively in tandem if their individual effectiveness is limited in the first place, collaboration itself is difficult, and boundaries present significant stumbling blocks? Crossing boundaries for collaboration has emerged as something of a last resort. This is because collaborative approaches are understood to be more complicated and burdensome than unitary approaches. Crossing boundaries implies extra time and effort to get to know "others"—other individuals and other organizations, probably with other agendas and other interests. Crossing boundaries as well as collaboration thus impose demands in their own right, constituting tasks to be managed on top of the technical details of conservation and development or any other substantive goal.

There is, however, reason to believe that despite the complications, crossing boundaries for collaboration offers significant benefits. Here the management literature provides valuable insights. While arguments for the advantages of collaboration for businesses are not new (Kraus 1980), they have attained greater currency in recent years. Perhaps as another reflection of a shared zeitgeist in support of crossing boundaries for the achievement of shared goals, work on business administration has heralded the importance of collaboration, whether among firms (Austin 2000; Cropper, et al. 2008), between for-profit and nonprofit entities (Cropper, et al. 2008; Reuer 2004), or among divisions of large companies (Austin 2000; Hansen 2009; Heckscher and Adler 2006).

In these instances and others, single-firm approaches to competitive advantage are giving way to collaborative approaches, summed up in literature on "collaborative advantage" (Huxham and Vangen 2005; Lank 2006). The notion of collaborative advantage posits that organizations with shared goals but complementary capacities and experience can achieve more by working together than by going it alone. The benefits of collaborative advantage are specified as applying to joint ventures pursued for complex, large-scale problems of strategic importance rather than simple routine tasks. Insofar as strategic challenges have ill-defined solutions due to the complexity of the dynamics involved, collaborative advantage asserts that there are benefits to collaborating across boundaries.

Collaboration involves non-competitive behavior among partners, which implies relations of trust and mutual accountability (Gray 1989). This is crucial in any case, but especially when crossing boundaries to work with various "others" who may not be well understood. Crossing boundaries for collaboration therefore implies a structural shift in relationships from emphasizing vertical hierarchies with chains of command to a focus on more horizontal networks with complementary partners. Collaborating across boundaries is only worthwhile if everybody brings something different to the party and thereby contributes. Collaboration across boundaries thus involves the allocation of human resources based less on job titles or other forms of authority, but instead on functional needs and capacities among collaborators.

Beyond the general philosophy, there are specific assertions regarding the advantages offered by collaboration across boundaries. One advantage is that by collaborating, partners gain access to a larger pool of human, financial, and other resources. To the extent that resources are complementary and shared, partners have greater capacity to act than they would otherwise. A second advantage is that collaboration permits the spreading of risks to pursue a strategic enterprise. Organizations may be unwilling to take solo risks on new endeavors, but if they only provide a share of the resources necessary, the risks are thus reduced. Third, crossing boundaries involves social learning, in which the two-way transmission of information among partners can be extremely valuable. To the extent that partners have different experiences and knowledge funds, spanning divides is extremely valuable for adopting new means to pursue shared goals. Failing all of those advantages, there is a fourth argument, the moral argument. If all other options to achieve a shared goal have proven inadequate, and the goal is vital in its ramifications, then collaboration across boundaries becomes a moral imperative.

Collaboration across boundaries requires a long-term orientation. A substantial investment of time and effort is required to build and sustain relations of trust among partners. That said, the benefits of collaborative advantage follow an investment logic: while it is costly to develop strong collaborative ties, you get what you pay for. Strong loyalties are necessary to withstand the stresses of the shared challenges involved. Moreover, relations of trust are crucial to pursuing capacity building over the long haul.

A final point merits mention about collaborative advantage in the specific context of conservation and development. The great secret to the logic of benefits from collaborative advantage arises from a paradox. Collaborative advantage proposes to take the unitary approaches to conservation and development and use their assets to offset their limitations. Crossing boundaries, such as among types of organizations, stands to increase the potential complementarities. If governments have capacity and resources but communities have local knowledge, the complementary advantages of the two ought to be combined. Collaborative advantage asserts that the unitary approaches to conservation and development are important precisely because they invoke social actors with distinct capacities for complementary contributions. The question then is to identify the issues and consequent tasks, challenges, and strategic practices in order to combine the capacities of different social actors in order to capture their collaborative advantage.

This book offers an insider's perspective on the exigencies of collaboration across boundaries. It is not a research report, nor the result of interviews with management consultants. Instead, the empirical basis for the book stems from firsthand practice of crossing boundaries, complete with successes and failures along the way. I thus offer reflections on direct experience, effectively immersed in the melee as it unfolded. Such a perspective permits a richer and more in-depth rendering of the relationships among partners who must overcome the difficulties presented by boundaries in order to collaborate effectively. The inside view also conveys the lived experience of the shared anxieties as well as the collective joys of collaborating across boundaries. Hence, rather than dispassionately report research findings about somebody else's experience in a typical academic mode, I seek to span the academic/non-academic divide as a scholar-practitioner in order to provide a series of reflections as a fully involved participant. This is not to say there is no framework or design behind the cases chosen or the issues highlighted. Instead, it is to point out that the empirics reported stem from firsthand practice and joint learning. What follows is thus a tract on crossing boundaries as experienced in the practice of collaboration.

Boundaries themselves are constituted for various reasons, including administrative tractability and political control. They separate distinct cultural worlds and institutional systems which define how things get done. Boundaries are thus largely beyond the power of collaborators to manipulate. The central question then becomes one of how collaborators manage to cross boundaries. Crossing boundaries means many things: recognizing a shared problem, identifying "others" across a divide with a common interest in dealing with said problem, reaching out to the other, and finding out who they are, just as they do with you. So begins the courtship, played out amidst the exigencies of cultural miscomprehension, administrative incompatibilities, and other challenges.

There is no shortage of previous work on crossing boundaries, though it has some limitations. To briefly mention a few prominent literatures, there is work on interdisciplinarity (e.g., Frodeman, et al. 2010; Klein 1990; Weingart and Stehr 2000), inter-organizational collaboration (e.g., Poncelet 2004; Robinson, et al. 2000; Wondolleck and Yaffee 2000), and international cooperation (e.g., Gaillard 1994), whether via international scientific networks (e.g., Newman 2001), transnational advocacy networks (e.g., Keck and Sikkink 1998), and other forms of alliances. A limitation in most such work is the tendency to focus on crossing one specific type of boundary. The result is a set of distinct literatures on crossing boundaries, which are fragmented because they focus on different types of divides. These literatures deserve to be considered together under the broader banner of "crossing boundaries." For one thing, the challenges in crossing a specific type of boundary are often similar. At the same time, there are distinct insights in the different literatures, so there are opportunities for learning from considering them together.

What is more, proposals for collaboration, to address various wicked problems, frequently underscore the importance of crossing multiple boundaries simultaneously, but then fail to recognize additional boundaries. One such example concerns work on transdisciplinarity, the crossing of disciplinary boundaries while also transcending academia to work with stakeholders. Then there is less emphasis on international cooperation. Likewise, work on collaborative conservation and multi-stakeholder planning processes like adaptive management are all salutary for their recognition of the importance of spanning divides among organizational types. But such efforts often pay much less attention to disciplinary or national boundaries. There remains a need to recognize the multiple dimensions of boundary crossing, and the implications for collaboration. Insofar as wicked problems require crossing multiple boundaries simultaneously, it is important to simultaneously consider multiple types of boundaries. Bringing together work on interdisciplinarity, inter-organizational collaboration, and international cooperation provides a means to address crossing boundaries in a more integrated fashion. This is especially important for considering the ramifications of crossing multiple boundaries simultaneously for the viability of collaboration. Each additional boundary crossed complicates collaboration itself. I submit that the compounded challenges of crossing multiple boundaries is a significant factor to explain what makes a problem wicked.

The explicit goal of this book is therefore to outline both the daunting challenges as well as facilitative practices, from courtships to breakups, in collaborating across multiple types of boundaries. The exigencies are indeed significant, but they are often viewed as intimidating without being understood. We must recognize that for any shared goal, collaborating across boundaries requires attention to an extensive set of tasks in and of itself, dependent on a specific skill set that few have well in hand. The requisite skill set for the substantive goal, be it conservation and development or community policing, differs from that for collaboration across boundaries. Failure to put adequate effort into either collaboration across boundaries or the substantive goal will similarly undermine achievement of the goal.

The requirements for collaboration across boundaries seem to be largely unknown outside of a small coterie of hard-working practitioners. They who know seldom write about how they do their work, likely because of the formidable level of effort required. By contrast, as an academic, I am beholden to the expectations to "publish or perish." I have thus had the privilege of working directly with practitioners who cross various types of boundaries, while at the same time observing proceedings from the institutional perch of an academic social scientist. By spanning the scholar-practitioner divide, this book seeks to clarify what is involved so that more people—from whatever discipline, type of organization, or country—can effectively collaborate across boundaries.

Consequently, this book is not written for specialist scholars in discipline X with intellectual commitments $Y_1 \ldots Y_n$ who specialize in the study of topic Z, including collaboration itself. Neither is this a book about conservation and development per se; nor is it a book about the Amazon. The central emphasis is squarely set on collaboration across boundaries. With that focus, I aspire to reach a mixed audience, and I hope this book can speak to different types of professionals as well as students seeking to collaborate across boundaries in various contexts. I do hope to reach academics, who I encourage to take a larger role than heretofore has been the case in collaborating across boundaries for conservation and development. At the same time, as an academic who has worked with practitioners, what follows includes many reflections on practice, non-academic organizations, and problem-based work. A key point is to issue a clarion call for more collaboration across boundaries, to both academics and non-academics, to argue that each needs the other more than they realize.

While this book draws on firsthand experience, it is not intended as an exercise in hand-waving and telling "war stories" about how awful collaboration across boundaries can be. I draw on frameworks to organize discussions of the issues of collaboration, the additional exigencies of crossing boundaries, the design of collaborative structures, and some persistent problems that defy easy solutions. All such frameworks were selected and organized based on previous work on collaboration and crossing boundaries as well as my own experience. Frameworks in turn provide a basis to organize the presentation of vignettes that recount noteworthy events in the practice of collaboration across boundaries. It is useful to have frameworks to organize the storytelling, for then the stories serve specific purposes and constitute concrete examples to motivate a discussion of topics applicable to collaboration across boundaries in general. Further, in the context of a framework, the vignettes offer not only cautionary tales of the problems that can arise but also lessons about how to avoid said problems. In particular, I present a framework for collaboration that identifies key issues, outlines the tasks required to address those issues, the challenges that impede performance of said tasks, and the strategic practices to address the challenges. Those practices are recommended to partners to manage the vagaries of collaboration. The vignettes thus provide illustrations of the issues, tasks, and challenges, as well as the strategic practices.

Having said all that, this is not just a "how-to" book. It should not be read as offering recipes for success or other forms of guarantees. This is just as much a "what to expect" book. The lessons learned came from specific experiences where collaborators and I made mistakes and had to learn the hard way how to do things better. Beyond learning from mistakes, we have repeatedly found ourselves in unexpected situations, precisely because we were collaborating across boundaries. There is much to learn from unintentionally causing problems not anticipated by standardized administrative systems or historically grounded cultures (Garfinkel 1967), in order to better understand how things work, and thereby pursue better practices.

A word on scope and generalizability is in order. The experiences on which this book are based come from collaborating across boundaries for conservation and development in the southwestern Amazon. This book therefore pursues an established strategy of managing the general and the specific in its treatment of collaboration across boundaries. On the one hand, it draws on previous work on collaboration, crossing boundaries, and experiences from around the world to make general arguments. To that end, I cite previous publications. Specialists should not, however, expect exhaustive reviews. The literature reviewed is suggestive rather than definitive, and references are provided for interested readers to find more details on particular points. The citations thus apply to topics of more general relevance, such as inequalities and politics. While such topics certainly have their context-specific vagaries, they are nonetheless issues of broad importance. Hence, I make general assertions concerning the importance of key topics. By implication, I avoid drawing specific conclusions about the details of those topics in various different contexts. For example, I will argue that inequality among partners is especially important when collaborating across boundaries, because such inequalities are likely to be greater when crossing boundaries. I do not, however, make more specific assertions about how exactly such inequalities may play out in a particular situation. On the other hand, this book features vignettes with concrete examples from specific cases of conservation and development projects in the southwestern Amazon. This amounts to employment of the "illustrative method" from qualitative social science, which highlights documentation of general concepts in specific cases (cf. Bonnell 1980; Skocpol 1984). The vignettes thus serve as illustrations of how general issues played out in a defined context where collaborators crossed boundaries. The vignettes are thus illustrative rather than definitive. Readers unfamiliar with the Amazon can decide if the specifics of the cases presented offer lessons generalizable to other contexts.

On that note, it is worth reflecting on the central issue that collaboration across boundaries presumes that one does not know enough to complete a task by oneself. Collaboration across boundaries necessitates an effort to increase one's awareness of one's ignorance. This certainly applies to ignorance of what is on the other side of a given boundary. One's training (and thus one's intellectual commitments), one's organization (and thus the administrative apparatus one inhabits), and one's country (and thus one's cultural background) all influence—and greatly limit—one's perceptions. But that is not all. Ignorance also applies to understandings of collaboration itself as well as what crossing boundaries means. The implication is that even practitioners with experience in collaborating in one context stand to learn from practitioners working in another. Furthermore, collaborators who have crossed one type of boundary can learn from counterparts who have crossed other types. One part of the learning stems from the differences in contexts, out of the recognition that context matters. Another part of the learning reflects the point that there are also shared issues in collaborating across boundaries, though there are differences in the details depending on the specific type of boundary crossed.

There are various definitions of collaboration across boundaries. For present purposes, I prefer to identify a few fundamental characteristics involving a key commitment, key decision points, and key outcomes. The key commitment is the resolution to metaphorically "get in the same boat" with the "others" with whom you propose to partner. This carries numerous implications, many of which won't be evident at the time of embarkation, which is exactly why the commitment is significant. Making the commitment to the "others" means stepping out from the familiar confines of one's home institution and building bridges with another. Issues of communication and logistics immediately arise. Bureaucratic procedures are likely to differ, stemming from distinct institutional histories and structures as they define roles. As a result, matching up personnel in one institution with those of another often turns the collaborators into diplomats who serve as the translators of "how things get done" in one institution to the other. Collaboration thus implies visits to and indeed working in the other's home institution. This entails the use of their office space, connecting to their servers, following their schedule, and drinking their water. While this is perhaps not "full immersion," it does offer significant opportunities to put oneself in the shoes of the partners, and provides more than a taste of what they face and how they operate. The commitment comes with the understanding that you will return, too.

The key decision points are what links the commitment to the outcomes. It is useful to think about the outcome as the result of a process of building confidence and trust that proceeds in stages. This is not to invoke a stage theory involving an inexorable progression, but rather stages defined by moments where the partners have to make decisions of strategic importance that affect the subsequent options for joint action. Not only that, those strategic decisions also speak volumes about the partners themselves and how they respond to pressure and ambiguity in difficult moments. The key moment is when one of the collaborators has to make a decision to go out of their way in support of their partners, as opposed to the decision to conduct business-as-usual in their self-interest. Decisions that valorize the partnership for the sake of the shared goal consummate the collaborative commitment; this is what generates trust and loyalty. A key decision point presents the opportunity to construct and fortify a collaborative bridge across a divide. This can be frightening to the individuals participating and threatening to the institutions at hand, since there are risks involved. But the trust and loyalty catalyzed by decisions in support of collaboration in turn modify relationships to make collaboration more productive and thus more advantageous.

The key outcome of collaboration across boundaries—setting aside the achievement of shared goals for a moment—is that the partners go from being mutual "others" to each other to becoming "known other persons." Do not underestimate the importance of this change. Stereotypes are overcome, mutual respect emerges, self-interests become joint interests, and loyalties grow. Partners go out of their way for each other, not only out of their common commitment to their shared goal, but out of loyalty to each other stemming from the understanding that they are both necessary to achieving said goal.

With this commitment-decision-outcome rendition of what collaboration involves, it is important to emphasize the structures in which collaboration occurs. At the end of the day, it is individuals who work together and thereby become collaborators. At the same time, those individuals also represent communities or organizations, which have internal structures. Hence one can readily imagine various types of collaborative relationships based on the institutional perches on which the collaborating individuals sit. Collaboration may occur between individuals of two divisions within one organization; among individuals representing two different organizations or communities; and an individual representing an organization with an individual representing a community.

In this book I am primarily interested in collaboration among individuals in different disciplines, types of organizations, and countries. I make two master arguments about collaboration across boundaries for conservation and development. First, *crossing boundaries augments the challenges of collaboration.* The boundaries being crossed are key to the difficulties of collaborating. Two scientists of the same discipline at the same university in the same country collaborating is a relatively simple matter. By comparison, collaboration becomes more complicated for two scientists of different disciplines at the same university, or of the same discipline at two different universities, or of the same discipline at universities in two different countries. In each of these cases, the scientists have to cross a boundary in order to collaborate. In this regard, the challenges of collaboration increase. Cultural assumptions about shared knowledge become more dubious, so investment in communication becomes more important. Frequency of communication may decline depending on ease of contact. A key corollary to the first master argument concerns the impact of crossing multiple boundaries. *When crossing more than one boundary simultaneously, the challenges stemming from boundaries are compounded.* Multiple cultural divides and institutional distinctions come into play, and the potential for miscommunication and bureaucratic snags increases. The complexities grow and the consequent difficulties multiply.

All that said, the second master argument is that *boundaries also magnify the complementarities of the partners, yielding greater collaborative advan-*

tage. Just as boundaries divide and impede, precisely because collaborators across divides are different, they are more likely to bring highly complementary skills, capacities, and experiences to a partnership. Rather than crossing a boundary by oneself and trying to learn the ropes on the other side the hard way over an extended period of time, learning is enormously facilitated by working with a counterpart. This is in some sense the quantitative magnification of collaborative advantage from crossing boundaries. But there is more. Crossing a divide involves trepidation even without detailed knowledge of the challenges that lie ahead, which leads to the result that many will not make the leap. This is a lost opportunity that forecloses on a possible future. In that regard, crossing boundaries brings the qualitative collaborative advantage of being able to achieve what would not be possible otherwise.

This then is the road. The challenges of crossing boundaries need to be better understood so they can more readily be met in order to realize the collaborative advantages that then become possible. Passing the trials of initial crossings of boundaries makes downstream collaboration more feasible. Consequently, weathering the rigors of crossing boundaries builds capacity to manage wicked problems in the long term. In sum, while collaborative advantage is hard to achieve, I submit that it is nevertheless worth working for. Collaborating across boundaries permits the achievement of what is not otherwise possible now, while making possible futures workable in the process.

The remainder of this book unpacks the exigencies and the blessings of crossing boundaries for collaboration. Chapter 2 focuses on a concrete study case with which I am familiar: the Amazon. The chapter begins by briefly reviewing the major conservation and development challenges of the basin writ large. I then focus on the southwestern Amazon and outline selected aspects of its biophysical environment, history of human settlement, and current conservation and development issues. That background serves as the context for a regional social movement seeking to support participatory environmental planning. The movement in turn generated proposals for conservation and development projects that sought to cross multiple types of boundaries. The remainder of the chapter describes several projects in which I have participated, noting certain details of the boundaries crossed for collaborative enterprise. In particular, I highlight the achievements of those projects for conservation and development. The discussion recounts how the projects crossed boundaries to gain collaborative advantage and thereby achieve objectives that would not have been feasible otherwise. Chapter 2 thus constitutes the "official story" of collaborative advantage from crossing boundaries for collaboration.

Chapter 3 takes a key step by pulling back the curtain on the work required to achieve collaborative advantage. To the old adage that law and sausage are more attractive if one does not know how they are made, I would add collaboration. I draw on previous work on collaboration in order to present a framework for understanding the complex suite of issues at play. The discussion then unpacks those issues in terms of sets of tasks entrained, challenges which complicate performance of said tasks, and strategic practices for managing said challenges. This chapter also introduces issues that will be featured in later chapters, namely politics and inequality as well as uncertainty and surprise.

Chapter 4 turns to a discussion of the significance of boundaries for collaboration. There I argue that crossing boundaries complicates collaboration beyond the issues of collaboration in general. I then offer discussions of disciplinary, organizational, and national boundaries, noting the additional challenges each poses to collaboration. I draw on the framework presented in Chapter 3 to make specific points about the particular issues of collaboration which are made more difficult to manage when crossing specific kinds of boundaries. This chapter then takes up the key matter of the compounding of difficulties when crossing multiple boundaries simultaneously.

Chapter 5 offers a structural perspective on collaboration across boundaries by focusing on issues pertaining to the design of structures of collaboration. Here I draw on literatures on social network analysis, business management, and organizational sociology, which make complementary contributions to understanding distinct types of collaborative structures. On that basis, this chapter offers a framework for collaborative design. The framework features several design considerations including scale, structural arrangements, functional roles, and modes of collaboration. I then introduce the issue of boundaries as they affect all of the other design considerations in the framework. Collaborative relationships that span boundaries are especially important because they are crucial to certain functions that support collaborative goals, and they are difficult to sustain. This motivates a discussion of a series of cases of different collaborative structures, noting their strengths and weaknesses. The cases include all of the projects in the southwestern Amazon that were introduced back in Chapter 2. Chapter 5 concludes with recommendations about the design of collaborative structures.

The next three chapters center on issues that in my experience have proven to be particularly salient when crossing boundaries for collaboration. Chapter 6 discusses inequalities and politics. I argue that spanning divides has the effect of widening inequalities and intensifying the politics of collaboration. Boundaries often exist for political reasons, making boundary crossing a political act. Crossing boundaries may thus be frowned on by authorities, with potentially difficult implications for collaborators. Further, whereas complementarities among collaborators can beget collaborative advantage, those complementarities may also reflect inequalities among partners. Hegemonic narratives, unpleasant histories with unexpressed grievances, and cultural differences involving unflattering stereotypes can all cause conflicts and thereby hamper the benefits of collaborative advantage. If not explicitly recognized and openly addressed, inequalities can become venomous and turn collaborators into combatants. This chapter therefore outlines various perils concerning inequalities and politics for collaboration across boundaries. I organize the discussion into four sections: the importance of boundaries for inequalities and the ability to manage them, the significance of boundaries for external politics as context may affect the viability of collaboration, the consequences of inequalities among partners for collaborative relationships, and politics among collaborators across boundaries. Each section comes with several vignettes from Amazon projects. The vignettes serve to illustrate how inequalities and politics affect collaboration across boundaries in a particular context. These discussions permit reflections on lessons learned as regards the clarity of shared vision, the lack of consciousness about inequalities, and the value of making oneself vulnerable in order to learn.

Chapter 7 turns to another central issue in collaboration across boundaries, uncertainty and surprise. I argue that crossing boundaries increases uncertainty for collaboration by multiplying the number of sources of surprise to which collaborators are exposed. Crossing boundaries becomes a matter of managing uncertainty and surprise, precisely because things work differently on different sides of a divide. The best-laid plans often encounter glitches that slow things down, or unexpected opportunities that present alternatives which cause dilemmas, or crises which may make an initiative impossible or suddenly raise the stakes by requiring the initiative to expand tenfold in scale immediately. Chapter 7 therefore presents illustrative vignettes from the Amazon projects on surprises, whether due to external politics, personnel changes, and biophysical phenomena like extreme climatic events. Collaboration across boundaries does not follow a predictable set of stages where planning of activities permits implementation as anticipated, yielding the expected outcomes, on schedule and at the expected cost. I then square up various approaches to management, including rational planning, adaptive management, scenario planning, and "muddling through," with the level of uncertainty due to surprises when crossing boundaries. Approaches to management that presume a low level of uncertainty are likely to be problematic, not to say absurd, under conditions of high uncertainty, as when crossing multiple boundaries. In the context of spanning many divides, Plan A is a useful point of departure, and little more. Further vignettes from the Amazon projects provide illustrations. There is a need to look beyond rational planning and adaptive management in order to appreciate the opportunities presented by crises, the vital role of hyper-flexibility, and the tortuous beauty of "muddling through."

Chapter 8 pursues issues of collaboration and the individual when crossing boundaries. Previous work on collaboration, including work on crossing boundaries, has a strong tendency to treat the players involved as categories rather than as individuals with distinct personalities. Personality traits, or behavioral dispositions, can greatly influence the effectiveness of the individual as a collaborator. Stated in positive terms, individuals with particular personality traits may more effectively cross boundaries. Put more negatively, collaboration across boundaries tests psychological mettle, and can thereby reveal behavioral disorders. Further, conflicting personalities can potentially allow conflicts to become personalized, which is destructive. This chapter therefore briefly reviews work on personality traits, emotional intelligence, and cultural cognition to illustrate aspects of personalities as they may affect collaboration across boundaries. Crossing boundaries imposes the hard knocks of experience on the self, making collaboration an effective if sometimes uncomfortable teaching tool. I therefore highlight behavioral dispositions especially relevant to collaboration across boundaries, derived from prior work and firsthand experience: a collective orientation, tolerance of ambiguity, and equanimity. Several vignettes from the Amazon projects illustrate the importance of these dispositions. I conclude Chapter 8 with comments on strategic practices for self-management when collaborating across boundaries.

Chapter 9 concludes the book by offering reflections on the "seasons" of collaboration across boundaries. Collaborative relationships begin with organizational courtships and continue to the work required to sustain collaboration, especially in moments of crisis, which may lead to the cessation or termination of collaborative relationships. Whereas there has been considerable attention to forming and maintaining collaborative partnerships, there has been less to managing crises and ending collaboration. I argue that boundaries make crises more frequent and harder to manage, with the consequence that crises and breakups become more serious menaces. This chapter therefore focuses on crisis management and unpleasant endings when collaborating across boundaries. I draw on complex systems theory, particularly thought on bifurcations. In the context of collaborative relationships, I interpret bifurcations as processes that result in functional breakdowns and structural breakups which may be hostile. A final selection of vignettes from the Amazon projects provides context-specific illustrations. The chapter and the book conclude, however, with more positive observations on the strategic importance of collaboration, including the manner of collaborative breakups. Just as collaboration across boundaries imposes a steep learning curve, there is much to be gained in learning from breakups.

Chapter Two

The Official Story: Examples from the Amazon

> The MAP frontier is the center of the world.
>
> —Various MAPienses

In this chapter I focus on collaboration across various types of boundaries in the southwestern Amazon. I offer this background of experience not because I think it is special or because I think the Amazon is somehow fundamentally different from other places with regard to the challenges facing would-be collaborators and boundary-crossers. Rather, I draw on my experiences in the southwestern Amazon because they serve to illustrate a specific instance of how issues of more general importance unfold for collaboration across boundaries.

The southwestern Amazon serves as a useful illustrative case for various reasons. With regard to conservation and development, the socio-environmental issues in the southwestern Amazon constitute something of a microcosm of the basin as a whole, and beyond. Development policy, infrastructure projects, deforestation, resource-based livelihoods, agrarian conflicts, biodiversity loss, watershed management, and climate change are all major issues in the southwestern Amazon, the rest of the basin, and many other parts of the world. With regard to collaboration across boundaries, the southwestern Amazon includes three countries and spawned a social movement advocating for participatory environmental planning, thus spanning academic disciplines and different types of organizations. This makes reporting on conservation and development work there useful as an illustration of collaboration across boundaries. Finally, the southwestern Amazon is useful because it is a stimulating study case with no shortage of colorful if occasionally bewildering and even frightening stories. Discussing collaboration in the southwestern Amazon often provokes rousing comparisons by colleagues collaborating across boundaries in other places.

The Amazon is a demanding place in which to work. It is a huge developing region where travel and communications are sometimes difficult. It is changing very rapidly, and the pace at which new exigencies appear is rigorous. The Amazon is the subject of no small body of writings including adventure novels and travel tales, replete with often-embellished accounts of killer snakes, hostile Indians, and so forth (e.g., Grann 2005; Millard 2005; Smith 1990; Wallace 2011). Capturing the region in a few summary words misses the enormous differences, biologically and politically, that one finds among different places. Local histories are invariably colorful and adumbrated if often fragmentary. There are stories within stories, and different people have different pieces.

In this context, collaboration permits some measure of compilation, yielding composites which do not fit together neatly. One acquires greater understanding of the story, but "the story" is actually several mutually inconsistent accounts. Collaboration is also generative of stories, especially when projects are well funded, high profile, and stand to make an impact. These are examples of the "official story" of what collaboration makes possible, the discourse that seeks to confirm the advantages of collaboration. However, collaboration also generates stories behind the official story. These are also important, since they pertain to the process of collaboration itself and permit understanding of the official story.

This chapter presents the "official story" of my experiences with collaboration across boundaries in the southwestern Amazon. The official story focuses on "what got done." The point is to first show what collaboration permits, and thereby illustrate the power of collaborative advantage. This provides a more digestible appetizer for the longer and heavier exposition about "the story behind the story" of "how things got done," which comes in later chapters.

I begin by offering a brief outline of the southwestern Amazon's biophysical characteristics and its history, concluding with a review of contemporary challenges to environmental conservation and sustainable development. I then turn to an innovative polycentric social movement, the "MAP Initiative," which emerged in the southwestern Amazon around the turn of the new millennium to address those challenges. The MAP Initiative in turn produced a suite of proposals for collaborative projects, which is where outsiders such as myself became involved. These opening sections thus provide a concrete case out of which collaboration across various types of boundaries came to be seen as necessary and important.

The rest of the chapter then provides "official stories" for a suite of collaborative projects that involved crossing multiple types of boundaries. This includes a collaborative scientific effort to evaluate social-ecological resilience to paving of highways, a consortium of organizations that sought to implement the regional environmental planning model elaborated by the MAP Initiative, the two projects that followed these initial efforts in trying to take the next steps scientifically and in terms of environmental planning, and a recent collaboration focusing on university capacity building for biodiversity conservation. In each case, I note the development process, governance structures, and goals, but I pay particular attention to the achievements in each case.

AMAZONIA: AN OVERVIEW

The Amazon basin is both a massive watershed and a forest biome (Figure 2.1). It has a humid tropical climate with seasons marked by changes in precipitation (Dickinson, 1987; McClain, et al. 2001; Prance and Lovejoy 1985). The interactions between land and climate are a fundamental biophysical interface in Amazonia (Keller, et al. 2009; Malhi, et al. 2008). Climate, vegetation, and soils are mutually engaged in cycles, as of carbon and water (Davidson, et al. 2012; Kim, et al. 2012). Moisture transport and precipitation are the main features of the earth-climate system that maintains the Amazon (Arraut, et al. 2012). Evapotranspiration of vegetation moisture contributes to atmospheric moisture that accumulates as it flows westward, resulting in precipitation that feeds the rivers that carry water to the east.

The Amazon has a range of natural variability in its climate system and therefore in its vegetation, soils, and rivers. During the last decade, the Amazon experienced a series of extreme weather events including both droughts and floods (Lewis, et al. 2011; Marengo, et al. 2013). Such extreme events may indicate a change of climate from one regime to another (Davidson, et al. 2012). If climate change shifts the Amazon away from its historical range of variability with respect to temperature, humidity, and precipitation, it is likely that other biophysical components of the basin will also change (Davidson, et al. 2012). It is now evident that severe droughts reduce primary productivity of forest vegetation (Gatti, et al. 2014) and increase tree mortality (Aragão 2012; Brando, et al. 2008). There is debate as to whether the future Amazon will die back, leading to "savannization" in the twenty-first century (Malhi, et al. 2008; Saatchi, et al. 2013).

The forests of the Amazon have considerable biological value, taking into account their ecosystem services and their biodiversity. In terms of ecosystem services, Amazonian forests constitute a key component in the Earth's climate system, in terms of water cycling and carbon storage (Davidson, et al. 2012; Keller, et al. 2009; Malhi, et al. 2008). This has fueled interest in the application of various proposals for programs of payments for environmental services (Hall 2008; Fearnside 2012). The Amazon forest is also

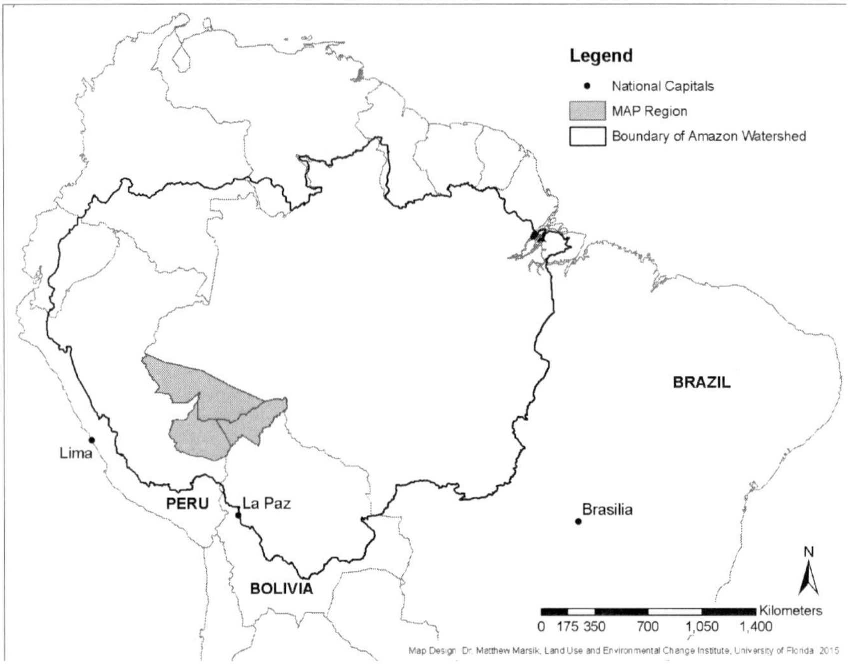

Figure 2.1. Map of the Amazon watershed and the states of the tri-national frontier in the southwestern Amazon.
Source: Map created by Matthew Marsik, Land Use and Environmental Change Institute, University of Florida, 2015. Used by permission.

valuable for its exceptional biodiversity (Laurance, et al. 2010; Phillips, et al. 1994; Prance 1994). Botanical inventories in the Amazon indicate high levels of species richness and endemism (Pitman, et al. 2008; Phillips, et al. 1994).

The Amazon has been marked by changes in human societies. There is evidence of human occupation dating back roughly 11,000 years (Roosevelt, et al. 1996). The indigenous population was severely impacted by the incursions of European explorers, mainly the Spanish and Portuguese, beginning in the 16th century. The colonial period in the Amazon was marked by a series of boom-bust cycles in the extraction of various raw materials, especially during the rubber boom (Santos 1980; Weinstein 1983).

In the twentieth century, the Amazon experienced accelerated human settlement and population growth. Especially important have been public policies of facilitating frontier migration, particularly along newly built roads where colonization was promoted (Goodland and Irwin 1975; Mahar 1989; Schmink and Wood 1992). In the 1960s and 1970s, governments of countries sharing the basin sought to occupy land and thereby secure their national borders. In turn, land settlement along the highways stimulated construction of secondary roads to facilitate access to land and other valuable natural resources (Perz, et al. 2007).

These changes in turn led to a rapid rise in non-traditional forms of natural resource use. In many heretofore remote communities, forest product extractivism was replaced by deforestation for agriculture. Logging, mining, crop cultivation, and especially ranching are the preeminent land uses along highway corridors in the Amazon (Faminow 1998; Walker, et al. 2009; Wood and Porro 2002). Initially, drivers of frontier settlement and deforestation included policy incentives, accessibility via infrastructure, cheap land, and limited opportunities in other places (Mahar 1989; Schmink and Wood 1992). More recently, deforestation has been driven by demand for commodities in regional, national, and global markets (Faminow 1998; Keller, et al. 2009; Walker, et al. 2009).

The switch from forest product extraction to a regime of mining, agriculture and livestock was not without complications, as newcomers encountered forest extractivists and indigenous groups. There are numerous accounts of rural conflicts and violence as an integral part of the change in the use of natural resources (Schmink and Wood 1992; Leonel 1992). This in turn led to social mobilization by indigenous peoples, forest extractivists, small farmers, large ranchers, mining associations, and environmental groups, all of whom sought to tilt the political playing field to serve their interests. Meanwhile, governments tried to manage this tumult by implementing new public policies, ranging from lines of credit for small producers to the establishment of protected areas, the creation of new models of land tenure, and other initiatives (Kainer, et al. 2003; Perz, et al. 2013a).

THE SOUTHWESTERN AMAZON

Like other national frontiers in South America, the southwestern Amazon has long been a crossroads (Paredes Pando 2015). When I speak of the "southwestern Amazon," I refer to the tri-national frontier where Bolivia, Brazil, and Peru meet (Mendoza, et al. 2008; Perz, et al. 2012). Specifically, I focus on the "MAP" frontier, named for the states that form the tri-national border: Madre de Dios (Peru), Acre (Brazil), and Pando (Bolivia) (Figure 2.2).

The southwestern Amazon falls within the altitudinal gradient that defines the ecological transition from the Amazonian lowlands to the Andean highlands. Because of the gradient, the southwestern Amazon is crossed by several rivers that originate in the highlands and drain eastward into the lowlands. The altitudinal gradient defines a series of ecological gradients which help delimit an array of biological communities. Consequently, the southwestern Amazon is highly biodiverse (Myers, et al. 2000).

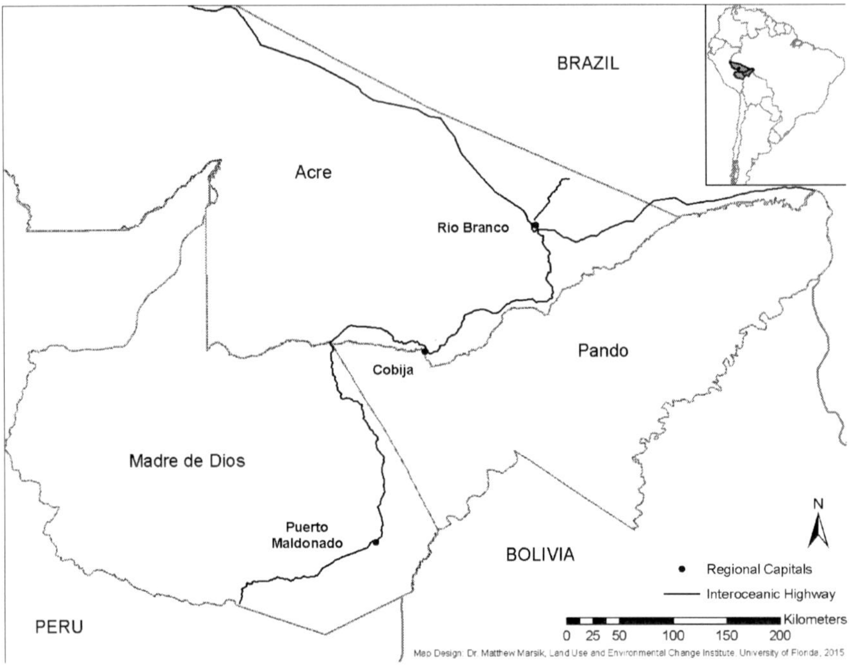

Figure 2.2. Map of the "MAP" frontier, Madre de Dios (Peru), Acre (Brazil), and Pando (Bolivia) in the southwestern Amazon.
Source: Map created by Matthew Marsik, Land Use and Environmental Change Institute, University of Florida, 2015. Used by permission.

As a reflection of the biodiversity of its forests, the southwestern Amazon is known for important timber and non-timber forest products. In the late nineteenth century, the southwestern Amazon became an epicenter for the rubber boom (Tocantins 1979). This led to the establishment of rubber estates controlled by rubber barons who attracted rubber tappers and then imposed labor relations based on debt peonage. These relations led to an economy dependent on rubber, a geography of settlement organized around river transport, and limited deforestation. When the rubber boom went bust in the 1910s, rubber barons abandoned their estates and left the rubber tappers to fend for themselves. As a result, a forest culture based on small-scale agriculture and forest extractivism emerged in the southwestern Amazon.

This changed in the 1960s when the governments of Brazil and other countries embarked on programs of national integration via the construction of new highways into the Amazon. Brazil opened the BR-364 highway into Acre, and Peru inaugurated what is now the Inter-Oceanic Highway from Cusco to Puerto Maldonado and from there to the border with Brazil. In both countries, road construction was accompanied by campaigns to attract migrants to the new frontiers for settlement in colonization projects directed by the state. At the same time, the government of Brazil sought to attract capitalist investment in cheap land, which prompted a wave of land speculation through establishment of cattle ranches (Bakx 1988). In similar fashion, in Peru, the Belaunde administration promoted colonization in the Amazon, and was followed by the Garcia presidency, which provided various policy incentives for agricultural land settlement (Chavez 2009). In parallel, during the 1980s and 1990s, the government of Bolivia opened a road from the capital in La Paz to Cobija in Pando (Stoian and Henkemanns 2000).

The expansion of new infrastructure prompted major changes in the MAP frontier. Populations in towns along roads grew rapidly. This occurred partly due to the arrival of migrants from other regions following the highways, and partly due to more localized shifts of populations from settlements along rivers to those along the roads. A second key shift occurred in natural resource use, from forest extractivism to agriculture. Under the new regime, forest was to be exploited (via logging) or removed outright (for farms and ranches). A third change was the emergence of land conflicts. New arrivals seeking to remove forests encountered forest extractivists seeking to retain forests, resulting in agrarian tensions. The murder of rubber tapper Chico Mendes in Acre in 1988 came after a year of record deforestation in the Amazon, and drew international media attention that linked the problem of environmental damage to human rights violations.

Rural violence stimulated social movements among rural peoples. Central to such mobilizations were proposals to recognize traditional claims on natural resources. This made land tenure a central question in the southwestern Amazon during the 1980s and 1990s. In many instances, social movements were successful in prompting governments to recognize new land tenure models that reflected traditional resource use. In Acre, the government of Brazil created the Chico Mendes Extractive Reserve (Ehringhaus 2005; Gomes 2009). In Pando, the government of Bolivia established communal lands for independent communities (Assies 2002; Stoian and Henkemanns 2000). In Madre de Dios, agricultural settlements were located along the roadsides with forest concessions in back (Chavez 2009; Rojas 2012). These are examples of a broader diversification in land tenure in the southwestern Amazon, where distinct lands have differing use rules, including deforestation limits. By 2000, the MAP frontier was a complex mosaic of lands types, and planning maps look like rainbows due to the numerous colors that have to be used to differentiate among types of lands (Perz, et al. 2013a).

In Acre, movements for social justice and forest conservation became allied with the Workers Party, which served as a platform for election campaigns. By the 1990s, movement activists were winning municipal and even state elections in Acre. Especially notable was the election of Jorge Viana, a forestry engineer, as Governor of Acre in 1998. This inaugurated the "Forest Government," which overhauled the state apparatus, cleaned up the accounts, obtained external funding, and embarked on a series of innovative policies in support of sustainable forest extractivism (Kainer, et al. 2003). Subsequent elections returned Viana as governor along with allies in local and state government positions, and resulted in numerous policy initiatives to support rubber extraction and processing, certify sustainable timber for export, and otherwise foster a culture of "forest citizenship" (Schmink and Cordeiro 2008).

Since 2000, the southwestern Amazon has incurred accelerating changes. One particular driver is new infrastructure. In 2000, the presidents of the countries of South America constituted IIRSA, the Initiative for the Integration of Regional Infrastructure of South America (CEPEI 2002; IIRSA 2008). IIRSA was conceptualized as a spatial strategy for development and trade along "axes of integration" targeted for infrastructure investments. One of IIRSA's first priorities was the Inter-Oceanic Highway (IOH), so named because it would link Atlantic ports in southern Brazil to Pacific ports in Peru. The key to the IOH is the "Peru-Brazil axis" of integration, where road paving in the southwestern Amazon was still needed to permit year-round transportation. The first phase of IIRSA involved a five-year implementation period from 2006 to 2010, during which the IOH was paved on the Peruvian side. In 2006, President Lula of Brazil and President Garcia of Peru met Governor Viana of Acre at the Integration Bridge over the Acre River to commemorate the IOH linking the two countries. In 2011, President Garcia came to the regional capital of Madre de Dios, Puerto Maldonado, to inaugurate another Integration Bridge over the Madre de Dios River. This officially inaugurated the IOH linking Madre de Dios to the rest of Peru. In both Brazil and Peru, the IOH runs right along the border with Pando in Bolivia, making the highway corridor important for all three sides of the tri-national frontier.

IIRSA and the IOH spurred contentious debate over the possible consequences (Killeen 2007; Dourojeanni, et al. 2010). There are multiple scientific literatures on the various impacts of infrastructure. For example, there is a body of work on "road ecology" (Coffin 2007; Forman, et al. 2003; Trombulak and Frissell 2000). This research emphasizes how roads encourage the degradation of hydrological networks, fragmentation of habitats, and the loss of endemic biodiversity along with the invasion of exotic species. In contrast, there is also an extensive literature on the positive economic effects of infrastructure investments (Calderón and Serven 2004; Straub 2008). Such work has consistently documented that infrastructure investments decrease transportation costs, which makes economic activity more profitable and improves access to urban services, which in turn generates economic growth and reduces poverty. In another contrast, in the Amazon and elsewhere, there is social research which shows that infrastructure often creates or exacerbates social inequalities and causes conflicts (Mahapa and Mahiri 2001; Schmink and Wood 1992; Shriar 2009).

The MAP frontier remains largely forested, but exhibits an accelerating deforestation rate. Forest loss before 2000 was slower than in 2000–2005 during the onset of paving of the IOH (Southworth, et al. 2011), which in turn was slower than in 2005–2010, when paving was actively advancing (Perz, et al. 2013b). Beyond deforestation, the paving of the IOH coincided with major economic changes in the MAP frontier. The global economic crisis of 2008 led to a rise in world commodity prices, notably for precious metals. A rise in the price of gold expanded migration and investment in gold mining, particularly in Madre de Dios (Swenson, et al. 2011). There, migrants from the Andean Highlands found easy access via the IOH to rivers and creeks where they could stake gold claims. As gold strikes proliferated, migration accelerated, multiplying the number of mining camps and attendant service enterprises. From 2009 to 2012, Guacamayo Creek in Madre de Dios went from a small mining camp to an open area many square kilometers in extent. As mining claims pressed deeper into the forest, reports of the environmental damage as well as human rights abuses emerged (CHS Alternativo 2012). Also expanding rapidly was the use of mercury to separate gold from silt, with various consequences including mercury pollution in rivers, mercury accumulation in fish, and public health concerns about human exposure to heavy metals. These problems prompted calls from environmentalists for state regulation of mining activity, enforcement of protected areas, and punishment of criminal activity in mining camps. The government of Peru has responded with military operations to detonate river dredges, disperse protesting miners, and inspect traffic on the IOH for contraband. Over time, as the informal mining sector has grown in Madre de Dios, political debate has intensified over the need to regularize mining in order to regulate it, especially in light of the large amount of state revenue otherwise being lost.

Beyond gold mining, paving of the IOH has stimulated other economic sectors, notably illegal logging and drug trafficking. Logging has an extended history in the southwestern Amazon, and the sector is characterized by informality (Cossio 2009). While there is legislation in Bolivia, Brazil, and Peru permitting logging in concessions according to management plans, such plans are rarely followed entirely. State capacity for enforcement remains limited, and while the most valuable tropical hardwoods such as mahogany and tropical cedar have become scarce, loggers have moved on to other species which have domestic or international markets. Over time, logging roads from the IOH have extended ever farther into the forests of the MAP frontier.

The profitability of illegal logging has generated its own complexities. Where there is risk of enforcement, log trucks move largely at night. Where papers can be fabricated, log trucks move openly along the IOH, though they sometimes do so in order to move timber illegally across borders. The IOH has unintentionally facilitated illegal cross-border commerce among criminal entities, who have proven themselves to be agile entrepreneurs. At the same time, such entrepreneurship has the obvious drawback of causing damage to forests by removing key species, damaging other trees, and so on. In addition, illegal logging has become associated with rural violence. In 2008, Julio Garcia, the mayor of the locality of Alerta in Madre de Dios, was shot and killed by loggers in broad daylight in front of the office of the governmental environmental agency INRENA for reporting an illegal shipment of mahogany leaving the town for Bolivia.

The IOH has similarly facilitated drug trafficking. In particular, the IOH has evidently made it easier to move cocaine from central Peru through Madre de Dios and out through Bolivia, and eventually to Brazil. By 2008, some of the small towns along the IOH in Madre de Dios had unused fields and pastures but exhibited a rise in the number of satellite dishes and other amenities. In August 2008, roughly 30 people died in and around Cobija due to gunfights over control of drug routes, evidently a result of new distributors moving into the region due to the easier access (La Razón 2008).

There is also a measure of cattle ranching, gas prospecting, and various tree crop "boomlets" occurring in one or another part of the MAP frontier. The result is a panorama of highly dynamic economic activities leaving a mosaic of impacts across the southwestern Amazon, such that local circumstances vary substantially from one place to another. It is thus not surprising to find a rural community pursuing agroforestry right next to a large cattle ranch, which in turn is a neighbor to an agricultural community, next to a forest concession for castaña harvesting. Depending on how tenure rights are defined, surface and subsoil rights may be independent, which causes new forms of land conflicts, such as between agriculturalists (with surface rights) and gold miners (with subsoil rights). All of this presumes that land claims have been surveyed and demarcated, which may or may not be the case, especially where there is a large influx of new migrants seeking land along the IOH.

If the IOH and its sundry impacts were not enough, the MAP frontier, like other parts of the Amazon, is experiencing the effects of climate change. The southwestern Amazon was located in the epicenter of the droughts of both 2005 and 2010 (Lewis, et al. 2011). In the 2005 drought, farmers set fires to clear vegetation, but the fires escaped control and resulted in large-scale burns of at least 400,000 hectares of mature forest in Acre (Pantoja and Brown 2009), with additional damage in other parts of the tri-national border.

Over the same time period, the MAP frontier also experienced major flood events, in 2006, 2009, 2010, 2011, 2012, 2014, and 2015. The 2012 flood was especially damaging, as the Acre River in Rio Branco reached a level just below the highest ever recorded up to that point in time. The damage in Rio Branco alone was estimated at around $100 million (Xavier 2012). The river crested at record levels upstream, and large areas of several cities along the Acre River were flooded, including 90 percent of the city of Iñapari, Peru, and 75 percent of Brasiléia, Acre (Lima 2013). In March 2015, the worst recorded flooding in the region's history unfolded, submerging these and other towns (Globo 2015).

Increasing deforestation along highways and incipient climate change bring the prospect of synergistic changes in the southwestern Amazon. Specifically, the effects of deforestation may magnify the effects of climate change. Deforestation fragments forests, yielding "edge effects" where forest edges exhibit greater tree mortality and other indicators of forest degradation (Laurance, et al. 2002). In turn, degraded forest edges and forest fragments may be more vulnerable to climate change, notably during drought events. On a larger scale, there is concern that deforestation can reduce evapotranspiration and reduce humidity and precipitation that support the Amazon (Spracklen, et al. 2012). Insofar as the IOH and economic dynamism accelerate deforestation in the southwestern Amazon, which has been an epicenter for droughts and floods, these synergies could be significant in the twenty-first century.

THE MAP INITIATIVE

The exceptional values of forest ecosystems and the unprecedented changes in the southwestern Amazon were not lost on local peoples, regional governments, or research organizations. As noises about IIRSA and the IOH reached the southwestern Amazon, and as talk of climate change and the savannization of the forest circulated, people took notice. National policy debates over the prospective impacts of the IOH were paralleled by local discussions of the potential benefits and problems of regional integration. Scientific argument about atmospheric carbon and climate change found confirmation in rural communities in the southwestern Amazon, where longtime residents noticed higher temperatures and more intense rainfall events.

Because drivers of change like the IOH and climate were shared across countries in the southwestern Amazon, residents and local leaders, as well as scholars and elected representatives, began to reach across borders (Brown, et al. 2002; Rioja 2005; van Oosten 2004). Recognition of common challenges in Madre de Dios, Acre, and Pando became explicit in the naming of MAP as a tri-national frontier and as a single arena for study and social action. What is more, "MAP" became a brand name associated with some key philosophical commitments. Two have stood out, and both stem from the UN Declaration on Human Rights (UN [1948] 2015). First is the Right to

Know: all citizens have a right to information about opportunities and threats that may affect them. Second is the Right to Participate: all citizens have a right to act on information in their individual and collective best interest. These commitments served as philosophical foundations for the "MAP Initiative," the name given to the tri-national social movement in the southwestern Amazon which served as the forum to pursue regional environmental planning via participatory processes. All stakeholders were welcome to contribute, and anybody in the tri-national frontier who thought about crossing boundaries was referred to as a "MAPiense," a name with the double meaning denoting people living in the MAP frontier and people thinking about shared tri-national issues (Reyes 2007).

The MAP Initiative is hard to characterize in organizational terms because it relies on a non-hierarchical structure and because it spans national boundaries. For lack of a better term, I describe the MAP Initiative as a polycentric movement (cf. Ostrom 2005) defined by a tri-national structure that features key individuals and organizations on all three sides of the frontier. Participation by collaborators in Madre de Dios, Acre, and Pando provided the basis for all activities of the MAP Initiative. Such partnering among organizers permitted cross-border coordination in planning of particular events as well as ongoing activities. However, boundary crossing did not only involve national borders. From early on, MAP Initiative organizers defined four priority foci for their activities: 1) environmental conservation, 2) economic development, 3) social equity, and 4) participation in public policies. The notion behind the first three foci was to inclusively conceive of the necessary ingredients for sustainable development. In the process, MAPienses thought about problems in a way that spanned academic disciplines. The focus on participation and public policies invoked yet another kind of boundary to be crossed, among types of organizations. The assertion was that broad public participation in discussions of the first three themes was necessary for adequate policies in the region. This required spanning divides among governmental and non-governmental organizations and community associations.

After an initial MAP forum in 1999, the MAP Initiative grew quickly (Mendoza, et al. 2008). During the 2000s, organizers held forums roughly once per year. Invited speakers from the southwestern Amazon and elsewhere provided presentations on conservation, development, and equity. Elected representatives and agency directors provided presentations on public policy initiatives. Working groups formed around specific themes, including conservation priorities like biodiversity and development opportunities such as castaña and ecotourism. Attendance rose from 220 participants in the 2002 MAP forum to 1,200 in 2004. Similarly, the number of organizations represented multiplied, and increasingly included delegates from multilateral banks, national government agencies, international environmental organizations, and other entities interested in the MAP Initiative's boundary-crossing approach to environmental planning.

The working groups meeting at the MAP forums, which came to be known as the "mini-MAPs," took on lives of their own. While some proved to be short-lived, others gained more members and began to produce proposals for concrete action. Mini-MAPs caught the attention not only of banks and aid agencies, but also governments looking for opportunities to adopt viable ideas as policy. By 2007, the governors of Madre de Dios, Acre, and Pando were jointly attending the MAP forum, and they took the opportunity to cite cases where their administrations were adopting proposals originated in the MAP forums or mini-MAPs.

In the process, the MAP Initiative directly or indirectly fostered numerous conversations and contacts among people not only across the tri-national frontier of the southwestern Amazon, but other parts of Bolivia, Brazil, and Peru, and beyond. As participation and organizational representation grew, the dimensions of the MAP Initiative expanded. Consequently, the network became more diverse in terms of the complementarities among the participants, in terms of their knowledge funds, their experience with the issues at hand, and their organizational capacities. Participants began to realize that the complementarities presented opportunities to collaborate in order to implement activities that otherwise would not have been envisioned as feasible. In due course, MAPienses produced proposals for projects to carry out participatory environmental planning activities based on ideas originated in MAP forums and mini-MAP working groups.

Hereinafter I report on a selection of projects that resulted from the MAP Initiative, including one especially large project that encompassed a suite of priority activities identified by MAP Initiative organizers. The remainder of this chapter describes six collaborative projects that resulted directly or indirectly from the MAP Initiative, in which I participated and thus have detailed firsthand information. The projects do not include everything that resulted from the MAP Initiative, and I make no claims about universal representativeness. Nonetheless, all six projects are useful for present purposes because all involve crossing of two or more types of boundaries. As a result, they serve as illustrations of efforts at collaboration across boundaries. Consequently, the projects provide the basis for vignettes later in the book on numerous issues that arose in the process of collaborating across boundaries.

PROJECT #1: "NSF ROADS"

Concern about impacts of paving the IOH prompted mini-MAP working groups to engage in discussions on highways and regional environmental planning. Such discussions in turn catalyzed demand for scientific information on road impacts. In particular, given its structure, the MAP Initiative came looking for information about the ecological, economic, and social ramifications of paving the IOH, as these corresponded to the emphasis on conservation, development, and equity.

As it happened, there was a parallel discussion already under way among faculty and graduate students in various departments and colleges at the University of Florida. Operating as their own working group, which came to be denominated "the Roadies," UF personnel were debating how to articulate a framework for analysis of road impacts that could include the ecological, economic, and social dimensions. The Roadies constituted a interdisciplinary group, with participants ranging from systems ecologists to wildlife biologists to geographers—physical and social—and sociologists and anthropologists. The focus of the Roadies fell on identifying insights from various conceptual frameworks previously used to study road impacts in different disciplinary literatures, and then seeking to integrate those contributions in a theoretical synthesis. Because some UF Roadies participants were also working in the southwestern Amazon, MAP Initiative organizers extended an invitation to UF personnel to offer presentations at MAP forums as a means to engage local stakeholders. UF personnel participated in MAP Initiative activities beginning in 2002.

The result was a series of animated dialogues between UF personnel and MAP Initiative organizers that sought to relate regionally specific issues in the MAP frontier with the broader conceptual issues at play in discussions at UF. Out of those dialogues came proposals for research projects focusing on the modeling of road impacts in the MAP frontier. Here I focus on the project led by UF via partnerships with three regional universities, one from each side of the tri-national MAP frontier.

The Roadies produced several proposals for collaborative research projects. In 2005, the Roadies sent a proposal to the US National Science Foundation's Human and Social Dynamics (HSD) program for work on road impacts in a dense road network (north Florida) and a sparse road network (the MAP frontier). Reviews came back requesting a focus on the Amazon site with a comparative design taking advantage of the tri-national MAP frontier. We agreed, and NSF funded the project.

The reviews also highlighted interest in our conceptual framework, which underscored two key concepts that had survived an extensive interdisciplinary triage at UF: connectivity and resilience (Cumming, et al. 2005). We understood highways as exogenous drivers of change, insofar as governments and banks planned roads with limited local consultations. We also viewed highways as catalytic of a wide array of change dynamics including the ecological, economic, and social, which raised questions about how the shocks wrought by paving the IOH would impact these different aspects of the affected region. We thus conceptualized road impacts in terms of social-ecological resilience and viewed affected regions as complex dynamic systems, where highway paving would lead to socioeconomic and biophysical changes that would in turn likely affect each other.

Connectivity served as a means of identifying key linkages among processes in a complex system, which provides a means of relating the ecological to the economic to the social. We found that connectivity has been used in distinct ways by systems ecologists (to denote how tightly processes are linked; Gunderson and Holling 2002) and geographers (to denote the importance of interactions among different locations in space; Haggett, et al. 1977). We exploited these multiple usages by invoking them both under the common banner of the same word as a means of stitching together a model of road impacts, whether spatial or systemic. Hence, we framed the project in terms of how highway paving affected spatial and system connectivity, and how they in turn influenced the resilience of the frontier in social and ecological terms.

Meanwhile, conversations with partners in the MAP frontier got under way. One key challenge was to confirm the participation of collaborators; another key challenge was to explain the theoretical framework in understandable terms. With those issues lined out, we moved on to identifying student research assistants. In the end, we constituted a management team of 21 faculty at UF and the three partner universities, and identified 69 participating students (Table 2.1). The management team was large because we sought a structure that permitted balanced representation of partners among universities across the three countries. Further, we divided the project up methodologically in terms of key data collection efforts under botanical, socioeconomic, and remote sensing components. The design permitted identification of collaborating counterparts, whether among peers within a project component or across countries. The project came to be called "NSF Roads" by the partners (in Spanish it was *NSF carreteras* and in Portuguese *NSF estradas*).

The goal of the NSF Roads project was to compile comparable botanical, socioeconomic, and remote sensing data across the MAP frontier in order to permit comparative analysis. At the time of the project, we had the good fortune of a natural experiment: the IOH was completely paved in Acre/Brazil, paving was under way in Madre de Dios/Peru, and paving remained incipient in Pando/Bolivia. While there were of course other differences, the distinction in paving status was important in the MAP frontier and served as a point of analytical departure.

We therefore drew from previous studies of road impacts and in the Amazon to develop our methodological protocols. Detailed descriptions of those protocols are available elsewhere (Perz, et al. 2012, 2014; Baraloto, et al. 2014; Southworth, et al. 2011). Project coordinators sent field teams out during the Amazon dry seasons from 2007 to 2009. The socioeconomic

Table 2.1. Faculty and Student Participants in the "NSF Roads" Project, 2005–2010.

University	*Total*	*Botanical Component*	*Socio-economic Component*	*Remote Sensing Component*
University of Florida (UF), USA	5 / 14	2 / 2	2 / 9	1 / 5
Federal University of Acre (UFAC), Brazil	4 / 17	1 / 5	1 / 10	1 / 2
Amazonian University of Pando (UAP), Bolivia	7 / 15	2 / 8	2 / 5	1 / 2
National Amazonian University of Madre de Dios (UNAMAD), Peru	6 / 21	1 / 10	1 / 9	1 / 2
TOTAL	21 / 67	6 / 25	6 / 31	4 / 11

Note. For each case, the first value indicates the number of faculty coordinators and the second indicates student research assistants.
Source: Perz (2010).

teams pursued their fieldwork in two stages. First, they first visited 103 rural communities along highways in the region and conducted key informant interviews with 292 leaders about their communities. Second, teams revisited 27 communities known to differ in order to conduct more detailed surveys of 742 households. The remote sensing teams made field visits to obtain GPS points of land cover in order to train classification algorithms. This supported development of mosaics of eight Landsat images necessary to cover the MAP frontier for each of six different years: 1986, 1991, 1996, 2000, 2005, and later 2010. This period spans the onset and advancement of highway paving in the region. We then developed forest/non-forest classifications in order to quantify forest loss over time. The botanical teams visited forest sites along highways in order to obtain botanical inventories along the ecological gradient from the lowlands to the foothills of the Andes. This proved a challenging proposition due to access issues and the difficulty of identifying the numerous different plants down to the species level. The teams installed a total of 69 permanent botanical plots with each sampling 0.1 ha of area from a roughly 3 ha expanse, and left approximately 5000 physical samples in the herbaria of the regional universities.

Given societal demand for information about the topic of highway impacts, it was crucial to return initial findings to participating communities, local governments, and other interested constituencies. An initial activity of the project, in an alliance with other projects, was to pursue a tour of the

MAP frontier. Dubbed the "knowledge exchange train," the notion was to have representatives of multiple research projects travel together and jointly hold workshops with local stakeholders in order to disseminate previous findings as a basis for dialogues about road impacts (Mendoza, et al. 2007). As follow-up to the activities of the NSF Roads project, we developed a suite of products for local constituencies including six different reports (roughly 200 copies distributed), one poster (80 copies), and two pamphlets (1500 copies). Results from the project also found their way into presentations at local radio stations and other outlets.

For participants in the NSF Roads project, academic publications were also a priority. The team spent time to map out a strategic plan for publications, and developed a protocol to guide proposals for papers that named the co-authors and highlighted the focal question of the analysis, data to be used, and the target scientific journal. This allowed a more coordinated procedure for identifying first authors while recognizing appropriate co-authors among collaborators and students. The project yielded 18 publications in scholarly journals, all but one co-authored with in-country collaborators and students (with an average of 6.7 co-authors). Further, the project either financially supported or provided data that contributed to seven dissertations at UF as well as student theses and projects at the partner universities. Finally, the NSF Roads project provided the basis for subsequent collaborations on other projects and activities.

NSF Roads thus crossed disciplinary and national boundaries and permitted collaborative advantages by producing unique data products of academic and societal interest. Without the numerous meetings of faculty and students from many disciplines over several years at UF, the integrative theoretical framework on connectivity and resilience could not have been developed, and the project would not have been funded. Further, without the large coordination team that spanned countries and project components, development of shared methodological protocols for data collection would have been impossible. In the end, large amounts of comparable data were collected, and they in turn made it feasible to produce outputs of interest to stakeholders across the MAP frontier as well as journals in several disciplines.

PROJECT #2: THE G-MAP CONSORTIUM (AND MABE AND M-P)

If the NSF Roads project constituted a collaborative adventure into crossing boundaries for environmental science, the G-MAP Consortium involved a larger enterprise crossing boundaries for environmental management. The growth of the MAP Initiative and the attraction of governments, banks, NGOs, and other organizations led to discussions about the potential to leverage the networking functions of the movement into conservation and development projects. In this regard, the MAP Initiative was influential beyond the MAP frontier itself. The MAP Initiative came to be seen as a potential model for pursuing collaborative approaches to conservation and development, notably by crossing various kinds of boundaries. In particular, the MAP Initiative provided a concrete case of collaboration across disciplinary, institutional, and national boundaries.

In 2005, the US Agency for International Development (USAID) engaged in a consultative process to define an Amazon-wide program to provide development assistance for biodiversity conservation. The result was the Amazon Basin Conservation Initiative (ABCI), which in 2006 called for the creation of consortia to pursue activities in support of sustainable development and biodiversity conservation in the Amazon. ABCI bore many of the hallmarks of the MAP Initiative: it called for consortium proposals involving collaborating organizations; it required that consortia worked in at least three countries in the Amazon; the consortia were to support civil society and government actions; and the consortia were to address both conservation and development concerns (USAID 2006). In support of these aspirations, USAID brought its own expertise to bear as well as that of an Institutional Support Unit. In effect, ABCI called for the institutionalization of collaborative projects that would leverage the gains won by movements like the MAP Initiative.

That said, ABCI exhibited some important differences from the MAP Initiative. USAID required proponents of collaborative projects to navigate manifold federal guidelines about contractual obligations, mandatory standard provisions, budgetary rules and procedures, performance monitoring, and other requirements of project management that were foreign to the MAP Initiative. USAID also required that consortia identify a "prime" or lead organization, which imposed a more hierarchical structure than was characteristic of the MAP Initiative as a network. Similarly, USAID required identification of "key personnel" as representatives of consortia, also in contrast to the participatory spirit of the MAP Initiative. Finally, USAID sought not contracts but more broadly conceived cooperative agreements. Such agreements included "substantial involvement" by USAID in the administration and activities of ABCI consortia. Hence, ABCI consortia formalized collaborative relationships by imposing requirements and expectations outside the experience of many in the MAP Initiative.

While the ABCI call for proposals was open, the opportunity was hardly lost on the MAPienses, who initiated discussions to identify priority activities and partners to constitute a "G-MAP" ("Governance in MAP") Consortium. In classic MAPiense fashion, a diverse group of nine organizations participated in a series of meetings. The partners spanned both organizational types and national boundaries: four universities, four non-governmental organizations (NGOs), and one governmental agency (GA), in Bolivia, Brazil,

Peru, and the United States. Table 2.2 shows the distribution of G-MAP partners across organizational types and countries.

The strategy of the G-MAP Consortium was to leverage the MAP Initiative's networking functions in order to implement conservation and development activities in the MAP frontier. G-MAP would pursue its activities under the MAPiense model of participatory environmental planning, which had several key components, shown in Figure 2.3. The point of departure is the emphasis on collaboration among partner organizations with complementary capacities. Universities, NGOs, and GAs have rather different mandates and capacities, but the differences support complementary functions. By extension, individuals within different organizations had complementary training and experience, which G-MAP's architects were very careful to leverage. Some had experience working on road impacts; others had previous experience handling USAID funds. A second element in the planning model is that the consortium partners had networks of affiliated constituencies, including community associations and local governments. This is another aspect of the complementarities among partners: they serve as nodes in different networks. The planning process also followed hallmarks of MAP Initiative proposals. Emphasis went to the production of knowledge (as via scientific research and knowledge exchange), which supported capacity building (as via training, whether in classrooms or field outreach), which in turn would support participation in management planning activities (i.e., development of management plans, such as for sustainable resource use), which would lead to plan adoption and implementation by local communities and governments, in order to achieve the goal of sustainable resource use and thus biodiversity conservation. Key to all this was to work with local partners on the ground to implement a shared model that would permit replication in other sites and thus scaling up of the impacts.

Table 2.2. Collaborating Partner Organizations in G-MAP and Subsequent Consortia, by Organizational Type and Country.

	G-MAP Consortium			*MABE Consortium*			*M-P Consortium*		
	University	*NGO*	*GA*	*University*	*NGO*	*GA*	*University*	*NGO*	*GA*
Bolivia	UAP	Herencia					UAP	Herencia	
Brazil	UFAC	IPAM, SOS Amazonia			IPAM, SOS Amazonia				
Peru	UNAMAD		PEMD				UNAMAD		PEMD
USA	UF	WHRC		UF	WHRC		UF	WHRC	
Total	4	4	1	1	3	0	3	2	1

Sources: M-P (2011), MABE (2011).

A key question then concerned the substantive focus of the activities to be pursued, which required the MAPienses to identify their highest priorities. In the end, G-MAP invoked two key priorities, stemming from the major challenges then facing the MAP frontier. First, G-MAP observed the issue of regional integration under IIRSA via the IOH, and therefore proposed to pursue activities to mitigate the negative impacts of highway paving. Second, the MAPienses noted the onset of climate change in the form of extreme climatic events such as the drought of 2005 and the flood in early 2006 to argue for a focus on watershed management planning. But the MAPienses were not done: given the accelerating rate of change, and given USAID's priority on demonstrating sustainability of impacts beyond the lifetime of projects, it would also be necessary to invest in future capacity for conservation and development. G-MAP therefore proposed a capacity building program, which featured investments in regional universities, in order to multiply the pace of capacity building for capacity building itself. Heavy investment was programmed for student fellowships, training of community members and governmental technical staff, and workshops on planning tools, all

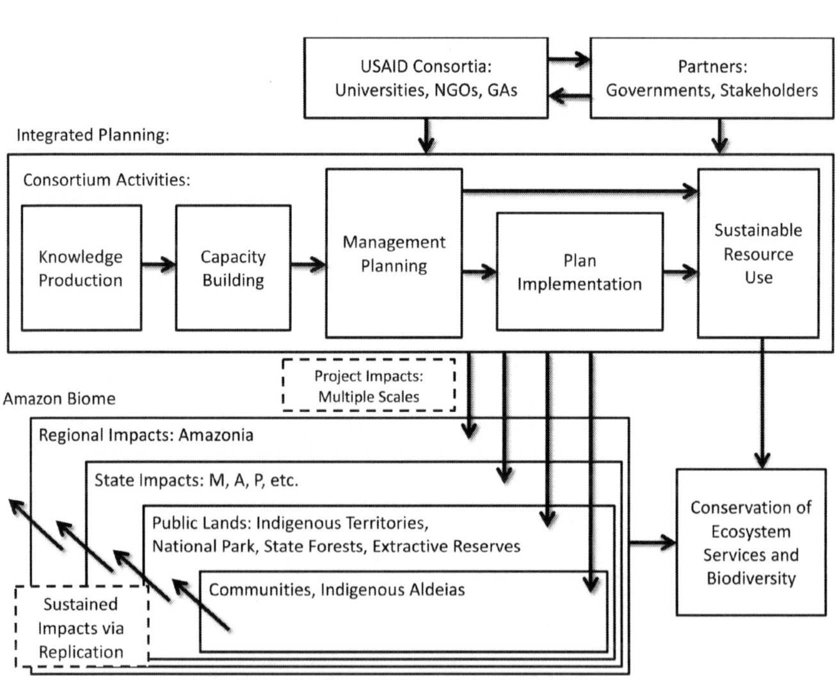

Figure 2.3. Regional Planning Model Underlying Environmental Management Consortia Stemming from the MAP Initiative.
Source: Created by author.

as means to support development of management plans for roads and rivers in the MAP frontier.

The design of the G-MAP Consortium thus exhibited many attractions: it leveraged the successful networking of a polycentric movement, it encompassed diverse organizations with complementary capacities, it responded to major questions of the hour in the Amazon, and it provided a roadmap for implementing conservation and development activities in a scientific and participatory manner.

However, the demands of USAID for an equally solid plan to manage the consortium itself left open some big questions. The G-MAP partners realized that nobody had sufficient organizational capacity to manage the contracts, accounts, reporting, and other requirements. The MAPienses therefore approached UF to serve as the prime. UF had the administrative capacity to manage a large USAID cooperative agreement. By 2006, UF personnel had participated in several MAP forums. This then led to that crucial phone call from Brazil in February 2006. I said yes, but that actually resolved very little, because there remained numerous other questions. For one thing, USAID required that the key personnel reside in the Amazon. G-MAP thus had a prime outside the region and key personnel not directly employed by the prime, an atypical arrangement. For another, UF as the prime officially received the agreement—and the money—from USAID. The challenge then was to manage the hierarchical relationships imposed by USAID. On the one hand, UF wrote sub-agreements to the MAPiense partner organizations in order to transfer funds, with the partners ensuring that work was done to be reported back to USAID. On the other, all partners sought to maintain their horizontal relations established via the MAP Initiative. The partners nonetheless agreed that this arrangement was the best option, and the collaborative adventure began. With double the number of partners and roughly ten times the funding as in the NSF Roads project, G-MAP embarked on an ambitious endeavor to implement the MAP Initiative's approach to conservation and development in the southwestern Amazon.

Along the way, ABCI was replaced by the Initiative for Conservation in the Andean Amazon (ICAA), with USAID activities in Brazil transferring to USAID-Brazil. Consequently, the G-MAP Consortium had to be divided, which prompted creation of two new consortia: MABE, which stood for "Environmental Management of Highways and Rivers" (in Acre/Brazil) and M-P, which stood for "Madre de Dios-Pando" (in Peru and Bolivia). Table 2.3 shows where the G-MAP partners ended up in MABE and M-P. I discuss the reasons for these changes, and their consequences, in Chapters 6 and 7. For present purposes, suffice it to say that both consortia incurred a series of adjustments but were nonetheless able to pursue implementation of conservation and development activities.

To give some notion of what the consortia did, Table 2.3 presents measures of the technical performance by project conclusion in 2011. We focus on USAID shared indicators, which are standard measures of technical performance under USAID cooperative agreements. The shared indicators employed in ABCI and ICAA include the number of people trained in the sustainable management of natural resources, the number of hectares under improved management, the number of policies, laws, agreements, and regulations for sustainable management of natural resources, the number of policy dialogue events for sustainable management of natural resources, and resource leveraging for other activities. USAID required setting targets in order to evaluate technical performance as well as cost management. Table 2.3 shows that the two consortia combined were able to train nearly 10,000 people, improve management on roughly 2.4 million ha, support promulgation of over 25 policies, organize at least 77 dialogue events, and leverage roughly $3.6 million for downstream activities.

This technical performance includes some putatively big numbers, but Table 2.3 begs questions about whether the consortia realized collaborative advantage. One basis for evaluation is to compare the targets set and the actual performance. Where comparable data are available, the comparisons are generally favorable: actual technical performance by the end was greater

Table 2.3. Technical Performance Indicators (and Targets) for the MABE and M-P Consortia, USAID Shared Indicators, 2006–2011.

Shared Indicator	*M-P*	*MABE*	*Total*
Number of people trained in sustainable management of natural resources	8653 (3233)	975 (N/A)	9618
Number of hectares under improved management	1,180,963 (169,000)	1,244,243 (150,000)	2,428,206
Number of policies, laws, agreements and regulations (PLAR) for sustainable management of natural resources	16 (9)	11 (N/A)	27
Number of policy dialogue events for sustainable management of natural resources	20 (+) (26)	57 (N/A)	77 (+)
Resource leveraging	$2.4 million ($2 million)	$1,261,485 (N/A)	$3.6+ million

Sources: M-P Consortium (2011) and MABE Consortium (2011).

than expected performance at the outset. While there are various explanations for this, a key reason was that demand for MABE and M-P services rose vertiginously during the project period. Governments and communities came seeking more information, and more support for environmental planning activities, as time went on. In the event, the consortia responded well, an indication of collaborative advantage in terms of flexibility and complementarity in the face of unexpected exigencies.

Table 2.3 provides some concrete numbers, but obscures what exactly the MABE and M-P Consortia were doing. Due to space limitations, I provide a brief summary, organized by the three main objectives: mitigation of highway impacts, watershed management planning, and capacity building for conservation.

MABE and M-P both engaged numerous stakeholders in watershed management planning. The consortia developed and disseminated technical reports outlining the planning process, which applied the MAP Initiative's participatory approach to environmental planning to the case of watersheds. M-P worked in Tahuamanu province in northern Madre de Dios, where rivers are smaller and water shortages greater during droughts. The first plan was for the Yaverija sub-basin watershed, and focused on conservation of riparian forests to ensure healthy aquatic habitats for biodiversity and water quality for local communities. This was so successful that the stakeholders insisted on developing a management plan for the much larger Tahuamanu watershed. In Pando, M-P developed capacity to monitor rainfall events and water levels in rivers. This supported an early warning system on which Civil Defense authorities relied in emergency management planning. Meanwhile, at the behest of the government of Acre, MABE worked in western Acre in critical sub-basins of the Juruá watershed. That experience was so successful that the government of Acre adopted not only the Juruá plan but also the MABE planning process as state policy (Governo do Acre 2012). The government of Acre's water management plan was the first of its kind in Brazil, and provided a model for Brazilian water management policy.

As for mitigation of the negative impacts of highways, both consortia focused on territorial management planning in roadside communities. M-P worked with the Bélgica indigenous community in Tahuamanu province. Together, they developed a management plan for both aquatic and terrestrial habitats. Because areas affected by fires in 2005 were usually agricultural communities along the highways, mitigation of road impacts also became a question of mitigation of drought impacts. M-P pursued emergency management planning by setting up communication networks via two-way radios in rural schools, by training firefighters to create firebreaks, and by organizing community emergency management exercises. MABE meanwhile leveraged previous organizational investments in dynamic simulation modeling of deforestation along roadsides (cf. Soares Filho, et al. 2004) to produce future scenarios of deforestation after highway paving in the MAP frontier to 2030. The resulting map became a policy tool for the government of Acre, which participated in side meetings at the 2009 climate meetings in Copenhagen. While there was no binding international agreement among countries, states from different countries, including Acre, signed MOUs. In particular, the government of Acre used MABE's deforestation map to argue for carbon payments agreements. As a result, Acre and the US state of California signed an MOU for a carbon payments program. MABE partners also signed agreements with the government of Acre to provide technical support for a carbon payment program in a new state Climate Institute.

While the watersheds and road impact mitigation programs led to concrete outcomes like hectares under improved management and new policies, MABE and M-P both spent considerable time and resources on capacity building activities. M-P established tree nurseries in rural communities and implemented training activities led by university students working with primary and secondary school children focusing on reforestation and agroforestry. In turn, M-P post-doctoral scientists worked with university faculty and students to establish or strengthen environmental monitoring programs, including geographic information system labs and certificate programs on deforestation monitoring and measuring carbon in forests for carbon payment program evaluation. M-P also implemented a "Children's Forest" activity in a rural community as an environmental education tool for primary schools. MABE meanwhile implemented a "Visiting Professionals" program that sent university students to work in various NGOs and GAs in Acre, many of which were involved in supporting or implementing the innovative conservation and development policies of the Forest Government. MABE also pursued three more Knowledge Exchange Trains (KETs) in Acre in order to expand dissemination and dialogue about environmental change in the region (Mendoza, et al. 2014). The new KETs in turn served as the basis for MABE partners to sign agreements with local governments and communities for downstream collaborative activities, notably municipal environmental management plans.

As with NSF Roads, G-MAP and its offspring provided evidence of collaborative advantage. Beyond the technical performance numbers, it was the administrative coordination between UF and the MAPiense partners which permitted the activities noted here—among many others—to proceed. This is a useful lesson: collaboration may be largely or purely administrative and still beget important advantages. Further, as in the case of NSF Roads, G-MAP (and MABE and M-P) also yielded latent benefits via the experience of working together. This built "collaborative capital" to make downstream collaborations more feasible and in that sense more likely. In that regard, initial collaborations not only involved institutional courtships, such as the incorporation of UF into G-MAP, but also constituted rigorous probationary periods to see if the partnerships would function effectively under the stress of funder requirements. Both NSF Roads and G-MAP (and MABE and M-P) exhibit this additional dimension of collaborative advantage by constituting valuable experiences that built up collaborative track records among the participating partners. This then made downstream projects more viable.

PROJECT #3: "NSF MODELING"

The success of the foregoing projects thus came not just in terms of the products delivered but in new collaborative opportunities. In the case of NSF Roads, the large-scale data collection effort led to discussions with agricultural and biological engineers about developing a modeling system with the data as inputs. To the cast of fieldwork and other "data" people, we added another breed of scholar, the modelers. Reminiscent of the old Roadies group but on a smaller scale, discussions were soon under way on the assets and liabilities of various modeling approaches and platforms. This led to a focus on issues of system representation in a model, which is related to debates over model complexity as it relates to model uncertainty.

Of particular interest was the question of how we could tell if the MAP frontier was in any measurable sense resilient to the impacts of paving the IOH. This stimulated an extended interrogation of work on resilience thought, model design and output, and previous ecological models. Most work on resilience invoked the term as an orienting heuristic rather than as an analytical goal (Strunz 2012), though some worthy examples did indeed appear (cf. Scheffer 2009).

We found ourselves proposing an audacious approach that amounted to a synthesis of insights from literature on ecological resilience (Holling 1996) and diagnostics of model uncertainty (Saltelli, et al. 2008). In brief, many renditions of ecological resilience highlight the possibility that a complex dynamic system can have more than one system state. This is often represented in "ball-and-cup" diagrams, with the system indicated by the ball and the cup in which the system resides indicating the system state. Global methods for analysis of model sensitivity and uncertainty generate probability distribution functions (PDFs). For more complex models, PDFs may have more than one mode. It became evident to us, in theory at least, that models of complex dynamic systems could yield multi-modal PDFs that in turn could indicate the probability that a system was in one state or another. In sum, a PDF flipped over looked a lot like a ball and cup diagram, and the peaks in a PDF indicated attractors in stability surfaces toward distinct system states. Hence, the use of global sensitivity and uncertainty analysis seemed a concrete and workable way to quantitatively evaluate ecological resilience (Perz, et al. 2013c).

With the data and collaborative relationships from the NSF Roads project and armed with engineers with tested methods for modeling and evaluating complex systems, we proposed our wild ideas to NSF via their Coupled Natural and Human Systems (CNH) program. The reviewers agreed that the ideas seemed crazy enough to work, and cited the strength of the project team, including its previous collaborative experience in the study region and publication record, and recommended funding. Thus was consecrated a new project on road impacts, this time called "NSF Modeling."

While this project is still ongoing, it has also proven fruitful. The personnel from NSF Roads found in NSF Modeling ample opportunities to conduct numerous analyses of the empirical data, generating an ongoing stream of publications, including those from graduate students. NSF Modeling has also overseen many new analyses, most notably those from the botanical component of NSF Roads (cf. Baraloto, et al. 2014), where data processing took the longest, and integrative analyses of data from multiple project components (cf. Perz, et al. 2013b).

NSF Modeling also led to a series of workshops with in-country collaborators in support of design of the simulation model. Workshops in 2013 prompted participants to design their own diagrams about key processes operating in the MAP frontier as a consequence of highway paving. Those diagrams in turn contributed to consolidated diagrams that then served as a point of departure for model design. That led to the first phase of model development with the engineers. This in turn led to a second wave of workshops, in 2014, to present initial model designs and output. The task in 2014 was to ask the participants a litany of very specific questions about particular parameters in the model for which we lacked published values in order to obtain a range of possible numerical estimates. This completed one full iteration in a cycle of model development, oscillating between consultations with locally knowledgeable experts and model development with the engineers. This manifested another instance of collaborative advantage: local experts from various countries and disciplines informing engineers with command of specialized software in order to permit model development.

PROJECT #4: THE 2020 CONSORTIUM (MABE II)

In parallel with the trajectory from NSF Roads to NSF Modeling, there also emerged paths of opportunity from MABE and M-P. In 2010, USAID issued another call for proposals for consortia working on environmental management for biodiversity conservation in the Andean Amazon under ICAA Phase II. Similarly, in 2011, the USAID Mission in Brazil issued a call for proposals for consortia working on the same issues in the Brazilian Amazon. Peruvian partners from M-P again asked UF to serve as the prime for a new

"Madre de Dios Consortium" (MDD Consortium). Brazilian partners from MABE invited UF on in a supporting role for a "2020 Consortium" (aka MABE II).

The need to follow parallel tracks was an eventuality that was not lost on the MAP Initiative organizers who participated in these consortia. It bears repeating that the whole point of the MAP Initiative was to cross various boundaries, notably national borders, in order to pursue conservation and development. At the same time, taking the opportunity to work with money from one government also led to complications with other governments that in turn resulted in the progressive division of MAP consortia, first into MABE and M-P, then from M-P to MDD as Bolivia went out. This posed challenges to the benefits of collaboration as the advantageous capacities of certain partners were lost.

MABE II is the case where another organization was initially tapped to serve as prime. Because that organization had not participated in G-MAP or MABE, there was not a shared history of experience. Consequently, there was some misunderstanding of how the consortium would be organized, particularly in terms of the allocation of responsibilities and funding. G-MAP and MABE were designed to have UF in a primarily administrative role with the partners spending most of the money for the in-country activities. MABE II was to follow that model, but it wasn't accepted by the proposed prime. This led to stressful days of reorganizing the consortium for UF to serve as prime instead. That required convincing not only UF administrators, but also new Brazilian partners. The established partners from G-MAP and MABE served as UF's references, and the proposal for MABE II was able to proceed.

That process served as a useful point of reflection on the MAP Initiative and G-MAP, particularly the conservation and development model outlined above in Figure 2.3, but also the model of collaboration underlying these consortia. The commitment to the second was instrumental to the viability of the first: without strong collaborative ties, the MAP Initiative's conservation and development model wasn't possible. What is more, the conservation and development model was evolving out of necessity due to the rapid change in the nature and scale of the issues on the ground. During the five years from 2006 to 2011, the IOH had been completely paved in the MAP frontier, and there had been not one but a series of droughts and flood events, making climate change a much more frightening prospect. As a result, the MABE II proposal declared 2020 as a key year for purposes of environmental planning, arguing for the need for long-range planning activities, and not just planning during project lifetimes.

MABE II thus highlighted the daunting challenges of climate change and proposed activities focusing on conservation of ecosystem services under instruments in development at the government of Acre. Via "integrated environmental planning," MABE II would leverage the network ties established during MABE to link road impact and watershed management planning in municipalities across Acre, many of which had county seats at road-river crossings. MABE II retained the emphasis on scientific data and public participation as key inputs in the planning process, but made the adjustments of unifying planning for highway impacts and climate change at road-river crossings, and scaling up from the community to the municipality level.

In the event, MABE II was not funded. It is nonetheless instructive to mention for the lessons learned about adjusting consortium structures, as well as modifying the conservation and development model in the process. The parties involved came out of the application process having suffered together through the changes in the partners at hand and the reorganization of the consortium structure. While this did not permit collaborative advantages in the sense of deliverables in a project, it did provide the basis to jointly seize opportunities later on.

PROJECT #5: THE MADRE DE DIOS CONSORTIUM

While MABE II failed to secure funding for activities in Brazil, the MDD Consortium ("MDD") did succeed with its proposal to USAID for activities in Peru under ICAA II. With Bolivia out of ICAA II, MDD had to organize around Peruvian partners, which led to discussions with Peruvian NGOs. As was the case with MABE II, partners served as references for each other, and pre-established relations of trust became the basis for accepting new partners into the consortium.

The larger adjustment, however, concerned the daunting scale and complexity of the challenges to conservation and development in Madre de Dios as of 2011. At that time, the gold mining sector was expanding vertiginously, paving of the IOH was coming to its conclusion, extreme climatic events were virtually an annual occurrence, and planning of the Inambari Dam was under way. We therefore audaciously proposed to address all of these threats via programs on gold mining, infrastructure, watersheds (dams), and climate change. We further elaborated on the MAP Initiative's conservation and development model by adding a second dimension of "transversal" activities that would support the four main programs (Figure 2.4). Transversal activities stemmed from lessons learned about project governance and the use of media and other tools for magnifying impacts of activities; we therefore included funds for communications, conflict management, and other activities. In turn, as a third dimension, which crossed both the programs and the transversal activities, we proposed the classic MAPiense activity of capacity building. Hence, activities under the programs to address threats would be supported by transversal activities, and everything would be supported by capacity building.

MDD proposed a positive agenda of activities in response to each of the four major threats. The positive agenda incorporated the standard hallmarks of the MAP Initiative, notably the use of scientific data and participatory stakeholder processes in environmental planning. This was more crucial than ever since some key threats, especially gold mining, were becoming politically sensitive issues about which any project was viewed with suspicion by miners. Activities to promote conservation and development with respect to gold mining thus included promulgation of clean mining technologies (which avoided the use of mercury), reforestation of lands degraded by gold mining, and support for alternative economic activities (agroforestry, honey production, etc.). The watersheds program employed the planning process developed under G-MAP and M-P, which had been applied successfully in Tahuamanu. The objective was to replicate the planning process for the case of the larger Tambopata watershed, which included the Inambari River. The infrastructure program similarly called for territorial management plans in road-

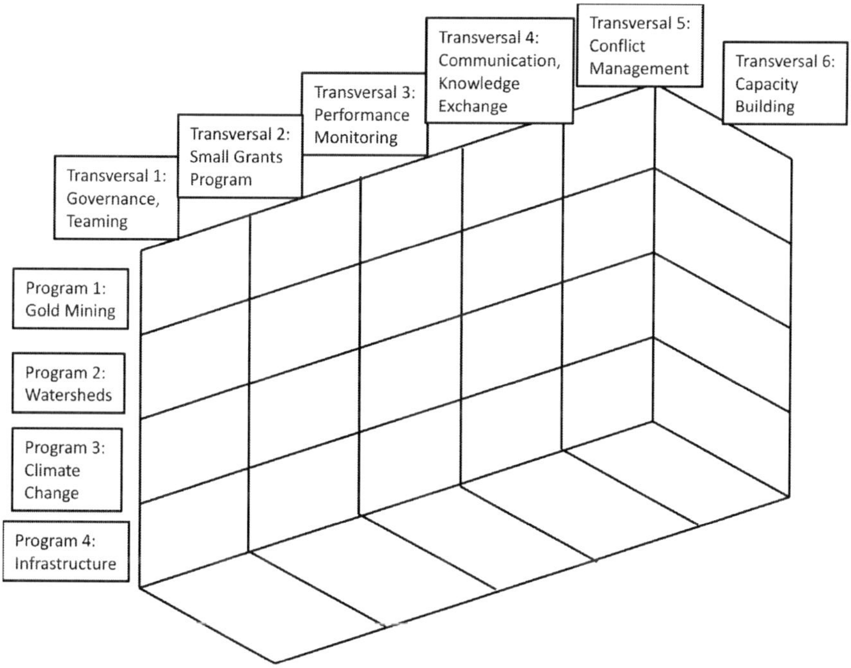

Figure 2.4. Three-Dimensional Conceptual Framework for Management of the Madre de Dios Consortium.
Source: Created by author.

side communities. In particular, MDD would focus on communities receiving large numbers of migrants coming from the Andean highlands along the IOH. In such locales, agrarian conflicts were becoming more common, so MDD proposed to work with lawyers to resolve tenure conflicts. The climate program under MDD had commonalities with activities proposed for MABE II. Specifically, MDD proposed expansion of the early warning system from Tahuamanu to all of Madre de Dios. This would be one step among others toward creation of a regional plan for adaptation to climate change.

Supporting the programs would be the transversal activities and capacity building. The conflict management activity contemplated workshops with stakeholders involved in various sorts of resource conflict, and training of conflict mediators. The communication program would produce videos for TV spots, audio products for radio spots, and printed media for distribution at regional cultural events and educational institutions. Capacity building under MDD refers to a suite of activities involving various trainings of community members, governmental technical staff, and other constituencies. Notable among the capacity building activities are training in management of nurseries for trees to be planted in agroforestry systems and reforestation plots. Communities would become producers of trees sold to local producers and firms to support agroforestry systems and reforestation. This exemplifies the underlying model of MDD's capacity building program: train people to take the lead in productive activities that generate income and also protect or restore the ecological functions of forests. Since then, the MDD Consortium has moved into capacity building for beekeeping as a means of honey production. Both raise agricultural incomes and constitute economically viable alternatives to gold mining. As with other consortia, MDD sought to leverage collaborative advantages of UF administration with the local knowledge and experience of Peruvian partners. The combination of the two made the activities proposed feasible. To date, MDD is arguably the only conservation and development project in Madre de Dios with positive relations with any miners, due in part to the positive agenda and scientific productivity, notably on reforestation of degraded lands (Román-Dañobeytia, et al. 2015).

PROJECT #6: ~~THE BOLIVIAN AMAZON CONSERVATION CONSORTIUM~~ THE ANDES AMAZON RIM CONSORTIUM

The final case study concerns another consortium project that had to change its name. Whereas G-MAP was split into MABE and M-P, the Bolivian Amazon Conservation Consortium (BACC) was suspended and reprogrammed as the Andes Amazon Rim Consortium (AARC). As with G-

MAP's fate, I elaborate on the reasons for the reprogram of BACC as AARC in later chapters.

For purposes of the official story, the BACC resulted from USAID providing additional funds under ICAA II for consortia of universities in the Andean Amazon to pursue university capacity building in support of biodiversity conservation. USAID contracted the American Council on Education (ACE), which served as the umbrella organization for the Higher Education for Development (HED), which oversaw a program of four consortia, each consisting of a US university partnering with one or more universities in Colombia, Ecuador, Peru, or Bolivia. UF faculty had contacts with Amazonian universities in each of those countries, but could only work in one country. We therefore consulted with our respective foreign colleagues. We settled on Bolivia, in part because those partners were especially enthusiastic, and in part because I had pre-existing ties with those partners in the ambit of M-P under ICAA I.

UF therefore engaged two Bolivian universities and two Bolivian NGOs with the proposition of accelerating the process of the Amazonian University of Pando (UAP) in building the capacity of its Area in Natural and Biological Sciences (ACBN). The argument was that via partnering, not only from a US university but from other Bolivian entities, ACBN/UAP would take major strides forward in its graduate training and applied research initiatives. We proposed to help ACBN/UAP build an MA program in ecology, manage a suite of collaborative research projects involving UAP and the other Bolivian partners, and use those activities as the basis for signing cooperative agreements for future activities.

The project could not have begun better: the government of Bolivia approved the project, the governance team was put into place, sub-agreements were signed, money was transferred, and the initial activities got under way. What is more, the initial meetings facilitated social learning among the Bolivian collaborators, who delineated important details for next steps in the key activities. We were on schedule. Then through no fault of the partners in BACC, the project was stopped.

To my surprise, HED and USAID indicated their interest in letting UF keep the funding by reprogramming the project for activities with partner universities outside of Bolivia. There was no way around swallowing the disappointment of having lost a promising expansion of our collaborative ties in Bolivia. At the same time, HED noted UF's ties to universities around the Andean Amazon and wanted to see a "regional approach" to university capacity building. A flurry of consultations with partners in Peru and Brazil ensued, and we listed partner priorities. We also consulted with the university consortia under HED/ICAA II in Colombia, Ecuador, and Peru regarding their interests in regional activities. We then sought to match up partner priorities with HED and USAID objectives under ICAA II.

Thus it was that out of the ashes of BACC, the AARC rose like the proverbial phoenix. We proposed some activities focusing on our long-time Peruvian MAPiense university partner, UNAMAD, some focusing on our long-time Brazilian MAPiense university partner, UFAC, and some with various universities from Colombia, Ecuador, Peru, and Brazil. The UNAMAD activity called for collaborative research projects on questions that the MDD Consortium didn't have funds to pursue. This was our attempt to adapt to rapidly growing demand for scientific data on priority topics like an experimental design for analysis of tree survival in reforestation plots, or evidence of heavy metal toxicity in medical reports in hospitals. It also provided the opportunity for UNAMAD faculty and students to work with specialists from other Peruvian research organizations and thereby strengthen ties as a basis for future collaboration.

The UFAC activity focused on two key MA programs relevant to the priorities of USAID under ICAA II: the MA in Ecology (MECO, which focuses on biodiversity conservation), and the MA in Regional Development (MDR, which highlights sustainability). In both cases, the MA programs faced difficulties due to low publication productivity, something observed in biennial evaluations by the government of Brazil. We therefore implemented a series of workshops on publishing in international scientific journals with MECO and MDR personnel. Participation by UF faculty supported production and revision of manuscripts by MECO and MDR personnel.

AARC also involved two regional activities, both of which permitted the participation of faculty and students from one or more universities in Brazil, Colombia, Ecuador, Peru, and the United States. The first was a "Study Visit" for students and faculty to come together in order to discuss shared themes of current importance in the Andean Amazon. Partners highlighted four priority themes: protected areas (including indigenous lands), extractive economies (like gold mining and hydrocarbons), infrastructure (highways and dams), and climate change (including extreme events). The Study Visit involved presentations by visitors from all participating Amazonian countries, and quickly revealed similarities and some big differences. The Study Visit also included field visits to sites in Madre de Dios relevant to the four themes, such as the proposed site of the Inambari Dam and gold mining areas. Participants from the Study Visit then wrote manuscripts on the four shared themes as they apply to their countries, which were published in a special issue of UNAMAD's journal, *Biodiversidad Amazónica.* The other regional activity was a short course on "Train the Trainers" offered by specialists in "scientific teaching" (Handelsman, et al. 2007). This responded to demands by governments of the countries sharing the Amazon for improvements in the quality of university teaching. There is by now an extensive scientific literature indicating that "active learning," "student-centered learning," and "collaborative learning," all of which are associated with scientific teaching, yield better student learning outcomes such as retention, critical thinking, and problem solving (Barkley, et al. 2005; Handelsman, et al. 2007). Visiting faculty thus brought syllabi for their courses and incorporated elements of scientific teaching, which they then implemented in their home institutions.

The AARC thus had a variety of ambitions tied to pedagogy and research, all in a highly collaborative mode that crossed various boundaries. AARC was possible precisely because of previous collaborative experiences with universities and other partners in the southwestern Amazon. AARC was also possible because of the consultations from UF, a known brand in networks of universities and NGOs in Latin America. Crucially, those consultations took seriously the priorities and suggestions of the other HED/ICAA II consortia, which ensured buy-in for the regional activities.

Together, the official stories of these collaborative projects offer some useful insights. For one thing, taking the risk of working long hours with the "others" can pay off. Collaborative advantage is real insofar as collaboration permits realization of research and management initiatives that otherwise would not be possible, especially those that span boundaries. The start-up costs in terms of the uncertainties can be formidable, but so are the benefits, both in terms of the learning that goes on and the opportunities to make an impact.

For another, making the investment in collaboration—sharing big ideas, taking the time, making the sacrifices—makes future collaboration more likely. Collaborative experience builds "collaborative capital" in terms of know-how and a record of past accomplishments, something funders want to see. If the start-up costs are large, paying those dues brings advantages over those who have not. What is more, collaborative experience constitutes the basis for recommendations by past partners to future partners. Failing all that, conservation threats in developing regions such as the Amazon are getting bigger, not smaller. Building and leveraging collaborative advantage is nearly priceless, if for no other reason than building ties and scaling up impact are necessary to keep up with the growing scale of the threats at play.

All that said, these "official stories" resound with timely opportunities seized, relations of trust established and then vindicated, innovative ideas implemented to good effect, and success stories to show the funders. Were it all so straightforward. The foregoing official stories cover something roughly on the order of 10 to 20 percent of what these projects involved in terms of tasks, time, and effort. The remainder of this book takes up the other 80 to 90 percent of the goings-on in collaboration across boundaries. The next chapter pulls back the curtain on the ruckus actually involved, by offering a framework that outlines the issues, tasks, challenges, and strategic practices for managing collaboration itself.

Chapter Three

The Challenges of Collaboration

If the previous chapter celebrated collaborative advantage, this chapter provides something of a reality check. For there is not just collaborative advantage, but also its nemesis, "collaborative inertia" (Huxham and Vangen 2005). Whereas collaborative advantage focuses on the benefits of collaboration, collaborative inertia results from failures to take seriously the many exigencies involved. Collaboration encompasses numerous tasks, and they are time-consuming to manage. If complementarities among partners can yield collaborative advantage, and if crossing boundaries magnifies those complementarities and thus the collaborative advantage, the same applies to the challenges at play and the potential for collaborative inertia.

This chapter therefore pursues a review of the tasks at hand in collaboration. Previous work on collaboration offers a variety of different task lists (Austin 2000; Gray 1989; Huxham and Vangen 2005). Here I employ a framework organized around groups of tasks I have found to be particularly salient (Perz, et al. 2010). I see the groups of tasks as being more or less independent of each other, in the sense that they must all be managed effectively to avoid collaborative inertia. At the same time, failure to address one set of tasks can indeed affect one's ability to address another.

I begin with a brief overview of the issues that structure the framework. Each issue entrains a family of specific tasks, with particular challenges, which may be addressed to varying degrees via certain strategic practices. The remainder of the chapter therefore devotes a section to each issue. Every section then spells out characteristic examples of tasks, and then discusses the challenges that are likely to arise to make the tasks more difficult to pursue. Each section therefore also devotes attention to strategic practices intended to help manage the challenges and complete the tasks in order to realize collaborative advantage.

Table 3.1. Issues, Tasks, Challenges, and Strategic Practices in Collaboration.

Issue	*Tasks (WHAT to do)*	*Challenges (WHY this can be hard)*	*Strategic Practices (HOW the tasks can get done)*
Introductions and communication	Sharing information, social learning, arrival at common terms, recognition of assumptions	Self-interest (sharing information can undermine advantage); assumptions are often cultural (implicit) or ideological (non-negotiable)	Engage in organizational and professional "courtships" to gain trust; net brokers can help to translate and negotiate
Inequalities and politics	Identification of shared histories, recognition of inequalities among partners, acknowledgment and evaluation of risks in the political context	More powerful partners seek to retain their hegemony; less powerful partners may make radical proposals; both may initially pay lip service to equality but later seek to renegotiate; partners may have complicated internal politics; external political forces may impede or threaten collaborative endeavors; political changes may modify partner relationships	Discuss shared histories (whether conflictive or not); spell out requirements for effective partnering among unequal partners (including non-negotiable issues set by weaker partners); call out assumptions and stereotypical perceptions; adopt non-aligned policies with respect to internal conflicts in other partner organizations; invest in capacity building in weaker partners; maintain regular contacts with authorities to share information; non-aligned policy toward authorities
Common interests and complementary capacities	In light of conservation threats, identify shared interests and complementary experiences and expertise	Statements of common interests may be vague and not clearly related to goals and objectives; complementarities may exist but there may remain shared limitations in some capacities	Critically evaluate claims in light of threats as opportunities for collaboration; critically compare options for responding to threats; identify gaps in pre-existing responses, especially by partner organizations; identify capacities in one partner that can be leveraged by another; note the consequences of not collaborating
Goals and objectives	Relate interests to specific goals, goals to particular objectives, objectives to activities, activities to outputs and outcomes, define targets for each, and set deadlines	Stating goals and objectives requires multiple iterations; partners may not agree on goals despite shared interests; partners may not relate goals to objectives the same way; inequalities among partners	Require joint learning in the process of developing logical frameworks; design logical frameworks to leverage complementary partner capacities; sequence activities in a logical order and clarify which partners contribute in each step of the process; incorporate capacity

Issue	Tasks (WHAT to do)	Challenges (WHY this can be hard)	Strategic Practices (HOW the tasks can get done)
		lead to dominance in articulating goals and objectives	building, especially of weaker partners
Leadership, governance and management	Constitute management structure; define leadership roles; identify key specialist roles (accounting, reporting, etc.); define responsibilities and allocate staff to serve support functions; secure support of organizational representatives; establish agreements	Partners may have differing models of leadership; some partners may not agree to or may refuse to engage in democratic governance practices; leadership may not understand or refuse to follow advice from technical specialists; partner staff may not uphold responsibilities; partners may resent or fail to uphold requirements of agreements	Servant leadership; decentralize and distribute leadership responsibilities (as via manager and board); clearly identify collaborating individuals and responsible authorities for each partner organization; define management processes for resolution of poor performance and internal conflicts; set clear deadlines
Expectations and benefits	Discuss expectations of and for partners; differentiate among types of benefits (monetary and non-monetary, short- and long-term, etc.); clarify benefits among partners	Spelling out expectations is often inadequate as to the details, which are often crucial; the relationship between compliance with responsibilities and benefits on offer is often insufficiently clear	Explicit sharing of benefits is mandatory (including salaries); workshops on responsibilities and benefits are crucial; relate short- and long-term benefits explicitly (the first should contribute to the second)
Surprises and uncertainty	Identify potential external and internal risks; establish protocols for managing internal risks; engage in scenario planning to manage external risks	Larger networks have more sources of uncertainty; rapidly changing regions have greater uncertainties	Explicitly recognize differences in organizational flexibility; identify capacities and other characteristics lacking among current partners to focus the search for additional partners; support net brokers in expanding networks; seek redundancies among partner organization capacities for network resilience; go beyond scenario planning to "muddling through" if necessary; respond to regime changes with communication

Along the way, I offer observations that foreshadow chapters later in the book. In that regard, this chapter and its framework serve as the organizing matrix for the discussions in subsequent chapters. Whereas this chapter will offer discussion of the issues, tasks, and challenges at play, I conclude by noting how crossing boundaries presents further complications in collaborative endeavors, especially if multiple boundaries are being crossed at the same time. This sets up Chapter 4, which discusses the travails of crossing various types of boundaries. I also raise issues about structures for collaboration, which are featured in Chapter 5. A key issue I highlight in this chapter concerns the complications stemming from inequalities and politics among partners, a theme which gets further play in Chapter 6. The same applies to the challenges of uncertainty and surprise, to which I return in Chapter 7. Near the end of the chapter I raise the issue of collaboration as it relates to the individual, which sets the stage for a discussion of collaboration and the self in Chapter 8.

ISSUES OF COLLABORATION IN PRACTICE: AN OVERVIEW

Collaboration is a means to various ends. Nonetheless, collaboration is an enterprise in its own right. This is because collaboration requires sustained attention to a set of issues which cannot be ignored if one expects to realize collaborative advantage and thereby achieve a goal. Whether seen as a celebration of what is possible by joining forces, or alternatively drudgery and donkeywork, collaboration involves an array of issues that entrain tasks with challenges. They in turn have prompted development of strategic practices in order to make headway. If the tasks and challenges of the collaborative undertaking seem daunting, it is worth keeping in mind the larger goals of the collaborative enterprise, and to recognize that there are established paths to follow.

Table 3.1 provides a framework in the format of a matrix which presents the key issues (first column), along with selected key tasks they entrain (second column), some of the challenges likely to come up along the way (third column), and strategic practices for managing the tasks and their challenges (fourth column). The issues provide the design for the table and orient us to a list of "matters that one should not neglect." The issues are unpacked in specific tasks which spell out "what needs to be done." The challenges include items that help account for "why the tasks can be difficult" if not managed attentively. And the strategic practices provide some "knowhow for managing the challenges." Together, these elements provide a roadmap—though not a guarantee of success—in the labor that is collaboration.

There are in total seven issues: 1) introductions and communication, 2) inequalities and politics, 3) common interests and complementary capacities,

4) goals and objectives, 5) leadership, governance, and management, 6) expectations and benefits, and 7) surprises and uncertainty. I suggest that each issue requires attention, or the benefits of collaborative advantage will be threatened. The issues are organized more or less in the order in which they are likely to arise in the process of initiating and sustaining collaborative endeavors. That said, issues tend to accumulate as collaboration advances. This is especially likely among greenhorn collaborators who do not thoroughly address an issue, leading to complications later. Issues also accumulate if one cannot be resolved via consensus, leaving a partner unhappy and a conflict to stew. This implies that addressing issue #1 does not mean that issue #1 will not require attention again in the future. On the contrary, issue #1 is likely to arise first, and it is likely to come up again when addressing issue #2. Further, when pursuing issue #3, issues #1 and #2 may crop up again. And when the meeting focuses on issue #4, issues #1, #2, and #3 are liable to become part of the conversation. The pattern continues to issue #7. The key implication is to honestly and thoroughly address the issues, but to expect them to arise again. Hopefully they will be less difficult to handle the second, fourth, and ninth times around.

Some issues could be unpacked further, notably goals and objectives as well as leadership, governance, and management. Both involve highly structured, relatively formal activities that operate through established protocols in organizations. Precisely because of those shared characteristics, I have lumped sets of activities together under those issues. This is not to belittle or simplify them, but rather to recognize that the means for addressing them are more or less similar. Further, I lump together issues on which much has already been written in order to highlight other issues that need more attention, particularly those that become especially complicated when crossing boundaries.

The first issue in Table 3.1 concerns introductions and communication. This is an issue that has often been viewed in narrow contractual terms. I agree that contracts serve to clarify formal relationships, which is very important, but I also view contracts as formalizations of pre-existing relationships that have been established over some period of time. This makes the issue of introductions and communication crucial beyond contractual documents. Collaborative relationships may start up quickly, whether out of urgency or happy circumstance when individuals or organizations suddenly realize they complement each other. In any event, introductions are an important moment because they frame the collaborative relationship. Eminently important in the process of introductions is effective communication. It is all too easy to misinterpret key terms or the cultural content of specific phrases and their underlying meaning. Introductions and communication stand the best chance of initiating and consolidating a viable collaborative relationship if there is the opportunity for a dry run, that is, a shared experience involving some unforeseen travail without gigantic consequences. This could be as brief as a shared task at a workshop; it need not be complicated. But if unexpected complications arise, each partner has to respond, and they both begin to learn together, which is the germ of a collaborative relationship. Without that shared experience, without that encounter with the inconvenient, it is hard to gauge how robust the commitment to collaboration will be.

Second in Table 3.1 is the issue of inequalities and politics. Previous formulations, as in the organizational behavior literature, often frame this question in terms of power (e.g., Ott, et al. 2003). That is a useful step, but the dynamics spelled out too often come across as ephemeral tiffs among organizations or departments. Hence, I reframe the questions surrounding conflicts in terms of more durable and difficult questions of inequalities and politics. I invoke inequalities to call attention to the differences in organizational knowledge, experience, and capacities. Whereas collaborative advantage highlights in positive terms the complementarities among partners, those complementarities may also involve inequalities that can in turn hinder the spirit of collaboration, where there is often the presumption that everyone contributes and benefits equally. Beyond inequalities, politics may refer to the external context in which a collaborative effort exists or to internal matters pertaining to relationships among collaborators themselves. External politics can threaten the viability of collaboration. Joint work toward a goal may be valuable but that may be precisely due to the fact that certain powers-that-be benefit from controlling access to information or other resources. Insofar as collaboration involves the exchange of information and other resources, collaboration becomes a political act, and potentially subjects collaborators to various forms of external interventions. Further, internal politics may arise among collaborators, particularly unequal collaborators. Disagreements over decisions and benefits therefrom can result in durable conflicts. Here effective communication is golden. Otherwise, conflicts can become personalized, partners become alienated, and the feasibility of the collaborative venture will be threatened.

The third issue concerns common interests and complementary capacities. Collaborative advantage depends on both. However, it is naïve to suppose that potential partners automatically perceive either, or that they will agree as to what the shared interests and complementary capacities actually are. This requires time and effort to frame a problem of shared interest and then craft an approach that leverages complementarities in the capacities of the partners at hand. Framing of a problem is indeed a matter of careful semantics, but there is more. Agreeing on a shared interest also requires attention to the past histories of the partners, which partners often read as trajectories in which they situate the collaborative initiative being contemplated. Collaborators want to be seen as addressing a problem that is framed in a way so their contribution is clear. Identifying common interests and complementarities begins to reveal the accumulative nature of the issues at play. This requires prior attention not only to adequate introductions and effective communication, but also some navigation of inequalities among partners and recognition of the potential politics involved. The issue of complementary capacities can morph into the problem of unequal capacities, which can influence the effectiveness of communication of what each collaborator will contribute. Just because collaborators have complementarities does not mean that the unique characteristics of a partner are what it wishes to contribute, or what the other partners desire.

The fourth issue concerns establishing goals and selecting objectives. Logical frameworks (log frames) for projects possess hierarchies driven by goals which define objectives that entrain activities with indicators and targets (Coleman 1987; Dale 2003; Stem, et al. 2005). Clarity on goals and objectives is, however, rather more complicated in collaborative efforts. If there is excitement about the possibilities during the moments of introductions and identifying shared interests, it is daunting to bring aspirations down to earth and outline the goals, objectives, and so forth in very specific, concrete form. It is at this point that commitments have to be made, as time frames must be set and targets established. As with framing problems and approaches, the earlier issues of communication, inequalities, and capacities can arise again when identifying goals and setting objectives. Nonetheless, collaborations require the level of clarity demanded in log frames if partnering is to be productive: setting goals and defining objectives thus constitutes another necessary suite of tasks for successful collaboration.

Leadership, governance, and management together constitute a pre-eminent issue in discussions of collaboration. There has to be a structure in place with individuals assigned particular roles and responsibilities, which are attached to functions related to the stated goals of the partnership (e.g., Locke 2000; Ott, et al. 2003). This involves not a little attention to design. Any collaborative relationship requires delineating some form of leadership and identification of the leaders, which together constitute a chain of command. The issue of governance refers to the clarification of the responsibilities of leaders and their relationships with other personnel. Key in governance is the issue of accountability among leaders and personnel. Beyond those matters are the administrative and managerial tasks of relating the stated goals to specific responsibilities and thus necessary functions of the collaborative partnership. Collaborators must design the administrative machinery for implementation, indicating the particular technical specialists needed (accountants, evaluators, scientists, etc.) and lining up qualified personnel to fill those positions. This is by definition a process that requires coordination, even within an organization; among collaborating organizations, the effort required for effective coordination rises because the number of relationships involved is greater. In this regard, it is key to point out that collaborations involve action on multiple scales. Individuals initiate and take responsibility for specific functions in collaborative activities vis-à-vis the other partner organizations, but those individuals also operate from institutional perches within their home organizations, where they also have specific roles. Hence, collaborations require very careful analysis of the structure of relationships among individuals as well as within and between the participating organizations. The individuals involved must have some modicum of legitimacy not only to their collaborating counterparts, but also to their home institutions. Figuring out the coordination structure requires attention to both levels of scale. Of course, in organizing larger collaborative efforts, a third level of scale increasingly becomes important to recognize, such as networks. During the design of collaborative structures, issues discussed earlier again have the potential to arise, whether due to miscommunication, inequalities among partners, clarity about common interests, and questions about shared goals. Sorting out leadership, governance, and management among collaborators thus becomes a balancing act that requires time and care. There are multiple types of structures for collaboration, each with advantages and disadvantages, which constitute options for the design of how collaboration will be implemented.

At some point along the way, there is a need to address the issue of expectations and benefits. Defining goals and setting targets of course spells out expectations in terms of products to be delivered and deadlines for delivery. Similarly, designing the coordination structure for the management team requires explicit delineation of responsibilities and functions. But at the levels of individuals, organizations, and even the network, there is a need to explicate not only who is required to do what, but who gets what. Partners frequently come to collaborations with contrasting expectations about what they are willing to contribute and what they expect to receive in return. Leaving implicit the issue of expectations and benefits is to light a powder keg with a fuse of unknown length, which could detonate at an inopportune moment. I have placed the issue of expectations and benefits near the bottom of Table 3.1 because working on many other issues may in turn bring up this issue. Benefits may be understood in different ways, and the differences are consequential. Failure to clarify the specifics of benefits up front may not only lead to premature termination of collaboration, but also foreclose on collaborative opportunities in the future. Partners who can negotiate early on, or at least work through the disappointment of not having aspirations realized in a single collaborative effort, are more likely to stick together and leverage initial experiences into bigger endeavors. Managing expectations and benefits may reduce benefits initially, but yield greater benefits over time.

The seventh and final issue concerns surprises and uncertainties. Like expectations and benefits, this issue occurs late in Table 3.1 because many other issues can catalyze it. Surprises and uncertainty may stem from unforeseen external events, or from changes in collaborators themselves. Either way, surprises create uncertainties that complicate collaboration. There are indeed management protocols for dealing with surprises, but they tend to presume a certain level of ambient uncertainty. My experience, based on collaboration in a rapidly changing developing region, is that orthodox management models such as rational planning are insufficient in the face of a high level of uncertainty. When the pace of change is high and there are many sources of perturbations, they form something of a bewildering cascade, necessitating alternative management models. This is not to say that planning is not important, but instead that the rate at which surprises occur, and thus the ambient level of uncertainty, can have an enormous effect on all aspects of collaborative endeavors. At the same time, merely being in a collaborative network can be useful for rapidly distributing information and procuring means of adaptation to quickly changing circumstances. The old adage that crises are also opportunities is useful to bear in mind when some unanticipated catastrophe begins to unfold again this week.

The remainder of this chapter takes up each of the issues in Table 3.1. I unpack each issue by offering a review of the tasks and challenges they entrain. I also discuss strategic practices for addressing the tasks and their challenges. Each discussion is necessarily suggestive, for the details will depend on the cultural and institutional context. I call attention to elements of each issue that have in my experience at least been particularly salient, enduring, or difficult to manage. This carries us into the skunkworks of the labor that is collaboration. My goal with the tour that follows is not to deter prospective collaborators; on the contrary, it is intended to offer some helpful guidelines. For experienced practitioners, hopefully the tour will be confirmatory and prompt further comment on good collaborative practice.

INTRODUCTIONS AND COMMUNICATION

Introductions set the tone for collaborations, and that tone is set in large part based on communication practices. In a sense, the grand challenge for all would-be collaborators is to figure out what one knows—and more importantly, what one doesn't know—with regard to the prospective activity at hand, and to clearly impart both to potential partners. This is easier said than done. Learning to conduct helpful introductions and establish effective communication takes practice.

I unpack the issue of introductions and communication into a suite of four interrelated tasks, all of which come into play in any meeting of the prospective partners: 1) sharing information, 2) engaging in social learning, 3) arriving at the use of common terms, and 4) making assumptions explicit. Each item might seem obvious, and the list could be longer. But each is crucial and has its subtleties.

Sharing information is definitely the simplest of the introduction/communication tasks. Prospective partners provide each other publicly available information. They dialogue about that information, moving into experience and discovering more about each other's capacities. This is straightforward in most cases. Eventually, though, the discussion delves deeper into more sensitive information, such as the personalities involved, or past histories that may include conflict. This is where the process is likely to slow down; collaborators have to listen to each other and in the process set a mutually agreeable pace.

Social learning moves beyond mere sharing of information. Now the focus is on activities that involve incipient collaboration via shared experiences from which both partners are learning together. This can be joint development of a proposal for funding, an initial field visit, or other possibilities. The key is that the partners are both involved and interacting throughout the process. Social learning stems from common experiences on which all partners can reflect. In other words, the learning occurs not just due to the experience itself but from observing how the other partners respond and reflect. Partners with shared experiences learn about themselves by expressing themselves to each other and by listening (cf. Goffman 1959; Parker, et al. 1995). Ideally, they respond positively to each other by adjusting their goals and outlook on the shared experience in a way that aligns one partner with the other in a focus on common goals.

But in order to communicate effectively, partners must also learn to make use of terms each other understands. It is not enough to use the same words; there must also be a common understanding of what those words mean. This is not to say that partners must necessarily agree on meanings, but it does mean they have to be aware of any differences. This is a remarkably tricky business, as broad terms like "conservation" and "development" and "management" have considerable content. What is more, if partners have even minor differences as to the content, they have to find ways to speak with one meaning when interacting with other groups. Depending on how important meanings are to different partners, and how consequential any differences might be for pursuit of common goals, failure to arrive at a common terminology can become a nagging problem.

The most difficult task in communication involves the recognition of assumptions. Everyone carries assumptions around in their heads about how the world works, too many to spend time examining them all with any frequency or much depth. But collaboration requires such examination for effective communication, at least for assumptions that have ramifications for shared interests. Discussing assumptions is not easy, as it often takes substantial time to spell out what the assumption concerns, give it a name, outline what its consequences might be, and contrast it with alternative assumptions about the topic. I find that such discussions often irritate participants, who may not understand the point, or may not feel comfortable discussing the assumption in question, or may otherwise find the discussion to be a waste of time. Nonetheless, assumptions reveal much about how we think, and they render collaborators more comprehensible to each other. Once discussions about assumptions get under way, a key question then concerns how much discussion of assumptions is too much. Too little discussion may result in numerous little side discussions about assumptions which then fail to be vetted by the partners together. Too much discussion can slow down the progress of the partnership when it needs to focus on other priorities.

In the process of pursuing the tasks of introductions and communication, particular types of challenges are likely to surface. One involves self-interested behavior. Self-interest is the great nemesis to collaboration. Self-interested social actors by definition act in their own interest, which can potentially undermine a collaborative effort (e.g., Olson 1965; Ostrom 2000). Self-interest has long been associated with rationality in utilitarian behavioral models, on a strong assumption that social actors behaving out of self-interest yield optimal outcomes for all. This is not the place to delve into the minutiae of the long-standing theoretical debate stemming from such behavioral models (cf. Mansfield 1990; Sen 1977). In the context of collaboration, acting out of self-interest presents a challenge because it can hinder transparency in communication. Potential partners may withhold important information about themselves, communicate only when they see advantage for themselves, and so on. For present purposes, it is useful to note the distinction between unenlightened self-interest, described above, and enlightened self-interest, where multiple social actors identify means of action whereby they both benefit (cf. Keim 1978; Mansfield 1990). This is the requirement: a given social actor benefits, but so does the collaborator, on terms they value.

Additional challenges stem from the nature of assumptions. Two in particular often arise: assumptions may be implicitly held, or they may be explicit and thus ideological. Implicit assumptions are often cultural assumptions stemming from lifelong socialization as to how people are supposed to treat each other. There is of course a long list of problematic cultural assumptions, many of them tied to sources of social difference, including but hardly limited to racism, sexism, nationalism, and so on. They can all undermine collaboration in many ways, but because they are often implicitly held, it takes time and care to dredge them up for critical examination without alienating potential partners. By contrast, partners may make assumptions for very explicit reasons based on ideological commitments, often held for political reasons. The challenge when dealing with ideological assumptions is that they are often very strongly held, and those who hold them have prepared their defensive arguments to block discussion and prevent critical examination.

To deal with the tasks and challenges of introductions and communication, I suggest two strategic practices. The list could be longer, but I find that the notion of organizational and individual "courtships" and the role of "net brokers" to be crucial. Collaborative relationships arise in a variety of ways, often just by happenstance, and they may emerge slowly or quickly. In any event, it is useful to apply the model of a courtship to point out its value for managing those introductions and early interactions. Merely invoking the term "courtship" immediately highlights the provisional nature of an emerging collaborative relationship. It also calls to mind the demands and expectations at play, given that the courtship could lead to consummation in the form of binding legal and financial commitments. Courtships involve not only individuals, but larger social aggregates—in this case, the home institutions of the individuals courting collaboration—who will likely have some say in what is feasible. Courtships thus require extensive multilateral consultations among the courting collaborators and with the various gatekeepers and authorities in their home institutions. Such consultations tend to keep people honest as there is something of a structural enforcement mechanism as courting collaborators have to navigate their mutual aspirations for what might be in the context of the mandates and regulations of their organizations. Courtships require information sharing, social learning, common terms, and examination of assumptions. In my experience, it becomes very hard very quickly to remain opaque in communication during collaborative courtships. Further, as the potential benefits of collaboration become clearer, it is motivating to actively work to improve transparency in communication. This is not only to sort out potential stumbling blocks like assumptions but also to explain and justify the collaborative endeavor to the home institution—one's own and that of courted collaborators. Indeed, the collaborators can be powerful positive voices when speaking to one's home institution about the proposed benefits of the collaboration in question. Especially when organizations say they want to collaborate, when collaboration brings benefits and prestige, it quickly becomes hard to say no.

Another key strategic practice stems from the crucial role of network brokers (Manring 2007; Miles and Snow 1986). Network brokers are individuals who are particularly well-connected and/or talented at identifying collaborative opportunities due to their ability to recognize the potential advantages. Network brokers are agile at envisioning collaborative relationships and then facilitating the courtship and supporting steps toward consolidation. Collaborative courtships facilitated by network brokers are in effect chaperoned courtships. I hasten to note here that the chaperoning of potential collaborators focuses primarily on identification of the potential benefits, so the courting partners stay interested in each other. Of course, this assumes there is a network broker available for the job. My experience is that necessity becomes the mother of the network broker, and when the need becomes obvious or desperate enough, somebody in a leadership position with a high-level institutional perch—that is, somebody with a good view of the institutional landscape—will begin brokering activities. That said, it also helps to have more network brokers, including institutionally autonomous network brokers (consultants).

Beyond supporting courtships, network brokers can provide additional services relevant to addressing the challenges of communication. Network brokers may not only operate as matchmakers and midwives for collaborative endeavors, they also serve as diplomats, translators, and negotiators during discussions among courting collaborators. In particular, to the extent that network brokers are autonomous social actors, they exhibit considerable moral authority in questioning the arguments, motives, and assumptions of the potential partners. In this regard, network brokers become useful guidance counselors or peer reviewers. Brokers thus watch for signs of unenlightened self-interest and other communication snafus. They constitute guides whom the partners can accept or reject, but whom they would do well not to ignore. The key to the moral authority of network brokers, however, has more to do with the larger issues at play that are driving interest in the collaboration in the first place. Effective network brokers are by definition not self-interested, and generally have no direct interest in seeing the success of the proposed collaboration. This makes the questions of network brokers—about proposed collaborative relationships, assumptions, goals, and all other issues—a useful reality check, including when the collaborators themselves get in a hurry and don't reflect with sufficient care on their enterprise.

INEQUALITIES AND POLITICS

As noted earlier, complementarities among partners often come with inequalities. Inequalities create the risk that collaborative relationships will become politicized. As a result, collaborative relationships can become venomous, and mistrust may affect all aspects of relationships between partners. Having partners with politicized inequalities is like playing with fire during a drought: once set, the fire will spread quickly, burning everything in its path. Inequalities in partnerships thus have to be openly managed from early on lest they become dangerous to the entire collaborative enterprise.

At the same time, the external context of a collaborative relationship includes a suite of vested interests and powers-that-be. For various reasons, political interests may take exception to the goals of collaboration and thus a partnership itself. I have found this to be more likely the case for policy initiatives, as when a territorial management planning process led by non-governmental partners is interpreted as a political act against a government. Of course, even science can be politicized, especially when a collaborative effort takes up a politically sensitive topic like questioning the wisdom of a public policy or documenting the lack of implementation of development projects or enforcement of environmental regulations. Further, seemingly innocuous research activities can become delicate matters in election years if they involve funding to a partner with known political leanings.

In my experience, both internal and external politics arise repeatedly in collaborative enterprises, and they pervade discussions about interests, goals, and everything that follows. I therefore use this section to introduce the tasks entrained by inequalities and politics, the challenges that arise, and a number of strategic practices for managing this issue. The discussion that follows treats inequalities and politics in general terms for collaborative relationships. However, when crossing boundaries, matters of inequalities and politics intensify, so I return to these questions in Chapter 7.

With regard to inequalities among partners, there are two basic tasks at hand: 1) reflect on shared histories, and 2) recognize inequalities. The task of sharing histories acknowledges that prospective partners may in fact have worked together before, for better or worse. Shared histories can provide a strong foundation for later collaborations, particularly if previous collaborations yielded concrete collaborative advantages. But shared histories can just as easily be unpleasant—or have yielded collaborative advantages despite conflicts. Indeed, it is worth recognizing that many collaborations are marriages of convenience where the partners are being asked (or told) to work together, with some ambiguous modicum of choice.

Recognition of inequalities is crucial. Insofar as certain capacities or experiences are particularly prestigious, organizations who possess them in greater measure may dominate other partners, regardless of the advantages of those partners. It is in these instances that complementary capacities are easily reduced to inequalities, which creates the potential for dominance rather than collaboration. Recognition of inequalities thus demands that all advantages of different partners be acknowledged and valued. While some capacities or knowledge funds may be more prestigious, the notion behind collaboration remains that all assets are nonetheless necessary to the success of the collaborative venture. A key task then is to recognize inequalities by calling attention to the importance of various complementary capacities and to call out arguments to privilege one asset—and the partner who possesses it—over others. Prospective collaborators all need to frankly spell out where they are strong and where they are lacking.

The political context embodies potential risks to the collaborative endeavor, so an additional task for partners is to acknowledge and evaluate said risks. This necessitates some form of structural analysis of who the external political players are, what interests they have, pre-existing relationships with the collaborators, and how the collaboration might compete with or contravene vested interests. The ability to pursue this analysis in turn depends on pre-existing communication with the powers-that-be. Usually the lack of ties is itself problematic, as that often coincides with a lack of communication and thus more limited direct information. Similarly, the ability to properly analyze such risks is reflective of the wisdom in selection of partners. While any such evaluation is hardly going to be perfect, as interests and policies can be hidden and are subject to change, this analysis is necessary.

Inequalities and politics present challenges for management of collaboration. With regard to inequalities among partners, the most obvious challenge that arises is when the stronger partner engages in various behaviors to assert its dominance of the collaboration. Perhaps that organization has a strategically important asset, or is accustomed to being the dominant partner. At the individual level, similar issues will arise, typically based on highest degree attained, seniority, gender, race/ethnicity, and so on. The task of acknowledging inequalities can quickly become uncomfortable for collaborators accustomed to dominating conversations instead of listening. Indeed, the process can be even more problematic for everybody else. If dominant collaborators respond unpleasantly, collaborators outside the category of "old white educated male from a big organization based in a powerful country" may find it very hard to speak up. All of this is anathema to discussions in the spirit of "everyone has something to contribute" and "all contributions are necessary to collaborative advantage."

Another key challenge is that individuals or organizations may make radical proposals to overturn established hierarchies of authority, as by arguing that they should instead be the dominants. This occurs out of frustration from having occupied for too long a subordinate position. While there is a strong justice argument for such proposals, the threat of problems from dominance remains the same: collaborative relationships among unequals become politicized, undermining collaboration itself. The challenge remains the same: dominance by any partner is problematic for effective collaboration.

A third challenge to managing inequalities among collaborators is that they may initially pay lip service to the spirit of collaboration where everyone's contributions are valued, but then later seek to renegotiate by exercising dominance. Some partners regard the initial meetings as formalities where pleasantries are to be exchanged, rather than serving as the actual onset of negotiations. Another version of this challenge is that the partners are in fact expressing authentic goodwill from the get-go, only to find later that one or more of the partners cannot keep up their end of the bargain. This circumstance then prompts other partners to assert the right to exercise greater dominance in the collaborative relationship. In any event, the issue of delayed manifestations of dominance is a key reason that the politics of inequalities among collaborators can repeatedly arise during the course of a collaborative relationship.

A final challenge concerning inequalities among partners concerns their own internal politics. Some partner organizations run smoothly; others are themselves political arenas, riven with conflict. When those organizations have individuals who seek other partner organizations for collaboration, the internal politics of the individual's home organization can affect the collaboration. Beyond the obvious observation that a dysfunctional organization may be less effective in getting things done, it is worth noting that political in-fighting in one organization can lead to criticism of the individual collaborating as well as of the partner organizations. This depends on how toxic the internal politics are.

Of course, the external political context presents its own challenges. Acknowledging and evaluating the political context may well reveal significant political risks to the collaborative endeavor or even the collaborators themselves. For the time being, I set aside the prospect of extreme instances such as imprisonment or violence. This nonetheless leaves open the possibility of various impediments in the form of public criticism or institutional pressures, such as via future funding or job prospects for collaborators, or restrictions on working in certain places (such as protected areas) or with certain groups (such as rural communities or indigenous peoples). Such measures can greatly hinder the viability and value of collaborative relationships.

Further, changes in the political context can modify political pressure on collaborators. Election years, big changes in the economy (particularly as they affect prices on goods in the region where collaboration occurs), and moments of major announcements for corporate investments are all potentially crucial moments. I emphasize that the changes can either increase or decrease political pressure on collaborators. In either event, there are likely to be significant implications in plans for joint action among collaborators. Increased impediments of course make joint actions harder, whereas reduced pressure can facilitate collaboration, though there is also the possibility that a partner may be sought by powerful players for a different collaborative activity. For example, hydrocarbon prospecting in the southwestern Amazon has motivated oil and gas companies to contract anthropologists to work with local communities to facilitate access for drilling.

The issue of inequalities and politics should not be underestimated. Chances are, the more important the conservation and development questions at hand, and thus the higher the stakes, the more difficult the tasks and challenges described above will be. I therefore outline a series of strategic practices with the proposition of offering an array of tools for managing inequalities and politics. As always, practices are not guarantees. I think of them not as conflict resolution, which I regard as a naïve conception of this issue. Instead, the practices that follow amount to elements of conflict management (cf. Burton and Drakes 1990), on the understanding that putting out the brush fire now doesn't mean it won't reignite again.

To that end, I highlight a few strategic practices I have found particularly useful, notably at the outset of collaborative relationships, anticipating a longer discussion in Chapter 7. In cases of prospective collaboration between past combatants, time must be dedicated to the open venting of grievances. The potential list of grievances can be long and the content of said grievances diverse: who insisted on being in charge, who ignored concerns of the other partner, who wouldn't reimburse those expenses that weren't foreseen, who said that unfortunate phrase with that particular colloquialism, and so on. A network broker or other moderator is often required for this exercise, as it must come to a conclusion involving changes in practices if the future collaboration being contemplated is going to work any less badly than that which came before.

Much the same applies to cases of highly unequal or incipiently contentious partners that do not share a common history. Here network brokers can serve the general role of discussion moderator, and as need be the more specific role of conflict mediator. For one thing, they can oversee an exercise wherein prospective partners explicitly outline what they see as their unique capacities in the context of the collaboration. This permits open discussion of what the capacities are, whether they are unique, and how they are related to the proposed collaborative effort. To the extent that this merely leads to debates and disagreements, moderators can turn the discussion instead in the direction of spelling out requirements for effective partnering. This puts the onus on the partners themselves to articulate practices to avoid dominance and value complementarities. This also opens the door to non-negotiable issues—for the weaker partners—as a means of offsetting inequalities (cf. Edmunds and Wollenberg 2002).

The other function of network brokers in this context is to serve as a moral authority, especially if discussions remain conflictive and partners are not moving toward some type of agreement. In particular, moral authorities can "call out" partners—especially the more dominant ones—on various assumptions and stereotypes. Assumptions and stereotypes cannot go unchallenged. Moral authorities can be very effective in questioning assertions by picking apart the underlying assumptions, offering examples that run contrary to stereotypes, and so forth. Such tactics can be usefully applied to questioning assertions about who is good at what, who should be in charge of something, whether a responsibility should be shared, and who stands to learn from whom. Questioning assertions is instrumental not only to the venting of grievances but also to rethinking of assumptions, both of which facilitate negotiations in order to collaborate. The key is to make sure any grievances are not left simmering. Simmering pots eventually boil.

The challenge of conflictive politics inside a partner organization is thorny. Collaborators are in no position to be scolding a partner organization, and yet there may be politicized social actors in that organization who are prepared to criticize the collaborators. Choosing who to work with can be a luxury if there is one interested, qualified person who fits the bill for the collaborative effort. Beyond the obvious step of making the benefit-cost calculation of whether the collaborative relationship is worth keeping, there is the strategic practice of maintaining a "non-aligned policy" with regard to the various factions within a partner organization that exhibits internal contention. Another strategic practice is to maintain regular ties with the authorities of the partner organization, regardless of who is in charge, so there is transparency as to the intent of the collaborative endeavor. These practices serve to incrementally isolate the less careful of the contentious parties as they make accusations, which are then apprehended as false or twisted by key parties with concrete information to the contrary.

Similar practices apply in the broader political context, though in many cases the ability of collaborators to engage governments or private vested interests may be more limited. Making the effort to proactively provide information, respond to questions in a timely fashion, and otherwise show respect all support transparency. This is effective if the issue is a lack of communication or misunderstanding. If there is instead a conflict and political hostility, they become evident quickly. Here it is worth noting the value of maintaining a non-aligned policy toward the powers-that-be, especially when collaborating across national boundaries. If the current administration is hostile, elections often lead to changes. Consequently, there is a tricky balance to strike as one weighs the current political context and future prospects. It is also necessary to admit that not all situations can be managed in the sense that in cases of hostility or violence, collaboration may well become impossible.

COMMON INTERESTS AND COMPLEMENTARY CAPACITIES

The third issue concerns the job of jointly defining—clearly—the parameters of the problem of shared concern, and the approach that the collaborators will pursue to address said problem. Whereas defining the problem leads to delineation of the elements of common interests, sorting out the approach to be pursued requires an articulation of the capacities which the collaborators bring to the table. To a great extent, the tasks at hand involve decisions about the selection of key terms and the content of what they denote. However, these tasks become more complicated to the extent that communication is not entirely effective, and inasmuch as inequalities among partners inhibit open discussion of interests and capacities.

In principle, sorting out common interests involves a rational planning process in multiple steps. First, prospective partners engage in some form of free listing of their individual interests and commitments. In effect, this involves defining the problem that motivates action. In the context of conservation and development work, problem definitions typically highlight conservation threats and development opportunities, and seek to identify strategies for sustainable development. Second, the courting collaborators identify items that appear on everyone's list. This may require some discussion as to whether similar-sounding terms are in fact denoting the same thing, and whether partners are using the same term in the same way. In effect, the second step amounts to construction of a Venn diagram that highlights the overlapping interests. Third, participants critically evaluate the common interests to decide if the collaborative activity will encompass them all, or only a subset.

Delineating complementary capacities follows a similar process with precisely the opposite logic. First, prospective partners free list their knowledge funds, experience, and capacities (including ties to other stakeholders). This involves itemizing assets that participants see as instrumental for pursuing shared goals. Second, participants identify differences among the knowledge, experience, and capacities noted. Prospective collaborators can construct matrixes with assets across the columns and partners down the rows, with comments in the cells of the matrix. This then motivates discussion of redundancies as well as complementarities, especially where assets are unique to one partner. Third, participants identify assets seen as crucial to the collaborative effort being contemplated. Whether that means access to information, having a powerful stakeholder ally, or significant administrative capacity, the discussion needs to hash out the assets that will be necessary for the collaboration to function. This step also squares up the assets seen as necessary with those seen as uniquely held by one partner.

In practice, both processes face significant challenges which can prolong the effort to achieve clarity. One challenge is that some partners will make vague statements about their interests, or provide optimistic evaluations of their capacity. Prospective partners come to the table for various reasons, and may avoid honest renditions of their auto-evaluations for fear that they may be left out. Another challenge is that participants may articulate interests or capacities not clearly related to each other. That is, they may identify problems for which they have limited capacities. And third, partners working to identify complementarities may find there are very few or that no one has a necessary asset. Limitations in complementarities may cast doubt on interest in collaboration, but a high degree of redundancy is not necessarily a bad thing, especially if one or more partners overestimate their capacity. The ideal is to have a mix of redundancy and complementarity, as will be discussed further in Chapter 5 on the design of collaborative structures.

Key to addressing the challenges of identifying common interests and complementary capacities is to have a robust critical discussion. A key strategic practice thus involves respectful but insistent critical evaluation of assertions made about interests and capacities in order to make them as clear as possible. This is not because partners necessarily doubt each other, but because it serves to strengthen their collaborative bond by emphasizing transparency as to what the contemplated partnership will be capable of achieving. This transforms risks into incentives. Further, any discussion of interests and assets must therefore clearly define and propose a response to the threats at hand. This translates threats into collaborative opportunities. The dynamic that can then emerge puts the onus on all prospective partners, and not one or another individually. It is likely that if one or another partner tries to exercise dominance, finger-pointing will follow and conflicts will ensue. Conflict management here takes the form of reminding prospective belligerents to focus on shared threats, and thus shared interests, as well as the value of both redundancy and complementarity. Negative energy can be turned positive by keeping the focus on what is held in common (as regards interests) and what is complementary (as regards assets).

Another dimension of the critical analysis at play in discussions of interests and capacities is the evaluation of options for responding to threats. Chances are high that partners will articulate distinct proposals for responses based on their particular organizational strengths. Critically evaluating said proposals is useful to permit broader discussion of various options by identifying the limitations in each case. This in turn allows for comparisons and perhaps incorporation of elements of multiple proposals. This is another opportunity to begin with proposals from individual partners stemming from their respective capacities, and then moving toward identification of complementarities in order to articulate a shared strategy.

Further, chances are that proposals offered by prospective partners are in fact activities they are already implementing, with varying degrees of success. Some partners will advocate for everyone to jump on the bandwagon of an ongoing activity, on the proposition that adding capacity scales up the impact. Critical analysis is useful here on the bet that pre-existing responses nonetheless have their liabilities. Identification of gaps and shortcomings provides the opening for a collaborative discussion for improvements in the pre-existing model, or more ambitious proposals involving complementary activities. Partners having their programs critiqued can of course become defensive, especially if the activity is public policy in an election year or the partner has committed a large proportion of its resources to said activity. The critical discussion thus requires careful communication and conflict management to avoid misunderstandings that can turn into catfights. But negative energy can be turned positive; the critiques need to come with articulations for improvements, especially via complementary capacities.

Here the discussion of redundancies and complementarities becomes important. As noted above, a mix of the two is useful; hence, one element of strategic practice is to highlight and value instances of both. A related strategic practice is to specifically articulate how redundancies and complementarities can be leveraged among partners. Having two partners with experience doing the accounting for a given funder is probably better than one, since the redundancy broadens the available capacity in the partnership for managing the accounting requirements. Conversely, having one partner with significant administrative capacity and another with a strong network of community stakeholders permits scaling up of a given activity.

If the courting collaborators seem unable to cease their squabbling, a final strategic practice concerning interests and capacities is to note the consequences of not collaborating. Where there are indeed shared problems, this usually ends the bickerfest, at least temporarily. The arguments posed tend to stem from attempts by one partner to exhibit dominance or a misunderstanding that leads to bad feelings and fear. There may well be disagreements about specific interests, which generally requires a narrowing of the scope of the problem as defined in order to permit collaboration. This is the heart of the negotiation: collaborate on what is viable, or nothing at all. While there is also the option of horse trading so individual partners get something they individually want, this tends to diffuse the focus and complicate the approach adopted, which can threaten the viability of the collaboration. If the critique of "we're doing too little" is important, it must be weighed against "we can't do more." It is also worth noting that collaborative efforts that start small and stay realistic can later scale up their ambitions by leveraging successful initial experiences.

GOALS AND OBJECTIVES

I once thought that discussions for collaboration began with identification of overarching goals and setting specific objectives. Hopefully by now it is evident that if issues of communication, politics and inequality, and interests and capacities are not addressed first, any discussion of goals and objectives is likely to return the discussion to the earlier issues, or lead to conflicts caused by those issues. Hence, the setting of goals and objectives falls in the middle of my list of seven issues; they must indeed be addressed, but this can only happen if other matters are already being actively managed.

Identification of goals and objectives occurs in the larger context of performance monitoring and evaluation. An overarching goal entrains specific objectives; achieving the objectives contributes to attaining the goal. In turn, an objective implies a suite of activities, such that pursuit of the activities permits achievement of the objective. Evaluation of activities and objectives requires identification of indicators and targets to permit monitoring of progress. Objectives generally imply an outcome where something changes in a way that contributes to the goal; particular activities have specific outputs that contribute to the outcome. And of course, all of this comes with deadlines. Hence, in practice, setting goals and objectives is the beginning of articulating a logical framework that thus delves into the details of activities and indicators, outcomes and outputs, targets and deadlines.

Drilling down into these details is no simple task, but effective collaboration requires this level of clarity and specificity. Without a results-based framework and implementation of monitoring and evaluation, it becomes very difficult to evaluate collaboration, whether for conservation, development, or any other goal. I confess my utter naïvete about this business at the outset of my collaborative experience, and it is prominent on my list of the things that have helped keep the learning curve steep ever since. That said, the business of goals and objectives is also crucial to the assessment of collaboration and its benefits, as well as conservation and development.

Sorting out goals and objectives, and all that they entail in a logical framework, does lead to challenges. Stating goals and objectives amounts to a test of clarity as to the definition of threats and shared interests, and likely will require multiple iterations. This is because statements of goals not only have to satisfy the partners, they also must correspond logically to threats posed and thus shared interests, as well as the approach taken to respond in light of partner capacities. Aligning all of this is not straightforward, as stating vague goals does not necessarily respond to particular threats or make clear the shared interests at hand. Further, stating ambitious goals may look dubious in light of the capacities possessed by the partners.

Even if the threats and opportunities are clear, the goal may nonetheless be complicated due to multiple priorities competing for recognition. Partners with distinct interests may rate various priorities differently. If the threat is illegal gold mining and the problem is regularization of mining in areas where it is legal, but the shared interest focuses on securing protected areas against mining, stating a goal that addresses both priorities will be tricky. In this case, it is key to determine how regularization aligns with management of protected areas. If one partner prioritizes protection over regularization on the argument that regularization provides no guarantee of protection, but another disagrees and argues the contrary, stating the goal becomes difficult. On the other hand, setting aside the regularization issue is problematic, as that is tantamount to ignoring a key dimension of the problem. One way to manage this type of complexity is to set the two priority themes of regularization and protection side by side and then delineate objectives explicitly tied to each.

Partners with clear goals may still encounter difficulties when they seek to articulate the objectives, or delineate the activities, indicators, targets, and deadlines. Partners may not logically relate goals to objectives in the same way, or objectives to activities. At each step in building the logical framework, there is the possibility of more disagreement and debate, as well as the need to go back a step and adjust. The bigger the log frame and the more numerous the partners, the longer this process is likely to take.

Similarly, the more problematic the issues discussed in previous sections, the harder the process of setting goals and objectives will be. In particular, the problem of dominance can greatly distort the process of identifying goals and setting objectives. Attempts at dominance undermine the process, as the dominant partner devotes energy to asserting its hegemony while the others engage in resistance. The result is that setting goals and objectives is hardly collaborative; on the contrary, collaboration is politicized, distracting from the task at hand. The log frame generated is likely to become the focus of subsequent contention.

Lest development of logical frameworks involve interminable travail, there is a series of strategic practices on hand. One is the obvious counsel of consulting with a monitoring and evaluation specialist. Here I focus on what collaborating partners, with or without specialists, can do. The first strategic practice is to engage in joint learning during the process. There is no shortage of resources on log frames, results-based monitoring, and established indicators (e.g., Hockings, et al. 2005; Saterson, et al. 1996). All partners need to go dig this stuff up and share the findings with each other. Learning is collaborative to the extent that all partners dig and share; learning becomes social as the dialogue proceeds. Crucial in this process is not whether all findings are adopted, but rather that the findings come from various partners and that decisions of what to use or adapt are taken jointly. This makes identifying goals and objectives an exercise in collaboration in the same stroke as the partners set joint goals and objectives.

Another strategic practice is to design log frames to specifically leverage partner capacities, especially where they are complementary. Obviously the log frame should reflect the portfolio of partner capacities overall. But if the log frame also highlights distinctive capacities of different partners, the framework itself becomes a document that explicitly records how each partner contributes to the goals and objectives of the partnership. This strategic practice can be especially enlightening to the extent that there is dispute about shared interests and complementary capacities. Identifying goals and objectives in a log frame that specifically leverage complementarities can throw in stark relief where the strategic opportunities lie. This also permits identification of individual commitments and proclivities that do not merit attention in a collaborative enterprise because they can be pursued outside the collaboration. At the same time, highlighting certain capacities for leveraging constitutes a commitment for collaborators, one where they will be evaluated before their partners. In this regard, the log frame also serves as a means of evaluating not only technical performance toward conservation and development goals, but also performance with regard to the collaboration itself.

A useful reality check is to engage in the strategic practice of testing the logic of the framework as to the sequencing of activities. There is a tendency to put the more exciting activities up front or otherwise emphasize them, without ensuring adequate attention to other activities which contribute to key objectives. In other words, even if one partner's unique capacity is up for show, the log frame also needs to include other activities which make contributions also necessary to the achievement of objectives. This often means that other partners will be doing heavy lifting in those other activities. In the process, the log frame can not only unpack goals and objectives in terms of logical sequences of activities and indicators, it can also spell out which partners contribute where in the process. If a given objective with a particular outcome requires the contributions of a suite of activities that yield particular outputs, a well-designed log frame will make clear which partners generate what outputs so that one or more partners will use those outputs to achieve the outcome. Of course, multiple partners may be contributing to a given activity. The point is that it be explicit in the log frame not only the order in which activities happen and when outputs are generated, but also which partners contribute when in the process.

Inequalities hound the process of developing goals and objectives into log frames. It is likely to be unavoidable that some partners will take on more responsibilities than others. A final strategic practice therefore concerns the inclusion of capacity building activities in the log frame as part of the objectives and thus the overarching goal. This not only recognizes the importance of addressing inequalities among partners, it also situates the current goal and objectives in a larger context of strategic investments to increase capacity for future activities. To be sure, there are devils in the details as regards the nature of the capacity building (whether individual or institutional, short or long term, etc.) and its relationship to other activities (whether direct or indirect, synergistic or independent, etc.). The ideal is to systematically pair training activities with implementation activities. Capacity building can of course be more ambitious, as when one partner supports another as the latter takes on tasks for which it has limited capacity but plans to use the collaboration precisely to gain experience. To the extent that the collaboration is robust and the support and commitment are real, this strategy can yield significant gains for the time and effort invested.

LEADERSHIP, GOVERNANCE, AND MANAGEMENT

Assuming goals and objectives have been set, the real fun begins with actual implementation. To that end, there are numerous details involved in establishing, maintaining, and adjusting the structure and personnel of the management team. This brings up several intertwined issues concerning leadership (who is in charge), governance (how strategic decisions get made), and management (how daily operations are organized). For present purposes, I focus on a limited selection of tasks, challenges, and strategic practices I have found to be salient in the context of conservation and development work.

The tasks of leadership, governance, and management can be understood as defining the overarching structure and various functions for the operations of a collaborative relationship. The complexity of these tasks depends on the number and capacities of the partners at hand, and the goals and objectives of the partnership. A key first task is to constitute the management structure. For small partnerships, as with two partner organizations, this is simple: indicate representatives from both partners who meet and jointly take decisions. For larger collaborative structures of five or more partner organizations, this becomes more complicated, and it is useful to distinguish between management and governance in terms of the leadership structure. Large consortia often hire a full-time manager and appoint a governing board. Whatever the structure adopted, there needs to be a commitment to representation of the partners in governance if not also in management, or the problems of dominance arise. The management structure is also likely to reflect the relative institutional capacities of the partners, where administratively stronger partners may take on larger roles in management. The issues of capacity and equity tend to run in the opposite directions here: greater capacity may justify a larger administrative role to ensure competent management, but equity concerns call for investing in weaker partners. In such debates, accountability and transparency become the measures of what is appropriate, regardless of the particular management structure adopted. In the end, the task of defining the management structure becomes one of ensuring that it can be changed if necessary.

Sorting out the management structure also involves the task of defining leadership roles. There can be complementary functions among distinct leaders with particular roles, especially in larger partnerships. As noted above, one arrangement is to differentiate between managers for daily operations and governance boards for strategic decisions. Beyond that, there are various values partners need to identify as priorities for what they want in leaders, particularly managers, as to the importance attached to consultations and meetings before taking decisions. Decisions about particular leadership roles are decisions about the values most important to the collaborators as to their means of operating jointly.

Beyond the question of leadership, a key task for implementation involves the definition of specialist functions and decisions about who fills those technical roles. Specialist functions include particular staff necessary to pursue the conservation and development activities at hand, whether a policy analyst, environmental educator, staff scientist, extension agent, and so forth. Specialist roles also pertain to management of collaboration itself, in the form of meeting moderator, conflict manager, financial officer, monitoring and evaluation specialist, and so on. Some or all of these functions may be necessary; others may be more important. Much of this depends on the goals and objectives of the partnership, as well as partnership size. Identifying the needed technical specialists can be informed by past experience and current difficulties among collaborators. This is a task that deserves care, and it benefits from a commitment to adaptive management, whereby the partners make adjustments in functions and personnel during implementation as needs emerge.

Having defined the specialist functions, there comes the task of staffing those positions. This becomes more complicated when several partners have candidates or when nobody has qualified staff. The first situation can easily lead to competition and perhaps conflict, especially if the collaboration has independent funding beyond core funds of the partner organizations in question. The ideal is to work for some sort of consensual sharing arrangement, such as rotating the position among partners. The second-best is some form of democratic process by the governing board involving a vote to achieve a decision, though this may leave some partners dissatisfied. The second situation necessitates a search for candidates, where the outcome then depends on market size and the ability of the partnership to attract applicants. Alternatively, hiring of consultants or efforts at individual capacity building in the desired function can be pursued.

Collaborative efforts do not exist in a vacuum. As noted earlier, collaborations among organizations are usually impelled forward by particular individuals attracted to join forces due to some combination of shared interests and complementary capacities. At the same time, the individuals doing the heavy lifting are often not the leaders, directors, or other "authorized organizational representatives." Hence another task concerning leadership, management, and governance is to confirm the blessing of those representatives. This may be as simple as a signed letter bestowing said blessing, or it may require a more elaborate series of procedures including meetings, various notarized documents, and so forth. If an organization has intense internal politics, they may well become evident and affect whether or when organizational support can be confirmed.

Stemming directly from the issue of organizational support, a final key task in establishing a management structure, particularly in the case of formal activities with budgets, is to establish written agreements. There is a range of possibilities, depending on the funding and the partner organizations, from agreements for cooperation (which may say nothing more than "we agree to cooperate" without spelling out how or when) to cooperative agreements and other types of formal contracts (which outline in detail who is responsible for what and by when, and often reference other legal documents, statutes, and codes). The more formal the collaboration and the more money involved, the more necessary the second sort of agreement becomes. Negotiating formal agreements can be a decidedly uncomfortable process, particularly the first time for heretofore informal collaborators. Highly legalistic language, references to other documents with more rules and regulations, and signature lines may be not only intimidating but insulting. The transition from informal friendships among collaborators to formal legal agreements is a momentous step, but often a necessary one to scale up activities, increase impacts, and make possible the achievement of long-term goals.

Those were the tasks, not the challenges. Make no mistake: collaboration requires significant effort dedicated to management, beyond setting goals tied to conservation and development. The foregoing tasks give rise to challenges to management of collaboration.

One challenge to organizing a management structure stems from different models of leadership. Different partners are likely to have divergent and possibly conflicting preferences about leadership styles. One basic distinction is between more dictatorial or democratic leadership styles. This reflects the classic debate between the two as applied to political systems (Moore 1966; O'Donnell, et al. 1986). For collaborative relationships, highly dictatorial styles tend to be inimical, but they are not without their advantages. Dictatorial leadership may be more time-efficient, at least at first. Of course, command-and-control leadership may lead to conflicts, which themselves cost time. By contrast, democratic leadership underscores inclusion but is often very time-consuming, and that may inhibit decision making in moments of crisis. Given these considerations, in the context of collaboration, a key challenge concerns the depth of commitment to dictatorial and democratic leadership styles. Balancing up efficiency and inclusiveness is a priority focus in defining keystone values and leadership roles. Once during discussions of how to constitute a management team, I had some collaborators pull me aside to advise me that I was being "very democratic," and asked that I just make some executive decisions in order to get to technical implementation. They seemed to sense that the other partners wouldn't have taken kindly to the proposal for a more dictatorial style.

Closely related to the challenge of models of leadership is whether those who disagree with decisions being made will accept those decisions, especially if they are made under a decision-making regime with which they also disagree. Hence, those who disdain democratic decision making may protest democratic decisions with which they disagree. Conversely, under a more dictatorial management regime, those who resent command-and-control may resist decisions by management. I've seen this go both ways; it seems the "grass is always greener on the other side of the fence." When moving into implementation, beyond the problem of dominance due to inequalities among partners, disagreement about management regimes can politicize decisions and poison collaborative relationships.

Another challenge in management concerns the relationship of technical staff with leadership. The charge of the first is to manage information and provide technical expertise and advice; the job of the second is to use that information and expertise in order to make decisions. Of course, there may be considerations beyond the specific technical data provided by staff, and there may be extenuating circumstances that influence decisions by leadership. A key moment thus arises when staff and leadership disagree about a decision, which complicates decision making if this happens repeatedly. There is also a greater likelihood of tensions over disagreements if specialty staff work for one partner organization but management is based at another.

Perhaps a more common challenge concerning staff arises when staff at a partner organization fail to meet their commitments. This can be an important consequence of unequal administrative capacities among partners. If the justice argument holds that responsibilities need to be spread among partners in a collaborative relationship, the efficiency argument responds that this exposes the partnership to risks if somebody fails at their end. Accounting, monitoring and evaluation, reporting, and other functions are all likely candidates, and in the context of formal agreements they generate tensions in the collaborative enterprise. Of course, in theory, stronger partners support newer or otherwise weaker partners, so a failure by one reflects failures by the others. Administrative problems are thus key moments that test the robustness of the collective commitment to collaboration; it is a teachable moment, for partners may start sniping or alternatively step up to support one another.

A final challenge refers to partner responses to the legalistic language of formal agreements. Partners accustomed to informal relationships, which are characterized by professional relationships embedded in social ties, sometimes react very negatively to the cold demands of a contractual document. In particular, clauses about rights to terminate agreements under specific circumstances, and declarations that the partner issuing the agreement is not responsible for a litany of possibilities (like funding cuts), may come across as very ugly sentiments, even betrayals, of pre-existing relationships. The challenge of emotional management of formal agreements only grows if a partner fails to comply with provisions in an agreement, particularly if that failure is grounds for termination. If ever there was a tool to ruin a collaborative relationship, it is one partner organization declaring an agreement null and void, for it not only means the loss of the formal relationship and its attendant mechanisms—including the movement of money—it is also likely to destroy the trust on which the formal agreement was originally based.

I have tendered a rather long litany of tasks and challenges involved in management of collaborative relationships. It would therefore be helpful to wheel out an equally formidable array of strategic practices to avoid the detonation of bureaucratic nuclear bombs. Fortunately, there is no shortage of strategic practices for managing collaboration.

Of the various styles of leadership, a particularly efficacious model for purposes of managing collaboration is "servant leadership" (Greenleaf 2002; Manring 2007). Perhaps ironically, this refers to leading by following, acting after listening, and consulting before making decisions. Key to servant leadership is that the leader in question puts the collaborative network and its partners first. Servant leaders act more as stewards than bosses, and in the context of collaboration, servant leadership by managers can build trust and strengthen ties among partners. Similarly, servant leaders who represent organizations but are committed to collaboration operate to reconcile potentially conflicting interests among the two. Network brokers often support collaborative efforts by acting as servant leaders.

If servant leadership defines the strategic practices of individual leaders, "distributed leadership" usefully characterizes the structure of leaders in collaborative networks (Spillane 2006). The larger the number of partners, the more important distributed leadership becomes. Not only does a governing board with broad representation become more vital to taking strategic decisions, but the distribution of functions among partners becomes more sophisticated. Consequently, while managers, net brokers, and other types of key individuals may still be important, larger collaborative groups require more decentralized management structures. This not only permits more efficient functioning via a broader division of labor, it also offsets centralism that can lead to problems of dominance and dictatorship. There are various means of distributing leadership. One can spread complementary responsibilities among partners with complementary capacities, or form working groups to manage key functions, and/or rotate leadership positions where multiple players want to perform key functions. The key is to ensure that different partners are represented among network leaders.

A third strategic practice for managing collaboration is to recognize and distinguish 1) the collaborating individuals and 2) the responsible authorities for each partner organization. The two should not be the same unless the organization is very small. Clarity on this question is crucial for purposes of efficiency in protocols, such as for calling meetings about the collaboration (with collaborating individuals) as opposed to signing agreements (with responsible authorities). Clarity also demands regular communication between the collaborating individuals and their responsible authority. Lack of clarity leads to confusion as to who is in charge and thus generates conflicts. Hence, while clarification of roles might seem obvious, it cannot be assumed or overlooked.

Also obvious enough but often overlooked is the definition of protocols for dealing with poor administrative performance. This needs to be spelled out up front. In some measure this is handled in the process of signing formal agreements. Partner organizations with more lawyers and experience in formal contracts generally have pre-established protocols, because they know that they must follow specific legal and accounting guidelines. While contracts and protocols may be offensive to partners accustomed to informal collaborative relationships, the underlying issue remains the same: there is a need for clarity about what constitutes inadequate administrative performance, and the consequences thereof. Ideally, management protocols will be set up to identify incipient administrative problems before they have to be brought to the attention of responsible authorities. This implies that there will be frequent communication between partners and managers, so little fires do not grow before being recognized. Protocols such as regular checks on accounts and time dedicated by personnel greatly assist in monitoring. This in turn permits more collaborative evaluations in order to jointly identify problems and make adjustments before more unpleasant and precipitous steps become necessary.

Much the same applies to effort up front to establish practices for conflict management. The earlier discussion of inequalities and politics made mention of conflict management. For purposes of managing collaboration, it bears repeating that some investment in conflict management may prove useful for downstream misunderstandings and disagreements that may occur. A typical protocol is to refer conflicts to the manager, and if they worsen, to the governing board. Network brokers may of course weigh in with their moral authority. The key is to avoid having conflicts rage out of control, whereupon partners leave the group and thereby undermine its capacity. That said, breakups can also be useful, as I discuss in Chapter 9.

A final and also fairly obvious strategic practice is to set clear deadlines and then hold partners to them. Missing one deadline sets a bad example that leads to missing more deadlines. It is certainly true that unexpected complications will arise. I will say more about surprises and uncertainty in Chapter 7. For present purposes, I am focusing on complications that arise due to factors internal to partners and collaborative relationships, as distinct from external factors over which collaborators may have no control. If partners begin missing deadlines due to internal conflicts or administrative failures, this is a problem that must be addressed. Action by managers and network brokers and servant leaders is crucial to clarify and improve protocols, and if necessary replace staff.

EXPECTATIONS AND BENEFITS

Some of the foregoing sections problematized the distribution of power and influence among collaborators. There is, however, more to this story, with regard to clarification of expectations and benefits. I provide a separate section on this issue, because it has multiple dimensions, and often involves implicit rather than explicit understandings. Because understandings about expectations and benefits are often implicitly held, failure to discuss them openly can lead to disappointments and worse.

After so much discussion of tasks and challenges involved in collaboration, the whole enterprise had better be worth all the trouble. This raises questions about expectations as to what exactly the partners will get out of the collaboration. By expectations I mean the level of hope or excitement partners have about what will result from working together. Expectations are likely to be high to the extent that the players like each other, engage in self-promotion, have complementary capacities, and offer an innovative approach for moving forward.

Benefits go hand in glove with expectations; partner expectations often refer to anticipated benefits. But benefits can refer to many things, which can become a source of conflict due to the potential for misunderstanding. Benefits may be monetary or non-monetary, accrue in the short or long term, and be directly instrumental to the goal or indirectly related subsidiary by-products. Further, benefits may accrue on multiple levels, from individual participants to partner organizations to the larger collaborative enterprise. The key question then is what kinds of benefits different partners have in mind. This entrains specific tasks to manage expectations and benefits.

The first task is to require that all partners engage in open discussion about expectations and benefits. Key here is to pair up expectations about contributions and performance ("expectations of" partners) with expectations about benefits ("expectations for" partners). After all, the whole point of the collaborative venture is that all of the partners jointly contribute and that benefits are consequently shared. Explicitly linking expectations about performance to benefits stems directly from discussions of common interests and complementary capacities, as well as goals and objectives. There is thus a degree of self-regulation at work in these discussions. Partners become monitors for each other. If everybody holds back, expectations and benefits remain implicit. But if partners are more transparent, opportunities become evident, along with the work required to achieve goals, and the fruits of those labors. Here network brokers can exercise their role as moral voice in calling attention to implicitly held expectations about benefits. This is especially important when partners hold back, whether due to self-interest or shyness.

Another key task is to differentiate among the types of benefits in order to clarify who wants what. Partners need to talk out their profiles of priority benefits. Partners with limited budgets may see collaborative opportunities as a means to expand their capacity, and may therefore think of benefits primarily in monetary terms. Partners with strong commitments to achieving shared goals may care less about the money and much more about achieving objectives that have an impact. There is thus a need for partners to talk out their expectations about monetary and non-monetary benefits. Similarly, benefits may be accrued in the short or long terms. Most partners primarily focus on short-term benefits, but long-term benefits can be more consequential on a strategic level. Capacity building and the formation of alliances are two examples. Weaker partners may have smaller budgets and roles, but may also receive investments via the collaborative effort that permit them to do more later on. Similarly, the collaboration itself allows new or younger partners to demonstrate their capacities, which can open doors to downstream opportunities.

A third task with regard to expectations and benefits is for partners to sort out the distribution question. First, it is crucial to recognize that because partners differ in various ways, expectations and benefits are also likely to differ. As has been noted in previous sections, partner inequalities in capacity and contributions to shared goals must be delineated clearly. Those discussions in turn become the basis for delineating the distribution of expectations about partner responsibilities. If the first is clear, the second is greatly facilitated. In turn, if the distribution of responsibilities is clear, the task of articulating the distribution of benefits has a solid foundation. But just as there are leaps from discussing interests and capacities to goals and objectives, and again to management, by the time the partners get to expectations and benefits, some adjustments may be in order. This reveals the reason for discussing expectations and benefits late in the process: various considerations affect the distribution of expectations and benefits, and they need to have been touched on already.

Of course, there are challenges in pursuing these tasks. Discussions of expectations and benefits can be uncomfortable, especially if organizations differ in administrative capacity and pay scales. Further, some partners may putatively come to collaborate, but may in fact act in their unenlightened self-interest. Hence, for a variety of reasons, partners may exhibit resistance to transparent negotiations or offer only bland official statements. A key challenge is thus that partners often provide insufficient details as to their expectations about contributions and benefits. Those details are, however, crucial. The trick then is to avoid coercing partners into admitting what they really want, which will likely prompt defensive reactions and negative feelings all around.

The other key challenge at play is that the link between "expectations of" partners as to their responsibilities and "expectations for" partners as to the benefits is complicated to define. The second depends on the first, and the collaborative enterprise depends on clarity on that point. Some partners grasp this immediately, but others may fail to take it to heart, which is perilous for all involved. What is more, there is great potential for tensions here. There is an "efficiency argument" that benefits should be distributed in proportion to contributions, but that can reinforce inequalities. Hence, there is also a "justice argument" that benefits to weaker partners ought to be greater relative to their contributions in order to expand the downstream breadth of capacity as a result of the collaboration. Assuming the partners have clarity on the need to link expectations about contributions and benefits in negotiations, they must arrive at some modicum of consensus as to the distributions of both, which need not be equivalent.

There are multiple strategic practices to avoid setting proverbial wildfires among collaborators over questions of expectations and benefits. The first is to be very explicit about budgets. The more partners there are, and the greater the inequalities among partners, the more important this becomes. It is invariably revelatory for partners to present the protocols of their organizations as regards their operating costs and their pay scales. To the extent that both are formally established, such presentations amount to sharing the rules of the game rather than a proposal from out of the blue. This can calm concerns and complaints, though it also reveals in concrete monetary terms the inequalities at hand. Despite the sensitivities about money, this is necessary. Better an up-front argument in the process of defining benefits than later once a budget has been set.

A second strategic practice is to hold workshops on "expectations of" and "expectations for" until everyone has the vital link between the two clear. This may require some lengthy discussions about different types of benefits. Once everyone has set out their expectations about contributions and benefits, the partners can focus the discussion by constructing a matrix that crosses the two. Ideally, the discussion will go well beyond short-term monetary benefits. While it might be very concrete to arrange agreements to require performance before payment, as in consulting contracts, collaboration goes well beyond consultancies. Monetary resources may need to be provided up front for start-up; other benefits may only arrive after the collaborative activities conclude. In all cases, the arrangements at hand should be designed and justified clearly, understanding that partners differ. The bottom line is that different kinds of benefits be made clear to all partners and that partners have a chance to discuss and if necessary critique the arrangements proposed.

Finally, it is important to underscore the relationship between short- and long-term benefits. There is an instrumentality between the two; the first is often necessary for the second to occur. Capacity building permits institutional capacity for downstream tasks on a larger scale or in a leadership role. Hence, it is often not enough to argue that some partners, especially younger or smaller organizations, only receive downstream benefits. A key strategic practice is therefore to carefully design the distribution of benefits by thinking on multiple timescales and viewing the current partnership as benefitting from but also contributing to the capacities of the partners. This fortifies the links of responsibilities to benefits while making evident that current commitments should bear fruit in the future. Collaboration is eminently strategic; discussions of benefits should be too.

SURPRISES AND UNCERTAINTY

The foregoing sections tended to assume that planning is efficacious, that the future can be predicted based on the past, and that changes of direction can be anticipated. These are optimistic assumptions. The contexts I know where conservation and development questions are of central importance can be characterized as surprising and uncertain. Institutions that incur regular changes in personnel, regional dependence on a few commodities with fluctuating prices, and even environmental change itself, all contribute to surprising events. Surprises can make plans unworkable, or suddenly much more urgent. Enough surprises and you have to plan for uncertainty, such that Plan A needs to be accompanied by Plans B and C and perhaps others. My colleagues in the MAP frontier have a refrain: "It is never dull here." My response is that it seems that Plan A never comes out; in practice we operate on the basis of Plan B or beyond. Hence, we plan, but we plan for multiple possibilities.

Surprise and uncertainty thus entrain tasks and challenges in collaboration, because unexpected turns of events require not only adjustments, but coordinated adjustments. Of course, coordination in moments of crisis is hardly more guaranteed than agreement in calmer moments when issues can be talked out. What is more, for those committed to democratic processes and consensus building, time may be of the essence and agreement may be impossible. I therefore outline some key tasks, challenges, and strategic practices as regards the issue of surprise and uncertainty in collaboration. This sets up a longer discussion on this issue in Chapter 7, which deals with surprise and uncertainty when crossing boundaries for collaboration.

The first task is to jointly identify risks to achieving the goals of the collaborative venture. Here it is very useful to distinguish between internal risks, posed by the collaborators themselves, and external risks, which stem from the broader context. Internal risks include institutional fragilities, which may come in various forms. Most notable here are the propensity for an organization to make personnel changes, and internal conflicts, both known and undisclosed. Organizations are understandably reticent to reveal instabilities in terms of personnel and politics, as well as administrative limitations or lack of collaborative experience, among other things. Subsequent discovery of these matters by partners amounts to an unanticipated event. Further, institutional changes, whether in mandate, budget, or personnel, can modify institutional capacity in ways that affect relationships with collaborators in unexpected ways. Even with the due diligence devoted to communication and the subsequent issues, it can be difficult to fully avoid surprises stemming from partner fragilities. External risks are harder to manage because by definition the sources of uncertainty are exogenous and thus often well beyond the ambit of influence of the partners. I have already noted some forms of external political risks, such as regime changes after elections, political instability, and in extreme cases political violence. Beyond politics, there are economic fluctuations which may entrain stresses on collaborating organizations in terms of budgets, especially for personnel. In addition, external changes may stem from biophysical alterations that conservationists are concerned about in the first place. As noted in Chapter 2, manifestations of incipient climate change, such as droughts and floods, constitute a significant form of uncertainty, whether due to nuisances such as temporary power outages or serious problems including bridge washouts and displaced populations. Such events undermine collaboration to the extent that collaborators rely on telecommunications and transport infrastructure. The results of extreme climatic events can of course change priorities for conservation and development work itself.

Institutional fragilities require attention to the task of adaptive governance to manage internal uncertainty in collaboration. If a key technical specialist leaves a partner organization, there is a need to institute a process for deciding who replaces that person in their key function. Management of internal uncertainty thus involves defining a protocol for that process in order to get from Plan A to Plan B. The key point is that collaborators need to define that process before personnel changes occur. If this is obvious at the outset of collaborative ventures, it avoids delays in technical implementation, and permits agile responses to changes in partner personnel.

Beyond internal uncertainty is the task of planning for external uncertainty. There are established planning methods that explicitly account for unexpected changes. Scenario planning has been used for many years as a means of managing uncertainty, especially due to exogenous factors over which partners have little or no control (Shoemaker 1991; Peterson, et al. 2003). In brief, scenario planning requires envisioning of multiple possible futures, and planning for each one. The envisioning process itself involves a critical review of possible futures in order to identify those that seem most probable. The planning process then focuses on three to five probable scenarios and details how the collaborators would collectively respond in each. The focus on a handful of scenarios permits detailed planning on the understanding that such plans will themselves likely have to be adapted due to the details of an external event.

Management of surprises and uncertainty involves challenges, because by definition it is hard to prepare for the unexpected. Earlier I asserted that internal uncertainty is easier to manage than external uncertainty. That mostly applies to smaller partnerships. The larger the partnership, the more complicated the internal uncertainties become, simply because there are more partners, and thus more sources of surprise. Managing a collaboration of two partners is one thing, since there are only two sources of internal uncertainty. With a consortium of eight partners, there are now eight sources, plus uncertainties stemming from relations among various pairs, trios, quads, and other permutations of partners, and how those relationships in turn affect one's home organization. The sources and pathways for propagation of uncertainty multiply rapidly.

Another challenge involving surprises is that many regions of importance for conservation and development are rapidly changing, which consequently involves large uncertainties. Rapidly changing regions exhibit alterations in the institutions present, their roles and functions, and likely their power relationships. Rapid change thus implies a cascade of adjustments and modifications—which may be for the better in some respects—but which require effort for collaborators to adapt. To the extent that there are institutional fragilities to begin with, regional changes can involve significant instabilities, such that how things function and how things get done may incur rapid modifications, even on the scale of weeks or days. Further, insofar as there are multiple independent sources of change, such as political instability, commodity booms, and extreme climatic events, the impact of one shift can magnify that of another. The last ten years in the MAP frontier have seen many changes that were not anticipated a decade ago, including a gold boom, a series of extreme climatic events, and migration from Haiti.

Given the tasks and challenges, there are strategic practices for managing surprise and uncertainty. One useful exercise in assessing internal risks among partner organizations is to engage in a critical evaluation of organizational flexibility. The next chapter will dive deeper into different organizational types. For now, suffice it to say that some organizations have heavier bureaucracies and more ossified personnel hierarchies than others. The degree of flexibility is useful to appraise because very high or low flexibility can be potentially problematic. Highly flexible organizations constitute sources of uncertainty if they change frequently; inflexible organizations may provide stability to collaborating partners but have difficulty responding to exogenous changes. Assessing organizational flexibility can support both the establishment of protocols for managing change among collaborators and scenario planning in anticipation of external surprises.

Another strategic practice for managing uncertainty and surprise concerns the possibility of seeking additional partners. An evaluation of the portfolio of partner assets may well reveal adequate capacities to manage planned activities, but surprises may reveal fragilities in other respects. This can drive the identification of missing functions to focus a search for additional partners to perform the functions for which there is otherwise inadequate capacity. The onset of conflicts among partners may entrain identification of specialists in conflict management; the inability to effectively communicate with stakeholders might prompt the search for public relations organizations.

Closely related to identification of additional partners is to support network brokers. After all, network brokers by definition specialize in the role of matchmaking, which involves identification and articulation of potential partners with functional complementarities to pre-existing networks of collaborators. Inasmuch as there are functional holes among collaborators, network brokers can be crucial in identifying complementary partners.

Despite the focus on complementarities, redundancy is also valuable in the context of surprises and uncertainty. Having multiple partners with specialists in a particular function bestows resilience on a network if one of those partners loses its specialist, shifts its focus away from that function, or leaves the network. At both the individual and organizational levels, redundancy begets resilience by backstopping key functions in a network. Hence, the search to expand a collaborative network need not only focus on complementarities, though they can be crucial; expansion for greater redundancy also has value.

The panorama of strategic practices for handling external uncertainty is unfortunately somewhat limited. This makes the available strategies especially valuable, if also frustrating. In cases of extremely rapid change, where instability is the norm and chaos is no longer surprising, planning becomes impotent and even scenario development can be inadequate. Under such circumstances, "muddling through" becomes the modus operandi (Lindblom 1959; Wollenberg, et al. 2007). The key to muddling through is to recognize the "spontaneous orders" that may arise in moments of especially rapid change. Muddling through involves hyperflexibility in managing collaborative relationships as well as ties with other stakeholders. Such hyperflexibility may demand retrenchment to relatively informal relationships. More proactively, hyperflexibility involves sustained engagement among collaborators with key stakeholders during times of rapid change. Reflections on those shared experiences can deepen relationships and build trust. Sustained engagement also permits expanded information sharing, which is itself valuable for reducing uncertainty and surprise, as well as reducing inequalities in knowledge funds (Mendoza, et al. 2014). The MAP Initiative was effective in these regards in the context of the paving of the Inter-Oceanic Highway (Mendoza, et al. 2007). That said, muddling through may be the best one can do, as high uncertainty may prevent efficient planning or coordination of action.

The importance of political change for uncertainty requires comment on relevant strategic practices by collaborators. Elections often mean change, and depending on the importance of patronage, elections may mean change not only in representation but also in staffing. Turnover often implies a lack of institutional memory. A strategic practice to manage political change is regular communication with authorities to share information. Transparency and proactivity in sharing information may indeed lead to criticism or a cold response, but they avoid ambiguity, which can readily lead to surprises. Further, transparency often leads to precisely the opposite response, to the extent that governments realize there are complementarities in what other organizations have on offer. While a positive response might just be for show, learning is nonetheless occurring. As with other sources of uncertainty, muddling through is useful for managing political uncertainty, and much of the effort to that end concerns sharing information and building trust with a broad profile of political actors.

In their book on collaborative advantage, Huxham and Vangen (2005) repeatedly issue the warning, "*don't do it* [collaborate] *unless you have to*" (pp. 13, 257, italics in original). This chapter offered one rendition of an explanation as to why. Against the calls for collaboration in various lines of work, and despite the advantages of collaboration, there are numerous tasks and significant challenges.

The length of this chapter bespeaks the order of magnitude of the tasks and challenges in collaboration. I have recognized many concerns enunciated in other work on collaboration, but I have emphasized issues that often become particularly salient in the context of conservation and development projects. In particular, the issues of politics and inequality, and surprise and uncertainty, are seemingly ever-present, because dealing with them now hardly means they will not arise again (Perz, et al. 2010). What is more, they tend to become entangled with other issues.

At the same time, while the list looks long and the tasks appear numerous and tiresome, there are also numerous strategic practices at hand. The strategic practices discussed above are all battle-tested, as they come from direct experience and/or work reported in prior publications. The larger point of noting the strategic practices is that the spirit of collaboration requires their application as a matter of course. Collaboration is a set of practices, which constitute their own skill set, have their own logic, and offer a variety of benefits.

Collaboration is ultimately a business like making laws or sausage, where the process can be protracted and messy. It should therefore not be surprising that some collaborators may disagree or even disengage. Among good-faith collaborators, threats of breakups tend to lead to more discussions and meetings. While they can be arduous, contentious discussions are eminently useful insofar as partners express their fears, which permits social learning to occur. If I seem to have overstated the perils and pitfalls at hand, I hasten to add that not everything that can go wrong will. Equally, if the strategic practices seem to hold out solutions, such practices offer no guarantees. Even setting aside the list of issues and tasks, the real work of collaboration is to manage the challenges in order to minimize the potential for complications to cascade and thereby grow.

This chapter made the simplifying assumption that collaboration occurs among professionals with the same training, working in the same kind of organization, in the same country. In other words, I said little to nothing about crossing boundaries. But collaboration across boundaries—whether disciplinary, organizational, national, or otherwise—is widely seen as necessary to pursue shared goals concerning wicked problems like conservation and development. To that assertion I will add that crossing boundaries complicates collaboration and expands its challenges. The remainder of this book therefore discusses the complexities of crossing boundaries for collaboration, and elaborates on the implications.

Chapter Four

Crossing Boundaries for Collaboration

> When an outsider says what needs to be done, they are like a saint away from home. (*Quando um estrangeiro diz o que precisa ser feito, ele é um santo fora de casa.*)
>
> —Foster Brown

The issues of collaboration stem in large measure from differences among collaborators. The pursuit of collaboration therefore involves managing those differences. Of course, this assumes that the differences are large and consequential. If collaborators come from the same discipline, institution, and country, the similarities rather than the contrasts are what stand out. This does not guarantee there will not be differences of opinion or conflicts. Nor does this rule out other sources of difference—gender, race/ethnicity, age, etc.—which can present their own difficulties. But the extent of similarities among collaborators is likely to simplify the difficulties to be negotiated in order for collaboration to proceed efficiently.

In this chapter I suggest that crossing boundaries complicates collaboration due to the greater differences at play among partners. Whether in terms of discipline, organization, country, or other sources of social difference, spanning divides not only constitutes additional challenges but can magnify the challenges of collaboration already discussed. The additional challenges start from the beginning: crossing boundaries complicates introductions and communication, along with every other issue that cascades therefrom, precisely due to the size and profundity of the differences among the partners. This is because crossing boundaries in effect means engaging collaborators who hold distinct cultural assumptions, whether those assumptions arise from their disciplinary training, the practices of their institution, the history of their country, and so on. Effectively engaging collaborators across boundaries thus takes more time and effort.

Fundamentally, collaboration is a process of "humanizing the other." When collaborators begin to engage, their introductions are effective to the extent that all participants convey information about themselves, both explicitly in what they say and implicitly in how they present themselves. As a result, each participant becomes a person with a background, interests, and capacities. Further, other participants begin to see the presenter as a person and not just an opportunity. Ideally, this prompts all of the participants to think not only of their own priorities but of shared interests and complementary capacities.

When crossing boundaries, because the partners have more background differences, humanizing the other takes more time and effort. The process of humanizing others across boundaries is thus more onerous but also utterly crucial. Even if the introductions are effective, they are likely to reveal significant differences among collaborators, notably inequalities, which will demand great attention and care.

All this is not, however, to suggest that crossing boundaries is not worth the trouble. On the contrary, if the differences are large, and the inequalities considerable, so are the complementarities and thus the potential collaborative advantages. "You get what you pay for" is an appropriate dictum in the context of collaborating across boundaries. If there are advantages to collaboration, they can be all the greater when crossing boundaries. A key aim of collaborating across boundaries is therefore to manage the challenges in order to achieve the advantages.

Success at crossing boundaries for collaboration requires an appreciation of boundaries themselves as they complicate the tasks and challenges of collaboration. This chapter therefore begins by discussing the significance of boundaries, lest anyone doubt their importance. I then elaborate on the nature of the challenges of crossing boundaries by offering a cultural interpretation of the distinct social worlds on the opposite sides of boundaries. The heart of this chapter provides discussions of three types of institutional boundaries: academic disciplines, organizational types, and countries. While these boundaries do not comprise an exhaustive list, they are particularly significant for conservation and development work. For each, I outline salient characteristics of the boundary in question, the complications it presents to collaboration when crossed, and strategic practices for managing the complications. I conclude by noting cases where the same strategic practices can be applied when crossing multiple boundaries simultaneously.

THE SIGNIFICANCE OF BOUNDARIES

Boundaries come in many forms and are understood in various ways (e.g., Ainlay, et al. 1986; Barth 1998; Brunet-Jailly 2005, 2011; Paulson and

Hernes 2003; Van Houtum 2005). One example concerns social norms and expectations that define what behaviors are appropriate in particular contexts. As sociologists often observe, transgressions get negative sanctions that are part and parcel of social control to keep everybody in line. Social boundaries based on sources of social difference, such as race and ethnicity, often serve to support social inequalities. Another example involves organizational protocols and rules which define categories of objectives and activities, and again delineate who gets to do what. Organizations have divisions, departments, and other units with specific mandates, each of which comes with particular job descriptions and responsibilities. A third example concerns government bureaucracies, which delineate various administrative divisions tied to geographic boundaries for the sake of territorial management. Sovereign states by definition control specifically demarcated territories, managed in sundry ways by various agencies charged with specified tasks, whether that means incorporated areas for infrastructure investment, postal codes for circulating mail, or voting districts for elected representation.

In these and other cases, boundaries exist for particular purposes and have multiple dimensions (e.g., Albert, et al. 2001; Baldersheim and Rose 2010; Newman 2003; Santos and Eisenhardt 2005). For one thing, boundaries are structural, in that they divide in order to organize. Boundaries thereby permit a degree of tractability for the sake of efficient management. Boundaries thus underlie the constitution of hierarchies that are characteristic of bureaucracies, and thereby facilitate administration. In addition, boundaries are political constructs. They reflect power relations and constitute a manifestation of who is in charge, rightly or not. Boundaries thus have histories that reflect chronologies characterized by interests, decisions, contests, conflicts, and some modicum of resolution. The histories behind boundaries often involve injustices and grievances, depending on who you ask. Whether the actual purposes of boundaries are publicly stated, or even remembered, varies considerably. In any event, boundaries as political constructs divide in order to permit control. Finally, boundaries are cultural. Boundaries define the cognitive limits that demarcate cultural worldviews, making the construction of meaning easier by being locally grounded. As such, boundaries support cultural assumptions. Consequently, boundaries hinder the flow of information among divided social groups. Further, boundaries hamper contextualized understanding of that information, should it get out. Whether viewed in structural, political, or cultural terms, the upshot is that boundaries influence how people on the opposite sides of a divide view each other. People on the "other side" are readily characterized as "others," which fosters stereotypes and misunderstandings.

Boundaries thus entrain a suite of consequences which are mutually reinforcing. The administrative advantages of partitioning for efficiency are attractive to governments which are themselves bureaucracies. Political control is facilitated by clear territorial demarcation legitimated in part by the validation of a national rendition of cultural identity. Cultural tensions are limited to the extent that groups which perceive each other as competitors or adversaries are separated. While boundaries may impede and control, they also order and legitimate.

Of course, things are not quite so simple. The advantages of boundaries also come with significant incentives for various types of boundary crossing (e.g., Bretell 2007; Ewing 1998; Klein 1996; Michaelsen and Johnson 1997). Whether legal or illegal, accepted or stigmatized, mutualistic or exploitative, crossing boundaries has many motives and takes many forms. For present purposes, where the focus is on collaboration for conservation and development, boundary crossing has some ambiguities. I note two relevant motivations for crossing boundaries for conservation and development. First, boundaries become problematic to the extent that they are applied differently depending on their purposes, and may not line up neatly. This is related to the problem of "scale mismatch" in the analysis and management of social-ecological systems (e.g., Cumming, et al. 2006; Schonewald-Cox 1988). Boundary overlaps and mismatches are perhaps most easily represented geographically, and they apply to many environmental problems. Whereas administrative units have demarcated dividing lines, ecosystems are defined more by gradients. Hence, political and administrative boundaries often divide ecosystems and watersheds. Boundary crossing thus becomes necessary for collaborators on the various sides in order to pursue coordinated study and management. Second, boundaries and their significance may change over time. The economic globalization literature includes debate over the question of "deterritorialization," that is, the issue of whether increased international flows of capital and products undermine national governments and elites (e.g., Elden 2005; Kearney 1988; Kennedy 2007; Robinson 2002). Similarly, major investments in trans-boundary infrastructure may erode the importance of borders insofar as crossing becomes easier. Under such circumstances, boundaries may become less important in practice.

CROSSING BOUNDARIES AS CROSS-CULTURAL EXCHANGE

Regardless of the boundary in question, crossing boundaries is in effect an exercise in moving from one culture to another. For present purposes, I invoke culture as an organizing concept to call attention to the differing assumptions underlying contrasting worldviews found on opposite sides of boundaries. Crossing boundaries amounts to cross-cultural exchange, the spanning of distinct cultures. Invoking culture in the context of crossing boundaries is thus useful for calling attention to the challenges of boundary crossing as being rooted in the task of operating simultaneously in multiple cultures.

There are numerous definitions of culture offered by anthropologists with varying intellectual commitments (e.g., Moore 1997). For purposes of a discussion of crossing boundaries for collaboration, I invoke a multifaceted definition of culture. When collaborating across boundaries, culture manifests itself on at least two levels. First is the relatively shallow, obvious aspect of culture evident in practices, which stem from accepted rules and institutional arrangements that define how things get done (e.g., Sahlins 2000). Culture as practice provides a set of acceptable guidelines for doing work in a particular social context. Second is the deeper and more implicit aspect of culture that concerns the unstated and often unconscious assumptions about how the world works (e.g., Peacock 2001). Culture as worldview generally arises from local history and long experience, whether direct or indirect.

Both aspects of culture have manifold implications for collaboration across boundaries. Crossing boundaries becomes a cultural adventure, because such crossings require recognition of, and learning as much as possible about, the shallow/operational and the deep/experiential aspects of culture (e.g., Shaules 2007). Without that recognition and understanding, collaboration across boundaries becomes a bewildering exercise prone to end quickly and ignominiously. This is because without cultural awareness, even the initial tasks of introductions and communication end up merely causing confusion and miscomprehension.

Consequently, crossing boundaries almost invariably requires close attention to cultural differences, especially the implicit aspects. Practices become evident quickly enough, as they are generally codified and explicit. It is still important to learn—quickly—about why institutions are arranged as they are. Grasping worldviews takes more time, as this necessitates some manner to dredge up implicitly held cultural assumptions. Unavoidable is the innocence of informally asking naïve questions of one's counterparts when the opportunity arises. Of course, there are well-worn methods for understanding cultures: immersion and shared experience, long part and parcel of ethnography, are also useful. For present purposes, the key point is that without cultural understanding, the "others" remain others, and many difficulties and limitations are likely to follow for advancing collaborative activities. Beyond communication, it will be difficult to discuss sensitive issues such as inequalities, identify complementarities, set goals, organize management teams, and so on.

Here it is useful to return to a metaphor employed earlier for collaboration, of getting in the same boat. Collaboration across boundaries means getting into a boat of different construction and handling than what one might know. Challenges arise due to the differences. By extension, using the new type of boat to navigate a river one does not know compounds the challenges. Letting the locals handle the boat facilitates things greatly. Of course, then it becomes crucial to know what their interests are in navigating the river, and what they expect from you out of the deal. There the adventure begins. If it is to be successful, then there has to be an ongoing dialogue about the boat, the river, and the interests at hand. This in turn entails questions and responses, admissions and explanations, story-telling and sharing histories. There will be no shortage of naïvete all around, so there will be rich learning opportunities.

But grasping another culture is only the point of departure. Understanding merely provides a basis for managing cultural practices and assumptions in order to collaborate. While collaboration becomes possible, there is still no guarantee of effectiveness. There remain the labors of managing the tasks and challenges of collaboration while crossing boundaries. Indeed, what one learns after an initial crossing may in fact reveal complications that pose dilemmas for the viability of collaboration. Strategic practices then have to be employed in order to proceed.

REFLECTIONS ON CROSSING BOUNDARIES

Once cultural navigation is under way, the work of collaborating across boundaries really begins. If I may be permitted another metaphor, crossing boundaries for collaboration is like climbing over a fence to see if the grass on the other side is actually greener. It might be; it might not. Indeed, the green stuff might not be grass. The point is that the trouble to cross boundaries and navigate cultures is all about what the other side actually has to offer to the shared enterprise.

The benefit of crossing boundaries is not merely a matter of resources and capacity. Learning also occurs in spades, and well beyond cultural understandings. Merely climbing over the proverbial fence is itself enlightening, whether the collaboration in question is successful. The view from the other side is itself invaluable; it teaches, profoundly. As one looks back, one gains new perspective from the other side on one's point of origin, which is illuminating about what one knows and the capacities of individuals and institutions in that origin point.

Even with cultural understanding, substantial reflection on crossing boundaries is also necessary. Following the classic renditions of the symbolic interactionists and social constructionists, collaborators co-create their relationships and in the process re-create themselves (e.g., Blumer 1969; Burr 2003; Goffman 1959; Parker 1998). This co-creation and re-creation is an iterative process as each collaborator responds to the other, and as each reflects on the other as well as oneself. All of this contributes to the construction of a collaborative relational space, a bridge among cultures and institutions, constituted by sharing experiences and histories, practices and assumptions, and of course common interests and complementary capacities. If that is too much jargon, here is a translation: Go. Listen. Then you learn. At the same time, you'll be teaching, including things you may not intend. All of that is useful. This is the point of departure for collaboration across boundaries.

But it only happens if one is committed to making the investment of time and indeed oneself to both the boundary crossing and the collaboration. Collaboration across boundaries is more challenging for some participants than others. Some are naturals: they are confident, gregarious, sympathetic, compassionate, and articulate. But whatever one's natural charm, there is more to the question than social skills. Crossing boundaries inevitably requires one to move out of one's comfort zone. This is never easy, but it is necessary, and what is more, it pays major dividends. The benefits of crossing boundaries stem from reciprocity and they increase progressively as one invests the time. The more one engages the "other" culture and humanizes collaborators, the more the collaborators open up about themselves and reflect on their culture. This can become a virtuous circle. As a result, the deeper elements of the collaborator's culture, and their problems and needs, become admissible for discussion. This permits deeper communication, which facilitates all of the collaborative tasks.

Everyone has their limit, living outside their comfort zone. A priority is to find that limit and negotiate it based on where it falls relative to the demands imposed by collaborators for crossing boundaries. Some prospective collaborators go out of their way to make the newcomers feel welcome; others may be indifferent to the cultural issues. Yet others may prefer a highly formal presentation that circumscribes the list of topics for conversation and thereby excludes discussion of culture. With time and effort, the delineation of comfort zones can be revised. The obvious hope is that comfort zones expand rather than harden or constrict.

Crossing boundaries and working outside one's comfort zone involve the experience of ambiguity, as cultural meanings attached to practices and words are often unclear. Ambiguity leads to revealing misinterpretations, various other mistakes, and hopefully enlightening corrections. These constitute teachable moments, not only about one's own handling of embarrassment, but also the tolerance of collaborators. This is extremely valuable, because teachable moments are often humanizing moments, when "others" become people we begin to understand and trust. In particular, there is considerable value to being the "visiting team" across a boundary, as mistakes and learning and thus humanization are all likely to transpire. Resolving ambiguities, even if that involves embarrassment, is instrumental for bridging the divide and thus taking steps toward acceptance by the home team. Of course, later ambiguities will arise, along with additional misunderstandings and clarifications, and you realize you're still a visitor with regard to some issues. But depth of understanding increases nonetheless.

Crossing boundaries thus takes some talent but especially commitment, the coin of the realm. Commitment and tolerance of ambiguity open the social space for learning by all those involved and permit progressively greater depth of mutual understanding. Comfort zones shift and hopefully expand. These changes build bridges that span gaps. This is the foundation for collaboration across boundaries.

THE IMPLICATIONS OF CROSSING BOUNDARIES FOR COLLABORATION

Crossing boundaries complicates the issues of collaboration. Insofar as cultures and sources of social difference are large, the tasks and challenges for every issue of collaboration become more complicated. From introductions and communication to surprises and uncertainty, the tasks and the challenges are bigger when crossing boundaries. The effort to cross boundaries must be sustained, because the dynamic of managing cultures in effect has no end, and because the cascade of issues from introductions to uncertainty is ongoing. Do not assume that just because the proverbial boundary has been bridged that the bridge will not fall apart tomorrow; on the contrary, it very likely will if not actively maintained.

This section therefore offers an overview of the complications likely to ensue when collaborating across boundaries, for each of the issues of collaboration, beginning with introductions and communication. One way to think of the implications of crossing boundaries for the tasks of introductions and communication is to invoke the linguistic element of culture. People in different cultures in effect speak different languages, insofar as words and grammar differ and the same words may have different meanings. Whereas collaboration requires a degree of effort to clarify the meaning of key terms, collaboration across boundaries is likely to have many more such instances than collaboration among partners operating in the same culture. Clarifications must begin in earnest with the introductions, where all sorts of cultural cues are being expressed, often non-verbally. Who shows up, where they sit, how everyone is dressed, who speaks loudest (and longest) all give indications beyond what is actually stated. Decoding all this is a significant task when crossing boundaries. Introductory meetings have a text and subtext, such that the text is officially offered in the meetings, whereas the subtext is culturally grounded and implicit. The amount of boundary crossing and informal discussions before the formal introductions makes the decoding of the subtext easier. What is more, it is most likely to be done among key players among the proposed collaborators.

Crossing boundaries significantly complicates issues of inequalities and politics. Insofar as certain boundaries exist out of structural and political considerations, they often permit and sustain inequalities and provide manageable institutional arrangements to secure the status of authorities and elites. Crossing boundaries is therefore very likely to result in collaboration among unequal partners, where the inequalities are eminently relevant to the collaboration itself. Whether due to individual training, institutional capacity, or other factors, inequalities among partners spanning boundaries can be large, making this question especially sensitive. Similarly, inasmuch as boundaries serve political purposes, crossing boundaries can be construed as a political act. How that political act is viewed by authorities and interest groups depends on who maintains the boundary, how much the relevant authorities benefit from said boundary, and whether the collaboration threatens said benefits. Boundary crossing may be warmly received if it permits a faster flow of information that brings benefits. By contrast, boundary crossing may be cause for suspicion if it is construed as undermining authority. To the extent that the boundary in question has a history involving inequalities and conflicts, boundary crossing for collaboration becomes more problematic, and at a minimum will require more time to sustain via strategic practices. At the limit, the boundary cannot be viably crossed, whether due to legal prohibitions or lack of safety. There is rather more to be said on this issue; I return to it in Chapter 6.

Boundaries also make discussions of shared interests and complementary capacities more challenging. Even if partners are meeting because they have already identified shared interests, when crossing boundaries, devils in the details underlying those interests often appear. Here language differences and inequalities cascade, casting doubt on shared understandings of the interests at play. As elsewhere, if boundaries map significant cultural differences, confirming in detail that interests are in fact held in common becomes more complicated. Instead it is more likely that while there may be shared interests, the commonality is vague or shallow, and there remain various topics that need to be recognized in order to better understand that the interests at hand are situated in distinct cultures. Two partner organizations may both be interested in biodiversity conservation, but whereas one wants to pursue biodiversity conservation to conserve biodiversity, the other wants to conserve biodiversity because there is funding for consultancies in doing so. When crossing boundaries, the issue of complementarity in capacities becomes a larger concern for intriguing reasons. It bears repeating that crossing boundaries is likely to heighten complementarities and thus collaborative advantage due to greater differences among partners. Nonetheless, as noted in the previous chapter, there is also value to redundancy. Crossing boundaries may result in complementarity in numerous respects but redundancy in few. As a result, crossing boundaries for collaboration may afford significant collaborative advantage, but at the risk of considerable vulnerability, such that the loss of one key person or one organization can cripple the whole enterprise. In sum, crossing boundaries can result in too much complementarity.

Cascading from the intricacies outlined above, crossing boundaries complicates the setting of goals and objectives. Due to the communication issue, clarification of terms being used in stating goals—and thus making commitments—is not trivial. Similarly, due to the issues of interests and complementarities, spelling out objectives requires careful attention to who will be responsible for what, and the consequences of failing to uphold responsibilities. To the extent that communication is not entirely clear and interests not wholly aligned, agreements on log frames can become contentious when some partners realize the targets were serious and not just suggestions. When crossing boundaries, setting joint goals and objectives becomes an exercise in asking the same questions multiple times to confirm that everyone understands that goals and objectives entrain commitments.

Crossing boundaries can radically increase the tasks of leadership, governance, and management. There are frequently very different assumptions in differing cultural milieus about appropriate leadership styles and how organizations are managed. The last chapter highlighted the importance of distributed leadership and servant leaders and network brokers. All are germane to horizontal management structures, but may in turn be inimical to cultures where highly vertical, hierarchical structures are the norm. Likewise, the relationship of leaders and managers to teams may differ considerably among collaborators across boundaries. Before long, crucial decision points appear, and if leadership styles differ across boundaries, decision-making processes must be negotiated. At minimum, this is awkward. This can become contentious in moments of tight deadlines and crises, when collaborating managers may revert to in-house leadership styles rather than negotiated protocols. This is not to suggest that vertical chains of command are not useful. There is more to this question than space in this chapter permits; further discussion awaits in Chapter 5.

Expectations and benefits can be a delicate topic among collaborators spanning cultural divides. Travails arising from issues of communication, inequalities, interests, and so forth all conspire to obfuscate discussions of expectations and benefits. This issue thus constitutes a significant risk for the best-laid plans to come unraveled, even if previous issues had already been addressed. Partners in different cultural settings often have very different and very particular expectations about collaboration, especially with regard to the benefits. Consequently, management of expectations needs considerable attention, and projecting spreadsheets of budget requirements and personnel salaries on the wall can provide a useful reality check for everyone involved. Even so, this only begins to address the question of monetary benefits, which stand among other types of benefits. If tolerance for ambiguity is crucial in crossing boundaries, dissipating ambiguity is equally important for managing expectations. The issue of expectations and benefits provides a stiff test of how far comfort zones extend among the collaborators. If people can politely discuss matters of monetary and other benefits, and engage in courteous negotiations over who gets what and when, you pass, at least for the moment. Be advised that due to boundaries, collaborators may not express discontent initially, and bring it up later. Such re-tests can come repeatedly.

Because crossing boundaries means operating in multiple cultural worlds, there are multiple sources of surprise and uncertainty to be managed. Do not underestimate the mischief that can be caused by delays on one side of a boundary-spanning collaborative endeavor, or the misunderstandings that can ensue as one partner tries to explain its circumstance to another, or the irritation felt at one's home organization when one has to explain why some task was not finished on time due to a problem somewhere else. In a globalizing world, it is a popular truism that what happens on one side of the planet may well affect the other. Nonetheless, there remain many instances of culture-specific hassles that only affect you if you are collaborating with someone in a specific culture. Of course, what comes around goes around. The point is that the more boundaries one crosses, the more one is exposed to various sources of glitches and inconveniences, catastrophes and disasters, and (hopefully) happy circumstances. All constitute surprises that can make crossing boundaries look like a fool's errand. But as collaborators become accustomed to the unexpected, they gain valuable experience in applying strategic practices, even if they amount to management by muddling through. These issues get further attention in Chapter 7.

The preceding discussion simplifies the exigencies at hand when crossing boundaries for collaboration. As noted above, some key topics will receive more attention in the chapters to follow. To this point, I have also assumed generic boundaries, and mostly highlighted their cultural aspects. There are in fact many kinds of boundaries, and they make distinct demands on collaborators. I have already mentioned various sources of social difference including gender and race/ethnicity. We could also discuss professional status (professor and student, manager and staff) or stakeholder category (outside organization or local community). I find that three types of boundaries are of particular importance when crossing boundaries for collaboration for conservation and development: disciplines, organizational types, and countries. The remainder of this chapter therefore provides discussions of the exigencies of crossing these boundaries.

DISCIPLINES AND INTERDISCIPLINARITY

There is a substantial literature on disciplines and interdisciplinarity (e.g., Becher and Trowler 1989; Frodeman, et al. 2010; Kagan 2009; Klein 2010), including work on crossing disciplinary boundaries for conservation and development (e.g., Hirsch Hadorn, et al. 2008; Kainer, et al. 2006; Lélé and Norgaard 2005). This work includes very good statements on the challenges involved: communication, jointly defining the problem, finding the crux of the synthesis to address said problem, and so on (e.g., on communication, Heemskerk, et al. 2003; Monteiro and Keating 2009; Pennington 2008). Thus the need and challenges are evident, and the strategic practices for the pursuit of interdisciplinarity have been set forth.

It therefore surprises me how hard crossing disciplinary boundaries remains. In particular, it seems a universal law that interdisciplinarity demands a long time to build the shared foundation of a common language, practice, and experience. Having been around researchers from many other disciplines, having supported and even led interdisciplinary efforts, I nonetheless retain a foolish assumption that past experience will somehow make the process easier and faster. But every single time I participate in a new interdisciplinary endeavor, that assumption is ruptured in short order. The words on the page are not the practice implemented to good effect. Nobody is exempt: crossing disciplinary boundaries complicates the tasks and challenges of collaboration.

One reason for the complications stems from the perils of a lack of a shared understanding as to what exactly interdisciplinarity is. The term is long and impressive, and it falls easily enough from the tongues of erudite academics. But it is one member of a family of terms for which I insist on some key distinctions in order to avoid confusion (Perz, et al. 2010). People generally agree that disciplinarity is not multidisciplinarity. The confusion begins when people use multidisciplinarity interchangeably with interdisciplinarity. The two are not the same. But the confusion does not end there, for there is also the term transdisciplinarity, which is something else again (e.g., Hirsch Hadorn, et al. 2008). The differences are important because the terms refer to different manners of crossing disciplinary boundaries, and thus contrasting demands in terms of the tasks, challenges, and strategic practices.

Multidisciplinarity is the "assembly line" approach to cross-disciplinary collaboration: each researcher does what he or she does in any event, and thus serves as a specialist who complements other researchers. This is what many people mean when they say interdisciplinarity. Whereas multidisciplinarity involves collaboration among researchers from different disciplines, the extent of actual boundary crossing by any individual researcher is likely to be limited since they continue to operate as specialists.

Interdisciplinarity goes beyond specialization in collaboration to feature integration. Interdisciplinarity is the "synthesis" approach to crossing disciplinary boundaries: researchers bring different insights to a research problem, but they engage in critical interrogation of those insights to figure out which ones are crucial to addressing the problem. Hence, the complementarities among different researchers are made to serve the response to the research problem via integration of relevant insights into a larger whole. The product is the synthesis of ideas which are creatively incorporated into a single framework. In turn, this requires integration at other stages of the research process, as in data collection and analysis. One can go farther, however.

Transdisciplinarity moves beyond merely crossing disciplinary boundaries to situate the academic research enterprise in a broader context where non-academic stakeholders participate more fully in the research process. Some people equate interdisciplinarity with transdisciplinarity, but in my view the latter sets a given discipline in the larger context not only of other disciplines but all of academia writ large in society. Hence, there is further boundary crossing in transdisciplinarity than interdisciplinarity.

Note that interdisciplinarity falls conceptually between multidisciplinarity and transdisciplinarity. This helps explain why there is such great potential for confusion as to what crossing disciplinary boundaries entails. People who confuse multidisciplinarity and interdisciplinarity will likely encounter difficulties over fundamental questions such as the framework to drive the research. In turn, they will likely also have conflicts over which discipline (if any) gets favored status in framing the terminology for the problem, setting the agenda for the research, and so on. In multidisciplinarity, it is often one discipline driving the process; by definition, this cannot be so in interdisciplinarity.

Similarly, people who confuse interdisciplinarity and transdisciplinarity are also likely to have trouble from the beginning. Whereas the interdisciplinarians will focus on which types of researchers should be participating, transdisciplinarians will ask about which stakeholders to invite. Because crossing boundaries complicates matters, the proposal by transdisciplinarians to not only cross disciplinary boundaries but also go beyond academia will probably be frustrating and seem needless to interdisciplinarians. By extension, the framing of the problem, the process for defining the framework, and setting the goals for the research will likewise run into fundamental and unavoidable questions as to the purpose of the whole endeavor. In sum, terms matter, and having clear definitions is crucial to avoid misunderstandings, exasperation, and indeed disappointment with the endeavor that is crossing disciplinary boundaries. Clarity of terminology is a challenge from the beginning in the tasks of collaboration: if introductions cannot resolve such questions, communication will be problematic, as will every task that follows.

Assuming that the issue of terminology can be settled, other questions arise. Not all disciplines are created equal. A review of the number of journals, articles, and citations in the medical sciences, other biological sciences, and the social sciences will reveal large differences. Similarly, while some disciplines are strong in the sense of having broadly (internally) accepted ontologies, epistemologies, and methodologies, others do not. Indeed, there is some tendency for disciplines in the former group to "trespass" on the subjects to be ostensibly covered by disciplines in the latter group (Portes 1995). Further, some disciplines are more akin than others. In the context of environmental science, crossing the gap between forestry and botany is not likely to be as time-consuming or harrowing as the voyage between cultural anthropology and chemical engineering.

Various implications derive from these observations. One concerns the politics of inequalities among disciplines as they may affect the possibilities for interdisciplinary syntheses (e.g., MacMynowski 2007; Oughton and Bracken 2009; Schoenberger 2001). Collaborator X from strong discipline Y may, consciously or unwittingly, push framework Z because it has worked so well in discipline Y. This may mean that collaborator A from weaker discipline B may, consciously or unwittingly, let X push Z, even if A's framework C offers an opportunity for an innovative synthesis. Collaboration becomes more difficult if X claims to be seeking an interdisciplinarity synthesis but in fact promulgates framework Y and thus multidisciplinary specialization. Further, inequalities among disciplines may undermine interdisciplinarity if X resists interrogation of framework Y and other efforts toward an integrative theoretical synthesis.

Another implication arises from the width of the intellectual and cultural gap between the disciplines being spanned (Kagan 2009). The challenge of crossing disciplinary boundaries is in many ways larger when trying to span the broader domains of academia. In the context of environmental science, including the social sciences along with the biological sciences has most often been seen as the focal challenge. But there is also the question of the contribution of the humanities to conservation and development. The bigger the gaps and in particular the more domains being spanned, the longer the introductions will likely need to be in order to achieve effective communication. But depending on the gaps to be spanned, establishing clarity of interests may well lead to tensions.

Part of what defines and differentiates disciplines are their contrasting intellectual commitments. The significance of said commitments should not be underestimated, including in the context of conservation and development. Systems ecologists concerned about "anthropogenic perturbations" of ecosystems may not respond well to cultural anthropologists proclaiming the importance of the intersubjectivities of the local stakeholders responsible for said perturbations (Schoenberger 2001; MacMynowski 2007).

In this regard, achieving interdisciplinarity requires critical reflections not only on the favored frameworks of one's home discipline, but also some shared interrogation of the underlying assumptions of the intellectual commitments at hand. The bigger the gap, the more likely there will be incommensurate commitments and assumptions in play. To the extent that the disciplines represented are unequal, and to the extent that the intellectual commitments among the collaborators stand in contrast (not to say conflict), the task of defining shared interests and setting common goals becomes more challenging. Consequently, more work will have to be devoted to spelling out—and then adjusting—the interests and complementarities of the proposed collaborators. Those committed to interdisciplinarity are willing and able to actively participate in questioning their own assumptions as part of a group discussion process. Some people take this as threatening to their disciplinary homes and indeed their professional identities. This fear is misplaced. Disciplines are obstacles if interdisciplinarity is required to respond to a problem, but that does not mean that we do away with disciplines in order to cross disciplinary boundaries. The goal of interdisciplinarity is to leverage and synthesize disciplinary insights, not to destroy disciplines. Interdisciplinarity thus requires and depends on the continued health of disciplines. At the same time, healthy interdisciplinarity involves long, careful discussions that critically evaluate the ideas and underlying commitments of all of the disciplines seen as relevant to addressing the shared research problem. To some this might feel like a protracted exercise in parading around the dirty intellectual laundry of the disciplines at hand. In my experience, this is in fact the key to unlocking the various disciplinary boxes available in order to then think outside of those boxes: then the opportunities for syntheses become evident.

In the process, the deeper ontological and epistemological differences among disciplines will be on display. Consequently, different concepts of what counts as science, theory, and evidence will therefore be subject to comparative evaluation. The differences and comparisons are grounds for misunderstandings and especially miscomprehension. In part this is due to the pluri-paradigmatic nature of the social sciences, which span the gamut from logico-deductive positivism to variations of phenomenology and constructionism. The "cultural turn" in the social sciences led to debates over post-modernism, which highlighted the production of knowledge as texts for interrogation, and included doubts about realist science and other "master narratives" (e.g., Rosenau 1992). Post-structural approaches thus departed from positivism for some but by no means all social scientists. This left some biophysical scientists confused and alienated from "those" social scientists who problematized the assumptions of positivism (e.g., Ludwig, et al. 2001). This constituted a challenging intellectual context for interdisciplinary initiatives for conservation and development. Social scientists in particular sometimes feel put on the defensive to explain themselves or their colleagues. The peril is for self-censorship to occur, whether one disavows constructionism and kindred approaches, or criticizes them in order to avoid them. This forecloses on the promise of interdisciplinarity as social learning as it permits broader understandings of the assets (and liabilities) of different disciplines. Self-censorship is thus arguably needless, insofar as the interdisciplinary endeavor requires interrogation of the assumptions of disciplinary commitments. Going back to the ecosystem perturbationists and intersubjectivities, bridging the gap between the systems ecologists and the cultural anthropologists would probably be served by a rollicking debate over the cultural turn and the implications of considering post-structural perspectives on ecosystem degradation. The point is not necessarily to accept post-modernism, but to better understand it, even if it is then discarded or only certain elements are retained.

There has been progress along these lines. Debates over the meaning and content of development confronted post-structural perspectives by the 1990s (e.g., Escobar 1995). There is also growing appreciation of the epistemological and theoretical ferment in the social sciences among biological scientists and conservationists (e.g., Moon and Blackman 2014). After all, a central point of constructionism is to understand how a given social actor sees an issue, regardless of how the scientist sees it. To the extent that disciplinary differences map on to contrasting epistemological commitments, interrogation of epistemologies may prove useful in the conduct of interdisciplinary research. Constructionist research may generate more valid conclusions about the behavioral assumptions of specific stakeholder groups, which can in turn influence the design of policies or institutional incentives concerning conservation and development.

It is also important to call attention to administrative issues that confront interdisciplinary science, at least in the university context (e.g., Pohl 2005; Hirsch Hadorn, et al. 2008). Even if the epistemological debates get sorted out among interdisciplinary collaborators, there remain institutional stumbling blocks. Whereas disciplinary boundaries present challenges to faculty and students, there are structural complications that stem from institutional arrangements defined farther up on the organizational chart of a university. There remain disincentives to dedicate human resources and financial resources to interdisciplinary efforts within universities. The first is true to the extent that faculty are evaluated on the basis of their disciplinary appointments and productivity; the second is true insofar as there is zero-sum competition among administrative units for resources. To the extent that interdisciplinary efforts bring in external funding, deans and directors are supportive. That said, inasmuch as interdisciplinary endeavors prove to be project-specific and thus fleeting, it becomes difficult to open physical space. Hence, whereas management of collaborative efforts entrains a suite of tasks and challenges, interdisciplinary collaboration involving universities can be hard to sustain. Of course, universities that take strategic decisions to support interdisciplinary efforts along specific themes or in particular buildings can obviate these administrative challenges. More difficult are issues of budget management across administrative units, especially if they have distinct protocols for faculty salaries and student fellowships. In sum, even where interdisciplinary collaboration manages to navigate the potential intellectual pitfalls, there remain administrative hassles that also require care and attention.

ORGANIZATIONAL TYPES AND INTER-ORGANIZATIONAL COLLABORATION

If interdisciplinarity is the endeavor to build bridges among disciplines in academia, inter-organizational collaboration is the modus operandi for incorporating academic with non-academic organizations in conservation and development (e.g., Cropper, et al. 2008; Manring 2007; Poncelet 2004; Wondolleck and Yaffee 2000). As with disciplines, different types of organizations exhibit important contrasts that pose their own complications for collaboration (Phillips, et al. 2000). Different types of organizations have different mandates, profiles of personnel, cultural assumptions, and other contrasts (Knoben and Oerlemans 2006). Consequently, joint work among different organizational types can present complications insofar as organizations have to learn to navigate those differences in order to work together (Easterby-Smith, et al. 2000). As with crossing disciplinary boundaries, the effort can be well worth the trouble due precisely to the complementarities.

This section therefore offers a systematic discussion of different organizational types in order to offer an analysis of both the complementary assets they present as well as the additional challenges they may pose in the pursuit of inter-organizational collaboration. I find it useful to identify five more or less distinct organizational types that come into play in collaboration for conservation and development. This is not to say my list is exhaustive; it could be expanded to fruitful effect. Here I focus on universities, NGOs, governmental agencies (especially at the local and regional levels), community associations, and funders. This leaves out potentially relevant players, notably for-profit firms (already the focus of a large management literature), multilateral organizations (such as UN agencies or international banks), and others. To an extent, this selection reflects my experience, which is also colored by cultural particularities of the United States and countries sharing the southwestern Amazon. I provide citations of relevant literature for those seeking substantiation from other contexts.

Table 4.1. Mandates, Strategic Assets, Structures, Cultural Assumptions, and Key Practices of Different Organizational Types: Universities, NGOs, GAs, community associations, and funders.

Organization Type	*Mandate*	*Strategic Assets (A) and liabilities (L)*	*Structures*	*Key Personnel*	*Cultural Assumptions*	*Practical Imperatives*
Universities	Student training, basic and applied research, extension	A: Considerable human resources, wide range of training backgrounds; L: limited flexibility, internal politics	Colleges, institutes/ centers, departments, programs	Administrators (deans, chairs), faculty, students, support staff	Education is an inviolable mandate; budgets and personnel should be stable or rising	Strong countervailing commitments to hierarchical authority, faculty autonomy, and student representation; heavily regulated bureaucratic protocols
NGOs	Applied research, policy input, outreach	A: High agility in allocating resources; L: highly dependent on "soft" money	Programs and projects	Directors, technical staff, support staff	GAs and other organizations require independent science and partners for policy implementation	Strongly oriented to securing external funding for projects and programs
GAs	Public service, technical data management, policy implementation	A: public mandate, resources for policy implementation; L: mandate and implementation subject to elections, politics	Ministries/ departments, agencies/ directorates, bureaus/units	Elected representatives, directors, technical staff, support staff	Public service is fundamental for effective governance in support of public well-being	Political patronage is central, heavily regulated bureaucratic protocols, emphasis on visible public works and programs

Organization Type	*Mandate*	*Strategic Assets (A) and liabilities (L)*	*Structures*	*Key Personnel*	*Cultural Assumptions*	*Practical Imperatives*
Community Associations	Local representation, community solidarity, social well-being	A: deep local knowledge; L: community may exhibit inequalities, conflicts	Committees, working groups, families	Representatives, former representatives, local professionals	Local peoples have particular interests that require collective mobilization to be heard by other organizations in order to receive necessary support	Reliance on social networks and local political ties, dependent on outside organizations for external resources
Funders	Identification and support of meritorious projects and other initiatives	A: financial resources, substantial expertise in program design, implementation, and management; L: very specific agenda and strong preferences that may or may not align with local priorities, subject to budgetary politics	As per organizational type of funder (usually NGO or GA)	Directors, program officers, contracts staff, accounting staff, performance evaluation professionals	Chosen funding priorities are fundamentally important as they fill gaps in needs for activities that would otherwise go unsupported, with costly consequences	As per organizational type (usually NGO or GA); GAs face external pressures from authorities with source funding

Table 4.1 provides selected details for different types of organizations. As a conceptual framework, I focus on several distinct dimensions of organizations in order to characterize them and thereby differentiate them. These dimensions include organizational mandates, strategic assets and liabilities, structures, personnel, cultural assumptions, and practical imperatives. Mandates refer to the central goal or focus of an organization. Strategic assets and liabilities provide implications of mandates for organizational strengths and limitations. By structures I mean the key units and relationships in organizational charts. Key personnel refer to titles for individuals with prominent roles in an organization. Cultural assumptions are the foundational arguments that highlight societal contributions which justify the existence of the organization in question. And by practical imperatives, I mean the primary commitments evident in key protocols that characterize how an organization's administration functions to accomplish its main tasks. While I recognize that there is considerable diversity within each organizational type, I highlight differences in order to characterize the organizational types in general terms. Insofar as the differences are valid, organizational types constitute distinct categories that require bridging as a form of boundary crossing.

Universities large and small share the common mandate of providing higher education via advanced training in a wide array of the arts, sciences, and professions. This bland sentence obscures more than it reveals, for universities also have other mandates, notably research and various forms of extension/outreach, which compete for resources with teaching. This complicates universities as organizations and partners. On the one hand, they offer the significant asset of their considerable human resources: numerous highly trained faculty in a range of areas of study, and even more students selected from large, competitive applicant pools. On the other hand, universities exhibit the liabilities of their organizational inflexibility, in part due to their multifaceted mandates, which have the effect of spreading faculty time thin. Universities are in some contexts also political spaces, notably in developing regions where there is often circulation between government and university positions and where university leaders are either elected internally among competing factions or selected externally by government authorities. University structures are in principle clear, with a central administration overseeing a hierarchy of colleges and departments. However, universities often create semi-autonomous units variously called schools, institutes, or centers with the purpose of pursuing new initiatives in a more agile fashion. This offers the potential for greater flexibility for would-be collaborators, especially if such units also cross disciplinary boundaries. However, such entities may be dependent on certain administrators and thus ephemeral, and their relations with other elements of the hierarchy may not be entirely clear or pleasant due to political questions related to resource allocation. The key personnel in universities are readily identifiable as they correspond to hierarchical structures: presidents or rectors oversee deans and directors responsible for faculty who evaluate students. Those characters of the cast have shared cultural assumptions but countervailing practical imperatives. Most everybody agrees that education is vital as universities provide the essential societal requirement of constituting human resource capacity for sufficing fundamental societal goals like regional development. Underlying that assumption is the belief that budgets should therefore be stable if not growing to meet societal demands. Universities with budget cuts often become sites of protests, strikes, and other resistance. At the same time, different parties in universities exhibit contrasting practical imperatives: administrators highlight their leadership roles when justifying their authority in identifying strategic goals and initiatives, faculty underscore the importance of academic freedom and thus their autonomy (including from administrators), and students frequently manifest their desire to have a stronger voice in university governance. Hence, beyond the influence of external politics and election of administrators, universities are sites of internal contention due to disagreements among players with different structural positions as to who contributes what to governance. A final item under practical imperatives deserves mention, as it is unavoidable in the context of a discussion of inter-organization collaboration. Universities, particularly public universities, have heavily regulated bureaucratic protocols. This also contributes to their relative inflexibility as compared to some other organizational types.

Non-governmental organizations (NGOs) are birds of a different feather. Here I primarily mean local, regional, or national social-environmental NGOs operating in developing regions where conservation and development questions are significant (e.g., Aldaba, et al. 2002; Bebbington and Thiele 1993; Jasanoff 1997). Such NGOs typically engage in some form of applied research, usually in the service of various types of community outreach and/or to provide technical support or critical evaluations of state policies. NGOs often have dense networks of contacts and often serve as bridge-builders between communities and governments. A particular organizational asset of NGOs is their high degree of agility, since they are typically incorporated as private, nonprofit entities. This no doubt helps support their bridge-building functions. NGO flexibility also permits rapid response to changing circumstances. NGOs thus offer greater flexibility than universities, which have more elaborate hierarchies. However, NGOs are often highly dependent on funders and donors for "soft" money that is generally tied to specific projects with time-limited objectives (Meyer 1999). Hence, the valuable flexibility of NGOs comes with the liability that a lack of resources can hinder or cripple their capacity in certain moments. NGOs generally count on a director (who may report to a board) and technical and support staff, plus some number of consultants. NGOs tend to hire staff and consultants tied to programs and projects, such that staff can be reallocated or scaled up (or down) as need be.

The underlying assumption behind the constitution of NGOs is that governments and other organizations are by themselves inadequate to fulfill some elements of their mandates. This is not necessarily a criticism: governments often actively seek partnerships with NGOs precisely because of their organizational complementarities. NGOs thus often serve as complements to governments by providing policy-relevant scientific data and offering outreach services. A key element of this complementarity concerns the importance of independence of NGOs from governments and other entities, which can serve to legitimate the data or outreach on offer. In turn, this supports agencies and communities in fulfilling their goals while avoiding potential criticisms such as conflicts of interest. The foregoing considerations mean that NGO independence is of pre-eminent importance, though it entrains a heavy commitment to the imperative of securing external funding for projects and programs. Consequently, most everything about an NGO—its flexibility, its independence, and its role as a bridging organization—are in sum measures to maximize the chances of securing funding from external sources.

Governmental agencies (GAs) occupy niches in complex bureaucracies that make up the apparatus of the state (e.g., Barzelay and Armajani 1992; Peters 2001; Tendler 1997; Wilson 1989). GAs have a wide range of mandates, even in local government, but they share a common obligation to serve the public interest in the territory of their jurisdiction. Whether their mandate concerns education, health, sanitation, defense, economic growth, trade promotion, natural resource management, watershed protection, species conservation, or something else, GAs have to manage technical data and follow legal frameworks in order to implement legislated policies. These are not small responsibilities. This and the mandate for public service make GAs extremely valuable partners for conservation and development work. Of course, these assets come with liabilities, as some GAs have more capacity and legitimacy than others. To the extent that there is limited capacity or the mandate of a GA is politicized, implementation is likely to be small-scale or ineffectual, if it occurs at all. Here it is worth noting that the term "GA" is a generic term for various units in rather byzantine hierarchical structures that characterize governments. Unpacking such hierarchies at the national level reveals a large range of units, from ministries and departments under which sit agencies and directorates, within which one finds bureaus and other entities. Similar if less complex hierarchies occur in regional governments (which lack the ministry level) and local governments (which usually also lack directorates). In classic hierarchical fashion, power and authority flow from the top downward. This has multiple implications: a given GA must generally do what it is told by authorities above it in the hierarchy, and partnering with a GA requires the blessings from said authorities. Of course, the farther up the hierarchy one proposes to go, the more difficult access becomes, and the less transparent the workings of GAs and their masters are likely to be. The cast of characters within GAs includes directors, various managers, and technical and support staff. Elected representatives and their staffs constitute additional players since executives and legislators dictate, directly or indirectly, what GAs are to do. This makes relations between representatives and directors especially important to understand, time-consuming though that is likely to be. Non-aligned policies along with proactive information provision constitute a useful point of departure. Underlying most any state apparatus is the assumption that the role of government is crucial for supporting numerous aspects of the public's well-being. Of course, how "the public" gets defined varies substantially among representatives and GAs with distinct mandates. This question in turn raises long-standing issues of government capacity, administrative effectiveness, and political legitimacy. These issues greatly influence the culture of GAs, which underscores an ethic of "CYA" (protect your posterior) and leads to cautious behavior in what is said and written. In turn, GA culture is tied to practical imperatives in government, of which there are several. Governments encompass complex systems of political patronage that strongly shape the distribution of public revenue and thus capacity among GAs, as well as staffing based on who gets elected or appointed to key positions. Periodic elections and changes in representation can mean wholesale staff replacements, which hampers institutional memory. At the same time, there is the question of administrative effectiveness. As entities subject to considerable scrutiny by the public as well as political interest groups, GAs face considerable bureaucratic regulations in their protocols, against which they must struggle for efficient implementation of their mandates. This is burdensome and stressful and can hamper collaboration. Finally, GAs face regular challenges to their legitimacy, whether due to media reports of poor performance or the vagaries of scandals revealed by political parties seeking to animate voters during election campaigns. This makes producing highly visible public works and programs a central imperative to GAs. A challenge then is to align that imperative with the priorities of collaborative ventures.

Community associations are again a different species of entity, though they too vary considerably (e.g., Berkes 1989; Pagdee, et al. 2006; Western and Wright 1994). While there are locally variable versions of what constitutes a "community" and thus a "community association," the unifying mandate of such organizations is to provide representation of people in a given locality to other organizations, usually starting with local governments. In turn, such representation is instrumental for communicating needs, venting grievances and issuing requests (or demands) related to community social well-being. Communities by definition are locally grounded, and as collaborators they thus bring the asset of deep knowledge of local history, natural resources, and management. Idealizations of communities have rendered them as homogeneous and harmonious; in reality, most have notable inequalities and simmering tensions (Agrawal and Gibson 1999). To the extent that prominent families look out for their own rather than their community, any attempt at collaboration can easily be undermined by fractures and conflicts. Relying on one point of contact in a given community is to run risks of unwittingly allying oneself with one group or family against others. Community associations typically have uncomplicated vertical structures, with one or more elected representatives (a president or council), though they often have complex horizontal structures in the form of working groups for particular issues as well as extended kin networks among various families. Whereas navigating the politics of competing families may amount to a long-term proposition, working with communities through their councils becomes a key avenue for collaboration. In this structure, representatives, former reps, elderly members, and local professionals (teachers, medical personnel) become key players. More formal community associations make clear who is in charge; less formal associations require much more time in sorting out the prominent families and personages who matter. In any event, the cultural assumption in community associations is that local peoples have particular interests that are routinely overlooked by other organizations. Collective mobilization is therefore required, as instituted via associations necessary to amplify voices that can shout louder in seeking external support for community well-being. This is what prompts community associations to seek collaborators: there is generally a pre-defined agenda, which prospective collaborators must understand. Correspondingly, the practical imperatives of community associations involve networking, whether via local social networks or via broader political ties, in order to open avenues and gain access to political actors who can channel resources. An eventuality is that community associations are dependent in greater or lesser degree on outside organizations for resources. The fundamental game is to find outside actors who can provide the resources needed.

It is important to recognize funders as relevant social actors in the context of conservation and development work since the work in question either does not involve income-generating activities, and/or seeks to support start-up but not management of such activities. Funders may be GAs such as aid agencies or NGOs such as private foundations; they may be individual donors or third parties that manage crowd-sourcing mechanisms. While these involve rather different organizational models, funders nonetheless constitute a useful organizational category since they have a common mandate: identify and support worthwhile projects and programs focused on specific topics and goals. The obvious asset funders possess is financial resources, but I hasten to add that funders also command considerable expertise in program design, implementation, and evaluation. I confess it took me time to recognize the importance of the administrative assets of funders, but they are indeed valuable and frequently facilitate learning with regard to collaboration across boundaries. Funders are uniquely perched to observe the travails of collaboration across boundaries from close by but outside collaborative relationships themselves, which makes them a useful source for insights, especially where the funder has long experience in overseeing collaborative projects. Indeed, wisdom from funders prompted reflection on certain aspects of all of the organizational types being discussed here. If there is a liability in dealing with funders, it resides with their clearly defined funding agenda and thus their strong preferences as to what they will support. A key question then is whether and how said preferences might align (or become aligned) with those of proponents of proposals. Here the legal constitution of funders becomes crucial, for that defines what they can and cannot support. An additional issue, notably with GA funders, is that the resources they can offer are subject to governmental budgetary politics. The structures of funders depend on the type of funder: GAs are situated in extensive governmental bureaucracies with the political implications that entails, whereas private foundations tend to be smaller and more agile as NGOs, if also capricious. The common denominator is that funders are in a position of power with regard to grantees: this relationship is unavoidably hierarchical. This does not mean grantees have no voice, just not the final say. The key players include directors, program officers, and staff tied to contracting, accounting, and evaluation. My experience is that there is considerable variation in how much grantees interact directly with the players beyond the program officers. In any event, all of the characters in the cast are relevant since they all review one aspect or another of a collaborative agreement. The manifest belief of funders is that they have picked funding priorities that are fundamentally important to meet needs for activities that would otherwise be impossible. Underlying this is a cultural assumption remarkably similar to the central contention of collaborative advantage: funders need other organizations with complementary capacities to get the necessary work done. The game then is for proponents of projects to demonstrate that they are appropriate grantees. These assumptions entrain a suite of practical imperatives, some of which stem from the type of organization funding the work, whether a GA or NGO. Similarly, the practical imperatives at play also stem from the type of organizations serving as grantees. There is thus a mix of imperatives, depending on the combination of organizational types involved.

Having outlined key dimensions that help differentiate organizational types, I want to draw out some implications of crossing organizational boundaries for collaboration. Whatever the permutation of organizations that join forces, the numerous differences among organizational types constitute boundaries that can present complications to collaboration. If collaborative advantage highlights the benefits of distinct mandates or complementary assets among organizational types, it is also important to recognize that differing liabilities, structures, personnel, assumptions, and practices can magnify the challenges of collaboration. Organizational differences themselves are not necessarily a problem: they can be sources of collaborative advantage or inertia. The issue is how to find ways to ensure that the differences among collaborating organizations can be leveraged to productive effect rather than becoming hindrances (e.g., Gulati and Gargiulo 1999; Manring 2007).

Mandates offer a useful example. Different types of organizations prioritize differing objectives. One implication is that the contrasts in mandates can pose problems to the extent that they entrain distinct calendars and thus timing for delivery of key products. Universities prioritize student training on an annual schedule; GAs often must pursue activities thinking of the next elections; and NGOs spend considerable time hunting for the next project. In each case, the organizational type has a mandate that imposes clocks ticking down to different deadlines with very distinct alarm bells (cf. Wood 2008). As a result, organizational boundary crossing complicates identification of shared goals and objectives. An additional task is therefore to map out a model with a flow of activities that are sequenced to permit integration of contrasting priorities and timelines.

Another complication concerns the specific reasons different organizations want to collaborate. All organizations have particular interests stemming from their mandates, structures, and cultures. Self-interest is a potential obstacle to effective collaboration, and it may be particularly pronounced among different organizational types. The question then concerns the extent to which organizations are prepared to adjust in light of other organizational interests. In the context of collaboration, this usually involves situating one's organization in light of broader collaborative goals. Such adjustments do not require one organizational type to necessarily become like another. Instead, the point is to find ways to leverage an organization's complementary assets in the service of other organizations. A university sending students to a GA for internships that support the GA's work would thus benefit the university students and the GA. Such an arrangement would parallel multidisciplinarity, where each organization does what it does best. But complementarity can also parallel interdisciplinarity, where organizations adjust their mandates and the rest in order to serve a larger goal that requires the adjustments in order to be feasible. This means reframing mandates, marrying structures, putting personnel together in working groups, and having them participate jointly in the same activities. The consortia described in Chapter 2 brought together university personnel who did research but then combined forces with GA and NGO personnel for outreach activities that in turn supported participatory planning. This required full collaboration of the multiple organizational types, including in activities beyond their typical routines.

If organizational assets may be complementary, they may also be highly unequal. Consciously or automatically, larger and more powerful organizations often seek to dominate proceedings, such as by imposing their mandate as the agenda (Hardy and Phillips 1998). This parallels the political problems among strong and weak disciplines when crossing disciplinary boundaries in academia. In the context of inter-organizational collaboration, one way to manage inequalities is to hire an autonomous manager. Often a network broker or servant leader with moral authority stemming from their independence, the manager can maintain a space for dialogue among collaborating organizations. Without an independent manager, disagreements among organizations can become personalized, such that the manager will be accused of working for their home institution and not the broader collaborative enterprise. Autonomy provides something of a structural safeguard against dominance by one organization as well as accusations of conflicting interests. Of course, autonomy can leave all of the various collaborating organizations unhappy at different moments insofar as the manager acts in the stated interests of the collaborative enterprise rather than in ways that correspond to the specific interests of an individual partner organization. My experience is that a generalized but low level of unhappiness all around is a good sign of effective management of a collaborative endeavor among unequal organizations.

While conflicts among different organizational types are a concern, there are arguably more ubiquitous complications. An enduring impediment to inter-organizational collaboration concerns contrasting practical imperatives. By this I mean incompatible bureaucratic practices. All organizations, if they had their way, would design rules and impose structures and practices to suit themselves and follow their mandates (Peters 2001). This of course follows the bureaucratic rationality of designing institutions to serve their stated goals, which is fair enough. That, however, carries costly implications for collaboration. Bureaucracies—even among organizations putatively of the same type—run differently. When they collaborate, it takes time for staff to figure out how to read each other's forms and so forth. This has motivated sundry initiatives for standardization, which typically face resistance, out of a preference for the "nightmare we know" rather than "nightmares unknown." The upshot is that organizational bureaucracies are designed to serve themselves with limited regard for other organizations unless forced to. Little snags and incompatibilities therefore arise and accumulate, increasing the need to translate and troubleshoot. Categories of personnel titles, contractual instruments, legal conventions, and cost items are thus often not the same, even if they sound the same or have the same names. It is therefore a significant task in inter-organizational collaboration to learn the collaborating organization's bureaucracy. As with academics learning the epistemological code of another discipline, appreciating another organization's bureaucracy may seem like a masochistic waste of time. I submit that it is not only worthwhile but revelatory. It not only constitutes an investment that attains to understanding, which will save time when snags arise later. I repeat the warning: bureaucratic differences lead to little snags and delays that gradually accumulate over time. Learning a bit about a partner's bureaucracy now can reduce the time lost due to snags downstream. Furthermore, making the effort to understand in turn establishes relationships and generates goodwill by partner staff on the recognition that they too have to put up with bureaucratic miscomprehension. The effort to bridge is thus appreciated, as knowledge gained establishes lines of two-way communication, and the bridge builder becomes a "go-to" person for questions about both sides.

I hasten to add that organizational self-interest and inequalities affect administration in collaborative endeavors. Self-interested and stronger partners may impose their wills on their partners in the form of unexplained bureaucratic protocols because it is administratively convenient. This disadvantages more open or weaker institutions, which can lead to downstream failures to complete tasks and deliver products. As a result, bureaucratic imperialism causes problems for all involved and thereby leads to collaborative inertia. Part of collaboration thus requires extensive communication among the key players operating on the bureaucratic level in support of practical imperatives. This is not to say that clarifying bureaucratic protocols resolves all administrative problems: indeed, communication may make frailties clear, including in ostensibly strong organizations. That is a good thing, because it forces critical interrogations over how things (do not) work, and therefore negotiations over how protocols might be modified. Figuring out how to translate among bureaucratic protocols permits greater efficiency of administration when collaborating across organizational types. Success means that people get hired and paid, invoices get processed, travel happens on time, work gets done, and reports get submitted correctly (and on time), and the partners look good to the funder, the government, and the stakeholders.

For these reasons, it is unwise for collaborators to complain to or threaten bureaucrats, particularly at their home institutions. Administrators and their staff might at first glance look like obstacles in the form of rule-following Nazis. What is more, they might even respond with fear upon receiving the first request for an exotic administrative procedure for interacting with another type of organization. But the up-front investment of being proactive with bureaucrats—getting to know them, learning their responsibilities, understanding their protocols—serves the larger strategic purposes of collaboration. Not only does it permit better collaboration within one's institution to get things done administratively, it affords the ability to explain what's going on to collaborators in another organization. This effort is extremely valuable to the partners, especially when the protocol in question involves the movement of money and/or deadlines. At the same time, being proactive provides the opportunity to explain the purpose of the collaboration to the administrators. In my experience, this is highly motivating to administrators and staff, which could be crucial the next time a glitch occurs, especially in moments of crisis when something must happen quickly. Engaging administrators and staff, both at one's home institution and in collaborating entities, not only motivates the bureaucrats, it opens more collaborative doors. Suddenly bureaucrats are no longer obstacles, but just the opposite: they become facilitators who in effect are part of the team. In the context of inter-organizational collaboration, this is virtually priceless, and is likely to bear beneficial ripple effects to partner organizations.

NATION-STATES AND INTERNATIONAL COOPERATION

Crossing disciplinary boundaries is increasingly common. Crossing organizational boundaries is more of a stretch, though there are strategic practices to manage the exigencies. Crossing national boundaries can be rather more complicated than either of the two previous cases, because national boundaries are political boundaries and often reflect long-standing divisions with profoundly different cultural histories (e.g., Chasek 2000; Gaillard 1994; Young 1989). If different disciplines and organizations have distinct cultures, they are not generally as deep and implicit as those of countries. As a result, there are more rigorous issues involved when crossing national boundaries, which are likely to take more time and effort to manage.

The additional complications run the gamut from minor inconveniences to outright prohibitions. One simple example concerns the logistics of travel among countries. There are invariably little logistical issues like passports and visas and overnight flights and new food. They all apply to any international travelers, and ideally they all fall at the "minor issues" end of the continuum of complications. But even these seemingly minor questions can become significant challenges depending on the countries in question and the content of the collaboration at hand. Moving toward the more complicated middle of the continuum, there are questions tied to authorities who regulate who gets to cross national boundaries. Doing conservation and development work may variously be construed as "research" or "business" or "other." Consulates make the decisions as to the category to be used; they also determine and implement the regulations pertaining to the extent of the paperwork required to secure the relevant entry document. Assuming the visa is issued, national police in ports of entry also get to review the document secured, which may entrain awkward Q&A sessions. Continuing toward the very complicated end of the continuum, politics can determine what happens. If diplomatic relations between two countries are strained, merely securing documents to travel between them can become particularly difficult or impossible. Requesting travel authorization can itself become grounds for suspicion by authorities.

There are other logistical issues that merit mention. Crossing national boundaries leads to complications in scheduling due to differing calendars. While days of business during the week are more or less standardized in many parts of the world, school terms and holidays vary considerably. Some countries also have days off when the national football (soccer) team is playing in the World Cup. The more countries one works in simultaneously, the more elaborate the scheduling gymnastics can become.

Similarly, understandings of time and deadlines vary among countries. Going from the United States to Latin America involved some education about "English time," where 3 PM meetings begin at 3 PM, as opposed to "Latin time," where 3 PM meetings begin at somewhere between 3:30 and 4. Also interesting is the effect of collaboration across boundaries on how partners view time. Meeting onset times with Latin American partners have increasingly been on English time. I'm told this is due to the influence of having Latin Americans visit the United States and finding it advantageous to hold meetings on English time. Here I must confess that the influence of partners has also gone the other direction. After grumbling to myself for years about the "late" onsets of meetings on Latin time, I now routinely start meetings in the United States up to thirty minutes after the scheduled time. Now I hear others grumbling.

Differing notions of time are but one among many cultural complexities that come into play when crossing national boundaries. Some are decidedly heavier than others in their origins, consequences, and mutability. There is pervasive selectivity in the transmission of information about different countries, whether due to economic and political relations, nationalist ideologies, media representations, or other reasons. The result is the formation of stereotypes with selectively positive or negative perceptions of other countries. Crossing national boundaries therefore requires management of those stereotypes by learning what they are and working by both spoken word and behavior to provide a more directly grounded basis for interpretation by both sides. How exactly this is accomplished involves many strategic practices. There is of course the traditional method of having a beer (or the national drink, if not beer) with the prospective foreign collaborators, as a means of facilitating a long evening's wide-ranging conversation. Workshops with substantial time dedicated to introductions, motivations, expectations, perceptions, and related issues can also help. I find that the larger the group and the more formal the setting, the less stereotyped perceptions can be addressed. In any event, a few alcoholic beverages or a day in a workshop are not themselves sufficient. Event-based approaches can only kick-start the process of undermining cultural barriers; they cannot obviate selective perceptions. Events provide collaborators the opportunity to begin to question stereotypes and thereby open a space for dialogue that can permit better understanding of why those perceptions exist and foster revisionism as to their validity. The key is to call attention to cultural perceptions beginning with introductions and to use every opportunity in the various tasks of collaboration to address stereotypes in order to maintain dialogues about cultural assumptions. While this requires extra effort, failure to attend to cultural perceptions is likely to lead to misunderstandings and conflicts later.

Crossing cultural boundaries among countries becomes more complicated when working simultaneously in multiple countries. Not surprisingly, collaborators among pairs of other countries often have cultural stereotypes of each other. This is particularly important for collaborators from neighboring countries with shared borders, as those countries have shared histories, including histories of how their borders came to be established. In the MAP frontier, Acre was once part of Bolivia. How exactly Brazil came to possess Acre is the subject of very different accounts depending on who you ask.

Beyond shared histories, inequalities among the countries represented can be extremely important for relations among collaborators. Representatives from larger, stronger, and thus historically more aggressive countries often dominate proceedings, whether consciously or not (Stewart and Gray 2006). This is likely to reproduce historical relationships defined by inequality and thus generate resentment. Crossing national boundaries among multiple countries thus requires a priori historical study and some careful consultations among potential partners. This can be quite extensive before the first meeting for a joint activity. Indeed, it may take time and effort just to get the key players to agree to come to the same table to talk.

In practice, crossing national boundaries extends the task of introductions and communication, which may be dominated by issues of politics and inequalities. Here the strategic practices of managing inequalities become crucial. There is a need to open spaces for dialogue about past histories in order to allow for the venting of grievances, even if they are not personal. Cultural perceptions mixed with shared political histories usually result in competing accounts with rather different interpretations, which deserve to be aired out and collectively interrogated. This is very similar to the critical interrogations of differing and often incommensurate assumptions found in different disciplines. Collaboration is not going to be viable if the prospective partners are unwilling to examine their views and have them examined by others, all the way around.

The cultural dialogues necessary to permit effective collaboration across national boundaries in turn reflect other aspects of divisions among countries. For simplicity's sake, I divide them into the macro and micro scales. At the macro scale, complications to collaboration stem from differences in national administrative systems as well as governmental policies. At the micro scale, crossing boundaries complicates collaboration due to differences among partner organizations as they work to navigate the cultural, institutional and political vagaries among countries. Hence, the first helps in understanding the second. In classic sociological fashion, one's context permits interpretation and hopefully identification of one's options and constraints.

While countries have broadly similar legal and administrative systems, there are various differences too, and the implications of the differences are often not evident up front. Focusing on one set of goals, such as conservation and development, hardly prompts one to dig into legal frameworks and administrative protocols as a matter of course. Instead, the tendency is to learn about these things only when necessary, which all too often means in a moment when deadlines loom and documents have to be reviewed and signed and notarized and approved, quickly. Not surprisingly, in such moments of crisis, legal frameworks and bureaucratic systems look like obstacles rather than systems with a rational logic. All of this is unfortunate and unnecessary.

It takes time to figure out how things work and why. This is necessary for a variety of procedures that are required to formalize collaborative relationships across countries in order to get joint work done. The establishment of contracts and other types of agreements, the payment of salaried and hourly personnel (and consultants), among many other key procedures, all require clarity within multiple legal and institutional frameworks. As noted earlier, bureaucracies were created with themselves in mind, not other bureaucracies. At the level of nation-states, the same phenomenon occurs, but on the deeper foundation of legal bedrock on which bureaucracies sit. Hence, working across countries means dealing with bureaucracies that may be very different in certain respects, which can lead to procedural gymnastics that may range from the awkward to the complex and occasionally to the unworkable.

Having spelled out the "nightmare unknown" magnified up to the formidable scale of the national level, it is crucial to recognize that all is not lost. For just as countries have histories with their own unique high and low points that help one understand their cultures, they also have particular legal frameworks that permit comprehension of their administrative regimes. They may not be the "nightmare we know," but this does not mean they are any better or worse. I also contest the cultural stereotype that it is always foreign country X that has "the worst administrative system in the world." This to me says as much about cultural perceptions inside the head of the person complaining as the administrative system in question. Legal frameworks and administrative systems have interpretable logics. What is more, the effort to understand them is not only necessary but also richly rewarding.

The key to the game is to recognize that institutions are by definition designed to accomplish specific tasks in a standardized fashion. Legal institutions in different countries share many of the same responsibilities in order to perform specific functions; the same applies to banks, businesses, universities, and so forth. In some cases, the administrative structures that perform a given function may be situated in different institutional structures, but once the administrative units responsible for a given function are located, the devils tend to be in the details. At that level, the devils follow specific orders.

Here the importance of collaborators takes on a whole new dimension. When crossing national boundaries, foreign collaborators become the "tour guides" to the legal frameworks and bureaucracies of relevant institutions in their country. This necessarily becomes a multidirectional dialogue, as the same applies to partners all the way around, who become each other's tour guides. The quality of the tours then depends on how well the tour guides have done their homework in cultivating bureaucrats in their home country institutions. Collaborators would therefore do well to heed the earlier orientation to proactively and positively engage administrators and staff in the relevant organizations, in order to understand how and why key protocols operate. The emendation when crossing national boundaries is that all collaborators need to be prepared to analyze institutions in each other's countries insofar as they do not already have point people for dealing with procedures involving foreign organizations.

The upshot of the seemingly thankless quest to understand how a foreign bureaucracy functions is that one not only gains understanding of key procedures, one also earns other benefits in the process. Just as when crossing organizational boundaries, a collaborator becomes a bridge for facilitating navigation of procedures in bureaucracies in different countries. Do not underestimate the value of this knowledge; depending on one's institutional history, it may be rare. This permits reporting back to one's home institution with information, which can be vital if they cannot procure it themselves. While legal frameworks are relatively easily available for review by lawyers, bureaucratic protocols are not always posted on the web. Even if the information is available, engaging foreign organizations permits establishment of points of contact. This humanizes the organizations and especially the people in them who become contacts in moments when quick answers are crucial. In turn, the act of bringing back information is often greatly appreciated by the home institution bureaucrats. Even if the information is available, it is one less thing they have to hunt down themselves. My experience is that this builds appreciation and respect among administrators and staff, especially if they are not accustomed to international collaboration. Hopefully one's collaborators will be doing the same things in their home institutions. All of this in turn facilitates efficient administration across national boundaries, which will sooner or later be crucial in a future crisis involving a tight deadline.

Beyond the questions of institutions and administration, crossing national boundaries presents additional complications at the macro level in the ambit of public policies, state priorities, and diplomatic relations among countries.

Nation-states are by definition territories with sovereign governments. International collaboration on conservation and development may or may not square with the policy frameworks, priorities, and foreign alliances of the government where the proposed work is to be conducted. In countries where environmental regulatory institutions are weak, international collaboration may in some instances be a welcome complement, particularly if there is an emphasis on capacity building or supporting establishment of priority protected areas. That said, it is also possible that countries have weak environmental regulatory institutions precisely because effective regulation would inhibit economic growth. Where countries rely on natural resource exports for foreign exchange and state revenue, proposals for international collaboration to improve conservation outcomes are likely to be suspect. Arguing for alternative development pathways along more sustainable lines may lead to tensions insofar as vested interests feel threatened. Similarly, international collaboration between countries with diplomatic tensions can become something of a post-modern phenomenon. Election years, key international meetings, and other politically sensitive moments can lead to declarations by governments that can greatly affect the viability of international cooperation. Here I note that conservation and development funding across national borders often seeks to influence policy or policy implementation. This can itself be sensitive. Go with the understanding that if diplomatic ties are tense, the plug can be pulled at any moment.

The issues just described can be particularly tricky when collaborating across borders shared by countries with tense histories or diplomatic relationships. Even if the government of one's home country has positive relations with the countries where one is collaborating, those countries may not. Not surprisingly, to the extent that there are tensions among neighboring countries, spanning national boundaries may be frowned on for geopolitical reasons. This is because tensions among countries often become manifest on the ground in frontier areas around the borders between said countries. Even if the local populace doesn't care or doesn't share the preoccupations of their central governments, those governments nonetheless have various means at their disposal to restrict cross-border activities, including collaboration for conservation and development. National security agencies, various branches of the police and military, and of course consular services may impose restrictions or prohibit certain kinds of activities, even if their content has little to do with the specific issues of contention. This is a key point that bears repeating: restrictions on conservation and development work in frontier areas may not have anything to do with conservation and development itself. Diplomatic ties are dictated primarily by national and not regional or local interests. Governments "see like states" and by definition their top priority is management of their territory (Scott 1998). International cooperation is frequently valued, but it is unlikely to be a top priority.

At their limit, the complications presented by macro level questions of sovereignty, policy, and diplomatic relations are insurmountable. That said, these questions are rarely at their limit and thus the complications are rarely insurmountable. Management of political considerations across national borders in many ways parallels the strategic practices outlined above for administration. Proactively engaging consular staff on a positive tone can be helpful in humanizing the rationale for policies of another government as regards its handling of requests for entry. It can also provide useful information as to the challenges consulates face in implementing policy by making decisions about visas, entry, permits, and so forth. Further, engaging one's foreign collaborators on the historical and political context surrounding policy questions is often enlightening. This includes identification of what one's foreign collaborators do not know. A visit to their government's consulate in one's home country can provide complementary information, which can help connect some dots as to what is going on and why. Assuming one has gained entry but the issues instead reside with permitting to pursue certain kinds of activities, like collecting data in frontier areas, further consultations with colleagues and relevant agencies is also in order. This is where networks of in-country collaborators are priceless as sources of information from people who have their own contacts in government. To the extent that collaborators have taken the time to engage and humanize personnel in governmental agencies, information if not also direct access may be feasible. This of course hardly guarantees policy changes or rapid processing of permit applications, but it provides illumination as to the issues surrounding restrictions and the prospects for getting to work.

These observations lead to micro-level considerations involved in crossing national boundaries for collaboration. As noted earlier, the macro context in many ways sets the rules of the game for what happens at the micro level among partners. Indeed, many of the strategic practices suggested in response to exigencies of the macro context involved specific responsibilities on the part of foreign partners. Those responsibilities of course presumed that the foreign partners are in fact committed. This is where additional micro-level challenges and complications arise when crossing national boundaries for collaboration. Establishing personal rapport, gaining an appreciation of cultural sensibilities, and spending time with collaborators where they live and work are only the beginning. Similarly, letters for visas, signatures on applications for permits, finalizing agreements and other documents are just the start. Beyond the cultural and organizational issues already outlined, crossing national boundaries makes significant demands on all of the collaborators, which require particular attention.

Working across national borders becomes an exercise in responding to numerous sources of events and circumstances over which the partners typically have no control. Assuming that the cultural introductions have been navigated and not a few beer bottles drained, the macro-level exigencies will repeatedly test the commitment of collaborators from different countries. Every such test will present challenges as to how the collaborators respond, and how they respond to each other. How one partner chooses to respond becomes a possible occasion for the others to interpret the response in the frame of stereotypes stemming from cultural perceptions. Then there likely won't be time for a chat over a beer or a quick workshop.

Here the issues of language competency and effectiveness in translating become crucial. Even with solid efforts at introductions and the establishment of effective communication among partners, rapport becomes challenging to sustain when time-sensitive demands appear, whether they stem from complicated bureaucratic procedures or changing political circumstances. In both cases, the partners have to work hard to translate terms clearly and explain what is going on, and more importantly, how they are responding and why. This makes for long e-mails and longer phone calls, because the situations are often not simple and the responses may involve many steps. Patience is priceless, especially since there is a tendency for stuff to happen in one country and affect the other partner, and then the converse to occur with the roles reversed.

Extenuating circumstances in one country thus demand that partners make decisions that affect their foreign collaborators. Changes in bureaucratic protocols that modify procedures used by partner organizations, public political criticism of the collaborative venture, and replacements of key individuals are all examples of externally induced complications that test international partner commitment. The in-country partner has work to do on behalf of all of the collaborators, and may be unable to overcome the complication at hand, often due to no fault of its own. Alternatively, the in-country partner may have to make sacrifices in order to handle the complication, including engaging in conflict with others at its home institution to keep a key person involved in a collaborative project. This carries political costs to the in-country collaborator at its home institution.

In either event, these are moments that require extensive consultations among collaborators before making decisions. Any decision under duress can readily be interpreted in various ways which can cause confusion and mischief. Decisions may be interpreted in terms of partner interests. Such interpretations can lead to understandings of decisions as exercising the self-interest of the deciding partner, or the shared interests of collaborators, with exactly opposite implications for partner relations. Decisions can also be viewed in light of the structural context, where outcomes may be interpreted as the result of a lack of attractive options, or to incur some political cost in the name of the shared goal of the collaborative venture, also with very different impacts on partner relations. Because of the potential to reach very different interpretations, moments of decisions are moments of long e-mails and phone calls and impromptu meetings. All parties need to understand the situation and what is at stake in terms of the consequences of the options at hand. Will a partner go to the bureaucratic or political mat in an institutional wrestling match to uphold the collaboration in order to achieve a shared goal? Ideally, the collaborators should jointly take such decisions. In practice, the decision usually falls to the partner in the country with the complication. This is no small responsibility. What is more, in many instances, there is no time for consultations before, only explanations afterward. Whatever the decision and however it is made, the deciding partner needs to explain its rationale so the collaborators understand. Even in the case where a partner decides to go against its own organization or otherwise incur costs at home in order to serve interests shared with collaborators, this needs to be explained. The collaborators may not agree, but they must understand and respect the basis for the decision taken.

Such decisions are uniquely teachable moments: everyone involved learns a lot about themselves, and collaborators learn about each other. In many cases, decisions are straightforward: the in-country partner accedes to the demands of the structural context, because the costs of not doing so are very high. The consequence is that the partnership is in some way inconvenienced. In other cases, decisions are more difficult because the costs may not seem so high, though they are certainly not zero, and/or there is a unique opportunity to support a collaborative effort toward a key goal in a crucial moment. Both cases are teachable moments, but the second case is especially valuable. There, one partner is making a sacrifice in a crucial moment on behalf of another, indeed, on behalf of a shared goal. Partners need to recognize and remember those sacrifices.

Decisions to support or favor foreign collaborators, at a political cost to the local partner, stand to rapidly build political capital among collaborators. This is the silver lining to the daunting complications of crossing national boundaries for collaboration. For if the contextual exigencies of collaborating across national boundaries are formidable, so too are the loyalties that can be won in recognition of sacrifices made. Decisions involving sacrifices need to be recognized at the time and remembered thereafter by all of the collaborators. What is more, the collaborators had better be prepared to do the same thing themselves in similar circumstances. It bears repeating: what comes around goes around when collaborating across boundaries, including national boundaries. This is how the long bridges among collaborators in different countries get built and especially how they are maintained.

In sum, crossing boundaries, whether among the disciplines of academia, organizations with contrasting mandates, or countries with distinct cultures and institutions, substantially complicates the challenges of collaboration. The tasks become more difficult, starting with communication, due to the additional challenges of spanning divides. Fortunately, there are strategic practices for crossing boundaries, some of which are rather similar regardless of the type of boundary being crossed. Indeed, reflection on one's self-knowledge before even beginning is a significant asset, whether one is reflecting on one's home discipline, institution, or country. This is because crossing boundaries changes one's perspective, and looking back in from the outside is eminently useful preparation for the challenges to come, beginning with the introductions as to who one is.

The complexity of the challenges of course depends on which boundaries are being crossed, and especially on how many boundaries one proposes to cross simultaneously. Not surprisingly, the complexity compounds when crossing more than one type of boundary in a collaborative venture (Perz, et al. 2010). Insofar as the issues of boundary crossing are distinct when crossing more than one boundary, the complexities of crossing one type of boundary are compounded by the complications of crossing another. The exigencies of the tasks and challenges certainly become more entertaining, since the snags and impediments are multiplied. The employment of strategic practices in turn becomes heavier going.

In particular, the arrangements of collaboration become more difficult to navigate. Here the structural issues of how to design collaborative arrangements becomes an issue. When crossing multiple boundaries, it becomes vital to recognize that the differences among disciplines, organizational types, and nationalities make the formation and maintenance of collaborative relationships more complicated. Each task requires more effort due to the need to pay attention to epistemological and administrative and cultural differences, among others. One way to approach this complexity is to adopt an analytical perspective on the structures of collaboration, paying particular attention to the boundaries being spanned by collaborators. The next chapter offers such an analysis by discussing structures of collaboration, their advantages and disadvantages, and design principles to support crossing boundaries, especially when spanning multiple divides simultaneously.

Chapter Five

The Analytics of Crossing Boundaries

The previous chapter offered a discussion of the rigors of crossing boundaries for collaboration. I paid particular attention to what can happen due to complications that confront academics from different disciplines, personnel from different types of organizations, and citizens of different countries when they collaborate. The discussion frequently skirted into issues of relationships among collaborators. That raises a central question: how to arrange partners in specific structures to ensure that their collaboration is effective.

How collaborators are arranged can greatly influence how well they work together. I suggest that boundaries are important features to consider in the arrangement of collaborators. This is because relationships that span boundaries face additional challenges. Collaborators from different disciplines, types of organizations, and countries bring with them specific practices that constitute their established styles of working relationships. Those relationships in turn are defined by social structures in their home context that delineate who is in charge, how responsibilities are divided, and chains of command for managing those responsibilities. Collaborators generally come expecting more of the same from counterparts including those across divides, though they are also likely aware that there may be differences. The consequence is often confusion over specific aspects of those differences, followed by frustration when the differences cannot be readily resolved in favor of the operational preferences of one's home institution. Thus when crossing boundaries, practices must be reconciled, and said reconciliation reveals innumerable devils in the details.

The design of collaborative structures therefore requires particular attention to boundaries. Relationships that span boundaries often demand heavier investments by collaborators. This is not only to avoid confusion and frustration among pairs of collaborators on an operational level. It is also to array groups of collaborators in larger structures to capture cross-boundary collaborative advantage. Crossing boundaries thus prods one to take "the view from the other side" of one or more divides for purposes of designing collaborative structures with workable relationships. This implies that crossing boundaries is not merely an exercise in taking another's perspective in a cultural sense, but also in seeing the collaboration in structural terms as viewed from both sides of the bridge.

This chapter therefore pursues a systematic analysis of structures of collaboration, highlighting structures that span boundaries. I first offer a framework that accounts for specific design considerations that can influence the effectiveness of collaborative structures. These considerations include the scale of collaboration, structures that define relationships among collaborators, functional complementarity, and modes of collaboration. I then focus on how all of these design considerations are affected by boundaries. The second part of the chapter then portrays a series of cases with increasingly complex collaborative structures. These cases include the Amazonian projects introduced in Chapter 2, and provide a means of illustrating the challenges that arise in arranging collaborators, especially across boundaries. The cases serve as a point of departure for a discussion of how structures can lead to difficulties, which motivates concluding reflections on how effective design can mitigate problems.

In the process, this chapter draws insights from literature on social network analysis, business management, and organizational sociology. All three offer useful and complementary insights for the task of effective collaborative design. Whereas network analysis attends closely to identifying ties among agents and characterizing how ties permit flows of information (e.g., Freeman 2004; Harary 1969; Scott 1988), business management provides a thorough review of issues pertaining to the organization of collective effort toward a shared goal (e.g., Daft 2005; Gittell and Weiss 2004; Gittell, et al. 2010; Gomez-Mejia, et al. 2008; Heckscher and Adler 2006; Ott, et al. 2003), and organizational sociology offers a suite of perspectives on how horizontal as well as vertical ties operate in formal institutional contexts (e.g., Burton, et al. 2011; Courpasson, et al. 2012; Handel 2003; Reed 1992).

A FRAMEWORK FOR THE DESIGN OF COLLABORATIVE STRUCTURES

I begin by offering a framework for the design of collaborative structures. The framework highlights several key considerations in the design of collaborative structures: 1) the scale of collaboration, 2) arrangements that structurally define collaborative relationships, 3) the functions of personnel, 4) the modes of collaboration, and 5) the boundaries being crossed, particularly as they affect the previous four design considerations. The heart of the framework concerns the centrally important issue of how collaborative structures affect the performance of key functions to achieve shared goals. Any collaborative relationship is necessarily defined by its structure and functions. Whereas "structure" refers to the arrangement of collaborators that defines their relationships, "functions" refer to the responsibilities of each collaborator. The game then involves designing structures that permit the collaborators to effectively perform their functions. The influence of collaborative structure on performance of function can in turn be greatly affected by the scale, mode, and boundaries crossed.

Scales of Collaboration

Scale is a fundamental element in collaborative structures, as the scale of collaboration greatly affects who directly relates to whom. In invoking the term "scale" I am following the conventions of geographers, who employ the concept with regard to both spatial resolution ("grain") and extent (e.g., Herod 2011; Lloyd 2014). In designing collaborative structures, scale as grain refers to the human units involved which can form collaborative relationships. Such units range from individuals, which when grouped constitute departments and divisions, which in turn comprise organizations, which in their turn can form alliances and create networks. Scale as extent therefore refers to the number of individuals or other units who are collaborating. Small-scale collaboration may therefore involve a few individuals, a couple of organizations, or otherwise relatively few collaborating entities.

The issue of scale, however, requires further clarification concerning collaborative relationships in practice. Human units can be seen as "social actors" capable of "agency," the ability to act. Agency may be bound by structure, and the relationship of structure and agency has been the topic of longstanding theoretical debate in the social sciences (e.g., Giddens 1984; Long 2001; Sewell 1992). While there is no question that individuals can make decisions and act on them, there is debate as to whether organizations or other aggregate social entities do so (Hindess 1986). This is due to the potential for disagreements and conflicts within organizations and alliances. While it is useful as shorthand to say "organization X has decided" something, the statement reflects a decision by a lead individual within the organization, likely after consultations with colleagues. By extension, if collaboration among individuals is straightforward as to the identities of the agents involved, collaboration among organizations in fact usually involves ties between particular individuals within those organizations. It is therefore important to maintain the distinction between the entities being represented and the agents actually carrying out collaborative activities. Especially in large organizations, most individuals may not be involved in or even aware of the collaboration.

This has numerous implications. One is that the scale of collaboration is best analyzed at the level of individuals, even as they operate within divisions of organizations which may be components of broader alliances. This permits identification of the relationships of collaborating individuals with regard not only to their counterparts in other entities, but also with regard to other personnel in their home institution. In turn, that affords a systemic analysis of the collaborator's internal and external relationships, which is necessary insofar as each is likely to affect the other.

Another implication is that if the focus resides at the level of individuals in their broader context, the scale of collaboration primarily refers to the extent of collaborative ties. If scale as grain refers only to individuals, then scale varies only in terms of extent, that is, the number of ties from one collaborator to others. Hence, whereas small-scale collaboration is chiefly a reflection of having a few individuals working together, large-scale collaboration involves numerous individuals. Here social network analysis offers useful terminology, highlighting as it does the proximity of ties (direct or indirect), the centrality of a given individual in a network (where more central individuals have more ties to others), and other terms that define the extent of a network (e.g., Scott 2000).

The key caveat in this discussion of scale is that collaboration refers to partnering among individuals who would not normally be working together. Whether they be individuals from different divisions, organizations, or other entities, collaboration among individuals implies collaboration among individuals from different organizational contexts. The result is that large-scale collaboration among individuals implies collaboration of individuals from different departments, organizations, or alliances. This might or might not imply large-scale collaboration across disciplinary, organizational, or national boundaries. While many collaborating individuals within a university might indeed come from many different disciplines, partners from different organizations might still come from the same organizational type.

Structures of Collaboration

Discussing scale brings up the question of how collaborative relationships are structured. Here it is worth elaborating on a definition of structure: an organizational arrangement that sets roles and responsibilities, and thus authority and expectations, for groups of individuals. Analysis of collaborative structures thus requires not only attention to the arrangements and content of relationships, but also how both of these constitute authority for certain individuals for the sake of coordinating work and achieving goals.

Work on social networks, business management, and the sociology of organizations highlights two broad types of models for structures of collaboration: hierarchies and networks. Hierarchies are characterized by the centralization of authority at the top of a chain of command (e.g., Burton, et al. 2011; Daft 2005, Chap. 10; Ostrom 2005). Hierarchies are centralized due to the structure of a few authoritative individuals at the top who command more individuals below. The primary flow of information is thus top-down, with a weaker bottom-up feedback. The emphasis on top-down and bottom-up flows of information means that hierarchies are highly vertical in their structural orientation. Hierarchies are embodied in their purest form in military organizations, and for long also characterized bureaucracies in GAs, business firms, and many NGOs.

Here it is worth noting that there are variations among types of hierarchical structures (Daft 2005, Chap. 10; Gittell and Weiss 2004; Gomez-Mejia, et al. 2008, Chap. 9). Businesses may have "functional" structures with hierarchical tiers defined by distinct functions of a business (research, manufacturing, accounting, etc.), or "divisional" structures with tiers organized around distinct businesses and customer bases. Some business organizations have opted for "matrix" structures that combine functional and divisional hierarchies, and yet others pursuing outsourcing strategies have adopted network structures via contracts with other companies. All constitute hierarchical structures (or at least structural elements) with varying arrangements for centralizing authority.

Hierarchical structures of organization have established advantages and disadvantages. The key advantage stems from the vertical orientation of a hierarchy, which defines a clear chain of command and allocation of responsibilities. Decision making in hierarchies is thus clear and efficient. Similarly, once decisions are taken, underlings in a hierarchy have their marching orders, so implementation should not be a confusing ordeal. Because it is clear who is responsible for what, accountability is also clear. That said, hierarchies also have disadvantages. One concerns the relatively limited bottom-up flow of information. This is constrained due to the centralization of authority at the top, and the fear of underlings for saying something controversial and paying a price. Related to this problem is the issue that there may be limited horizontal flows of information. This can make hierarchies inflexible as to their latitude to respond to opportunities or crises. In both cases, information from below or from another functional group may be crucial to decision making and the well-being of the hierarchy itself. The variations on hierarchical structures noted above can be viewed as proposals to facilitate information flows and thus decision making without fundamentally modifying the vertical structure and undermining its advantages.

The other broad category of structure concerns networks. Long the focus of social network analysis, networks are usually characterized in terms of webs of connections among individuals or organizations. Social network analysis has a long intellectual lineage, from early anthropological studies of kinship and sociological research on communities in the 1920s to multiple applications in various disciplines since then (Scott 1988; Freeman 2004). Interest in the structural properties of networks has primarily focused on traits such as network density and centrality, as well as the strength of network ties and the direction of information flows among agents (e.g., Scott 2000).

Highly connected agents may command more knowledge than others, but in network analysis, this is not generally translated into authority per se. This is because a key characteristic of networks is their primarily horizontal orientation. Networks typically encompass webs of ties among agents who constitute each other's peers rather than their bosses or commanding officers. Hence, networks exist among friends, extended families, communities, businesses, and other organizations. They do not constitute command structures with a vertical orientation but instead are horizontal structures in which members benefit from more or less co-equal relationships. Network members work with one another not because they have to, but because they see mutual advantage in doing so. This is not to say that networks have no politics or conflicts, but rather that they emerge as more or less spontaneous orders that are sustained by the voluntary willingness of agents to continue participation.

The advantages of networks reside in their highly organic and flexible structures. Networks are organic in the sense that they permit the relatively free and multi-directional flows of information among members. Entry into a network and acceptance can open numerous lines of communication and thus opportunities. Networks are also highly flexible in size and arrangement, which allows modifications in extent and form over time. The organic and flexible nature of networks permits considerable autonomy among participating members. The disadvantage of networks concerns their limited capacity to efficiently support collective action. Without a command structure, it is far from clear how even central agents in networks can reliably motivate (much less order) other agents to act in a coordinated fashion. Networks thus emerge as rather inefficient structures for collaborative action.

As structures of collaboration, hierarchies and networks have highly complementary advantages and disadvantages. The organic flexibility of networks makes them especially useful for the constitution of collaborative arrangements, while the clear chain of authority makes hierarchies particularly effective at collective action toward shared goals. This has prompted numerous proposals to find ways to somehow combine the advantages of both. Beyond modifications to hierarchical structures, various collaborative structures have been identified that incorporate elements of hierarchies and networks.

One established model is a consortium, which can be described as a network among partner organizations, each with a hierarchical structure. This permits multi-directional information flows among partners. A consortium usually has a "lead" member organization with hierarchical authority over the others, so there remains centralism in decision making. The environmental management projects described in Chapter 2 include examples of consortia with a lead organization and members.

Similarly, confederations embody structures of organizations linked as in a network, with particular rules for membership. There is an umbrella organization over the members, as in sports leagues. Hence, a confederation differs from a consortium in two respects. First, consortia are led by one of the members while confederations are led by a uniquely supervisory entity. Second, confederations tend to be substantially larger than consortia. Both, however, embody networks among members with centralism in decision making.

Consortia and confederations are distinct from alliances, which exhibit a looser organizational structure. Alliances amount to networks of groups or organizations, but they lack a "lead" or supervisory entity. They do have recognizable leadership, usually in the form of an elected or selected or rotating individual, committee, or member organization. But no one group has unequivocal authority over the others, which results in polycentrism in decision making. The MAP Initiative can be viewed as an alliance with a tricentric organizational structure, with individuals from each side of the trinational frontier forming collaborative working groups. While the MAP Initiative had organizers, it lacked a vertical command structure and instead relied on shared interests to motivate participation.

Beyond such multi-organizational structures, there is also work on "heterarchies," which may describe organizations or multi-organizational structures. In either case, heterarchies retain vertical structures without the neat order of hierarchical authority (Beekman and Baden 2005; Birkinshaw and Morrison 1995; Ehrenreich, et al. 1995). Heterarchies may have specialist authorities for particular functions or divisions, but may still lack a single authority figure at the top. Alternatively, heterarchies may exhibit substantial horizontal ties along with vertical authority. Heterarchies thus permit coordinated action on specific activities, or substantial lateral information transmission, but not broadly coordinated action. Like the other "crossover" models that attempt to capture the benefits of both hierarchies and networks, there are limitations due to the significant tradeoffs at play. Managing the advantages of hierarchies and networks thus amounts to something of an organizational "holy grail" sought in collaborative endeavors.

Functions in Collaboration

As noted in Chapter 3, collaboration can be unpacked into a suite of issues, each of which in turn encompasses an array of tasks. Accomplishing those tasks requires organizational structures where the responsibilities of personnel are differentiated among a variety of key functions. Regardless of scale or structure, there are functions necessary to perform in order to support collaboration if the goals are to be met. Even on the smallest scales of collaboration, there is a fundamental necessity for some modicum of "functional differentiation" among partners. That is, collaboration works better if there is a division of labor wherein each partner takes primary responsibility for certain functions in order to address the shared tasks at hand.

Here it is crucial to differentiate structures from functions as they pertain to collaborating personnel. If structures describe the arrangement of personnel, functions delineate the tasks performed by personnel. In business management, the typical ideal is to organize structures to make performance of functions by personnel as efficient as possible.

It is crucial to recognize that performance and differentiation of functions applies not only to collaborators, but also to their organizations. Even small-scale collaboration among a handful of individuals still requires some degree of functional differentiation and coordination with administrative personnel in their home institutions. The time and effort involved are easily underestimated, even in cases where boundaries are not being crossed, because functional coordination within and among organizations is a significant task. In more ambitious multi-organizational collaborations with many participants, the extent of functional complementarity multiplies, and functional coordination becomes much more complex and demanding.

As there is no shortage of tasks, there is an array of key functions to be performed in any collaborative endeavor. Lists vary somewhat, so in the context of conservation and development work, let me note the following functions as being particularly salient: strategic planning, program development, operational management, financial management, task implementation, and performance reporting. This list could of course be further elaborated. Operational management could be broken out in terms of scientific oversight, human resources management, and conflict management. Performance reporting could be broken out in monitoring and evaluation, reporting to funders, public outreach, and communications. Task implementation could mean many things including field visits, data collection, stakeholder workshops, and so on. Whatever the list, a key point is that the focus of collaboration in conservation and development work is in practice embedded in a much larger array of other functions.

In any event, the performance of functions depends on optimizing among three key imperatives. First are individual characteristics. Personnel should have the requisite training, skills, and experience to perform the functions in question. Hence, there are characteristics of individuals that define their capacities to suffice specific functions (cf. Locke 2000; Ott, et al. 2003). Even as a PhD sociologist, I have come to greatly appreciate Human Resources departments for the various steps and hurdles they now require during the hiring process. This is to protect not only potential employees but also employers.

The second imperative concerns structural characteristics. Structures should be arranged to ensure efficient coordination among personnel with closely interdependent functions. This means that having qualified people is not enough; they need to be structurally arrayed so that everyone knows what goes to whom and who answers to whom. Further, the design of effective structures goes hand in hand with the design of protocols for accomplishing tasks, particularly with regard to how long tasks require. If the first step in handling an invoice to hire a consultant takes three days to complete, the protocols need to clarify this so personnel know what to expect. This avoids unnecessary conflicts before going to the next step. Among collaborators, much the same applies when arranging meetings to set goals, defining timeframes for travel for field visits, and so on.

The third imperative pertains to the issue of collaborative practices. Personnel have to work well together. This is affected by the first and second imperatives, but personality issues also play a significant role. This issue merits mention here despite the focus on structure; I return to individual personality issues in Chapter 8. For present purposes, the quality of relations among personnel sometimes does not become evident until they are in fact together attempting to work. Some people interview well but then exhibit antisocial behavior once hired. Here again HR people are golden. In sum, the hiring, organization, and management of personnel to perform functions required by collaboration is in fact a central issue. Unqualified personnel, inefficient structures, and personality conflicts require attention, or they spell doom for collaborative efforts.

The upshot is that functions affect relationships among collaborators. If collaborator X of organization Y has solid staff, structure, and HR protocols for performance of key functions in support of the partnering, the work with collaborator A from organization B is greatly facilitated. Accomplishment of tasks is likely to face fewer challenges. On the other hand, if collaborator X has a weak organizational home in Y, and staff, structure, and HR do not permit efficient performance of key functions, work with A from B will be hindered, and perhaps also the relationship of X with A. Personal, organizational, and relational capacities all bear big implications for performance of functions and thus what is possible in collaboration. This is not to say that one should never work with weak organizations. Rather, the point is that weaknesses constitute challenges that require greater reliance on strategic practices.

Modes of Collaboration

The foregoing discussion of functions and their imperatives motivates a further comment on "modes of collaboration." I refer to the strategy for organizing collaborative structures as they relate to functions. There are distinct modes of collaboration based on contrasting strategies for arranging personnel based on different functions. Figure 5.1 presents two general options among modes of collaboration, based on strategies that highlight "functional complementarity" and "functional redundancy."

Functional complementarity highlights functional differentiation. In the functional complementarity mode, everybody does something different. The typical organizational path is to define functions, identify a specialty for performance of each function, and allocate a specialist to each function. This builds on the classic argument that specialization permits efficiency in performance of tasks. What is more, functional differentiation into specialties

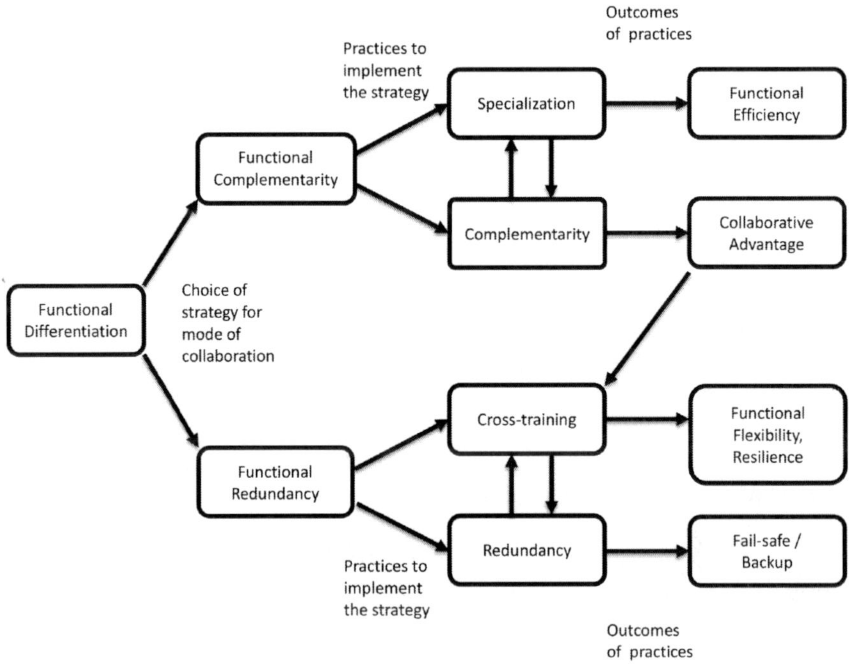

Figure 5.1. Modes of Collaboration.
Source: Created by author.

maximizes functional complementarity among collaborating personnel. In turn, functional complementarity promotes collaborative advantage. In practice, one need only add people to the enterprise if the newbie adds a necessary function that complements those already being performed.

However, efficiency is not everything, and functional complementarity can be viewed as one mode among others. Functional redundancy constitutes an alternative mode of collaboration. Here I draw on resilience thought as applied to management (Walker and Salt 2006). Resilience refers to the ability of a system to maintain its key functions in the event of a perturbation (Holling 1973; Gunderson and Holling 2002). A frequently cited factor that supports system resilience is redundancy among system components (Adger 2000; Gunderson 2000). In ecological literature on resilience, species diversity in an ecological niche is valued because the loss or removal of one species does not eliminate that niche in an ecosystem (Levin 1999). Consequently, managing ecosystems for resilience features the maintenance of a variety of habitats with many species and ecosystem services.

In the context of collaboration, collaborative structures can be viewed as systems. By extension, the collaborators constitute system components, their functions become key system functions, and their relationships system processes (Cumming, et al. 2005). Viewed from a resilience perspective, managing the collaborative structure to maximize specialization and complementarity is a recipe for disaster. Functional complementarity may indeed permit efficiency and collaborative advantage, but only in the absence of system shocks. In the presence of shocks, whether due to external or internal changes, all such bets are off. If something happens and one of the collaborators drops out, a key function is lost, various inputs to other functions cannot materialize, and the system grinds to a halt or crashes entirely. Hence, while efficiency and collaborative advantage are definitely valuable, they bring perils to the viability of collaboration itself.

Functional redundancy seeks to avoid the problem of a lack of resilience. Functional redundancy means managing collaborative structures so that more than one collaborator can perform a given function. This can be achieved either via cross-training among partners, whether via organic social learning or formal skills training, or by including more than one collaborator who can perform a function. Cross-training permits functional flexibility, so if one collaborator is unavailable to perform a task, someone else can temporarily step in. Similarly, having multiple collaborators who can perform a function provides a fail-safe in the event that someone is unable or unavailable. In either case, the loss of a collaborator does not mean the loss of a key function.

Of course, people are not just automatons and collaborations are not merely systems that occur naturally. If there are perils in managing collaborative structures for complementarity, management for redundancy has its own dilemmas. There are invariably devils in the details of how functions are performed, and switching among collaborators will not necessarily be smooth, especially since switching may occur in moments of crisis with deadlines looming. Further, collaborators are self-aware and have varying relationships with each other, which implies that some people may resent being asked to handle a function that is not their favorite or to work with people they dislike. Insofar as specialization simplifies tasks, functional redundancy brings with it the challenges of pursuing more tasks.

There are thus tradeoffs among modes of collaboration, such that managing for efficiency tends to undermine resilience. For a collaborative structure of a given scale, complementarity comes at the expense of redundancy, and vice versa. Despite the tradeoff, it is a false choice to manage for one mode or the other. There are multiple outlets for dealing with this tradeoff. One is to expand the size of the collaborative structure. Adding collaborators can build complementarity and redundancy, as when one adds two new collaborators, one of whom is a specialist in something new and the other has capacities similar to existing personnel. Of course, the larger the scale, the more complex the structure. Another outlet is to optimize on complementarity and redundancy by including collaborators with multiple specialties, some of which are unique but others which are similar to those of other collaborators. Then the devils reside in the details, for some specialties and capacities are easier to acquire or replace than others. A final outlet is to set explicit, mutually agreed upon expectations for cross-training or social learning among the collaborators. Pursuit of this strategy implies that regardless of the pre-existing collaborative structure, even one designed to maximize complementarity, redundancy can emerge over time as collaborators learn from each other. Among committed collaborators, complementarity with social learning can thus feed redundancy.

Boundaries and Collaborative Structures and Functions

To this point the discussion has assumed that collaborative structures do not cross disciplinary, organizational, or national boundaries. I skirted into issues of crossing disciplinary boundaries when discussing functions, such that different functions often require distinct capacities stemming from different training backgrounds. Now it is time to dive into the issue of boundaries.

Crossing boundaries fundamentally changes the game with regard to collaborative structures and functions as discussed so far. For one thing, crossing boundaries significantly increases the complementarities at play, and this bears implications for functional differentiation. The key issue that arises is that when crossing boundaries, differences in the knowledge bases, experiences, and capacities are significantly expanded. One consequence of this is that a given function may now have to be split into multiple functions because of the complementarities of the collaborators contributing to that function. Constructing collaborative structures among specialists in discipline X among universities A, B, and C is one thing. The specialists likely have similar knowledge bases, so they do not exhibit functional complementarities. Similarly, their home institutions likely have similar organizational charts, so constructing collaborative structures by matching up support staff is unlikely to be ambiguous or difficult. On the other hand, constructing collaborative structures among specialists in disciplines X, Y, and Z is another matter entirely, particularly if their home institutions also happen to be in different countries. There are now very different disciplinary knowledge bases as well as contrasting national contexts for the home institutions. Important functional complementarities in the technical content of the collaboration now appear due to the different disciplines represented. Even if the technical design of a research or management activity is still considered one function, the content of the technical design is now effectively a combination of functionally complementary knowledge bases. Similarly, even if the support staff from the partnering universities have similar organizational charts, it is very likely that they will take on functionally distinct roles such that staff at the university in country A will deal with travel to country B, where the support staff will be focusing on local logistics. Support staff in both places are still dealing with travel logistics tied to implementation, but they must coordinate with each other, effectively taking on distinct and complementary functional roles. To continue the example, staff at A need to know that staff at B have local logistics ready before staff at A make flight arrangements for their personnel.

This is just the beginning. While crossing boundaries increases the complementarities at hand, it also tends to reduce the redundancies. Just as the contrasts of knowledge bases, skills, and experiences across boundaries tend to differentiate functions, that same differentiation makes it harder to match collaborators on a given function. To an extent this is not a problem insofar as functional differentiation across boundaries may not require redundancy. Following the example above, if partners from multiple countries are all traveling to a country to work, there may only be a need for local logistics management by one partner in the destination country. But the expanded functional differentiation across boundaries can nevertheless create challenges with regard to management for redundancy. This becomes very difficult to ensure to the extent that the requirements for key functions become more specialized, organizational and individual capacities differ across boundaries, and functional redundancy is required across those boundaries. If a watershed management planning activity spans a national boundary, the activity is likely to require hydrological specialists based on both sides. If one is not available on one side, this creates risks to sustainable collaboration, and thus effective management planning. And if there are few hydrologists around on that side, finding one will be difficult and perhaps result in pursuit of alternatives that may be less cognizant of the local vagaries.

There is more. Whereas the expanded complementarities of crossing boundaries increase the potential for collaborative advantage, the opposite is likely to be true for the benefits of specialization in terms of efficiency. The great assumption behind the argument that specialization leads to greater efficiency is that tasks remain simple and relationships are standardized and understood. Crossing boundaries makes any task more complex due to the additional exigencies involved, starting with clear communication. Indeed, crossing boundaries inverts the effect of specialization on efficiency: the more boundaries crossed, the more specialization will lead to functional complementarity along with added complexity and thus operational challenges. Efficiencies in performing functions are likely to be reduced when crossing boundaries, even in cases where there would otherwise be functional redundancy. Two scientists from discipline X but from countries Y and Z will have to work harder to collaborate due to cultural and language differences than if they were from the same country. Further, scientists X and Y will face additional challenges if one is also working in a GA and the other is working in a university.

Even if the complementarities can be managed via some modicum of redundancy, further entertainment awaits. Boundaries undermine tie strength. This occurs for a variety of reasons, and bears a litany of implications for relationships among collaborators. For one thing, there is the challenge of communication. This implies that there is greater effort required to maintain ties across boundaries. For another, the magnification of complementarities when crossing boundaries is likely to be reflected in similarly magnified inequalities among partners. Inequality is reflected in unequal capacities regardless of structural position and function. Two NGOs collaborating in different countries such that one NGO is large and professionalized while the other is small and young are likely to exhibit significant differences in the capacities and experience of structurally and functionally equivalent personnel (Chapin 2004; Dourojeanni 2005). Beyond inequalities, ambiguities in structural positions across boundaries constitute another challenge that can weaken collaborative ties. In the parlance of social network analysis, "structural equivalence" of position becomes more difficult to establish when crossing boundaries (Borgatti and Everett 1992). Finally, due to the greater challenges and independent pressures on opposite sides of boundaries, there is likely to be more turnover in collaborators. I discuss this issue further in Chapter 7. For now, suffice it to say that when crossing boundaries, turnover is particularly likely to hamper resilience and thus threaten the viability of collaboration.

There are thus several structural threats to collaboration when crossing boundaries. These threats bear a final implication: a key functional requirement when crossing boundaries concerns bridging itself. Beyond the other functions noted earlier, bridging is a function in its own right that requires dedicated personnel when crossing boundaries. This can buffer against the effects of extreme complementarity, limited redundancy, and other threats to strong ties. Specialists in boundary spanning provide support to everybody else in order to maintain cross-boundary relationships. Hence, when crossing boundaries, it would be wise to design a collaborative structure that explicitly accounts for the bridging function.

THE ANALYTICS OF COLLABORATIVE STRUCTURES ACROSS BOUNDARIES

Scale, structure, functions, modes, and boundaries thus constitute elements of a framework for thinking about the design of collaborative structures when crossing boundaries. I now turn to specific cases of collaborative structures with illustrations to visually evaluate the strategies and opportunities as well as the risks and perils in various designs. While some cases are stylized, I also offer the projects outlined in Chapter 2 as concrete examples to elucidate the positive and negative consequences of specific collaborative structures when crossing boundaries.

For both hierarchies and networks, visualization in graphical form has long been a useful practice to communicate about structures. The same might also be said for identification of advantages and potential problems in the design of collaborative structures. I therefore employ a series of figures that illustrate collaborative structures, not only to support analytic discussion of their advantages and disadvantages, but also to concretely point out specific adjustments that could be made in particular designs to mitigate certain problems.

Graph theory provided a means of illustrating structures of relationships (e.g., Harary 1969), as applied in analysis of networks (e.g., Scott 2000) and organizations (e.g., Burton, et al. 2011). Illustrations of such structures typically render relationships using lines ("edges") that connect agents ("nodes") depicted as dots or circles or boxes. Such relational graphs in turn informed the "sociograms" of social network analysis, which visualize individuals and organizations as points and relationships as lines (Scott 2000). In this chapter, I adopt a version of this symbology to illustrate collaborators and organizations as labeled boxes and collaborative relationships as lines.

Collaborative Structures among Individuals

I proceed by starting with simple collaborative structures with few partners, little functional differentiation, and no boundary crossing; I then progressively add complexity. Table 5.1 provides a summary of these cases, noting the key design elements from the framework outlined above: scale, structural complexity, functional differentiation, collaborative modes, and boundaries. This permits direct comparisons among the cases.

The simplest case involves two individual collaborators who have the same capacities and perform the same functions when working together. This could involve two scientists seated side by side on the same lab bench. Here the structure is simple and the functions are undifferentiated. This case might be rendered as shown in Case 1 of Figure 5.2. Each box indicates an individual (A and B), and the line connecting them represents their collaborative relationship. The boxes appear the same size, shape, and color to denote that each agent is functionally equivalent. Case 1 of Figure 5.2 also portrays one

Table 5.1. Key Characteristics of Structures of Collaborating Individuals.

Case	*Scale*	*Structure*	*Functions*	*Modes*	*Boundaries Crossed*
1. Two individuals	Very small	Very simple (1 relationship)	Undifferentiated	Redundant	None
2. Two individuals	Very small	Very simple (1 relationship)	Differentiated	Complementary	Disciplinary
3. Two individuals	Very small	Very simple (1 relationship)	Some differentiated, others undifferentiated	Redundant and Complementary	Disciplinary
4. Three individuals	Small	Simple (3 relationships)	Undifferentiated	Redundant	None
5. Four individuals	Small	Moderate (6 relationships)	Undifferentiated	Redundant	None
6. Five individuals	Medium	Complex (10 relationships)	Undifferentiated	Redundant	None
7. Seven individuals	Medium	Highly Centralized (6 relationships)	One differentiated, most differentiated	Limited complementarity; largely redundant	None
8. Seven individuals	Medium	Moderately centralized (9 relationships)	Differentiated	Redundant and Complementary	None
9. Seven individuals	Medium	Moderately centralized (6 relationships)	Mostly differentiated	Redundant and Complementary	None
10. Ten individuals	Medium-large	Moderately decentralized (12 relationships)	Mostly differentiated	Redundant and Complementary	None
11. Nine individuals	Medium-large	Moderately decentralized (12 relationships)	Mostly differentiated	Redundant and Complementary	None
12. Nine individuals	Medium-large	Moderately decentralized (12 relationships)	Mostly differentiated	Redundant and Complementary	National

line with arrows at both ends, indicating bidirectional communication with the same technical content. This line could also be rendered with two lines pointing in the opposite directions. In this case there are no boundaries to be crossed.

Having the two scientists permits each to check the other's operations to ensure everything is done as intended. In other words, there is redundancy rather than complementarity in the design of this collaborative structure as regards the functions performed. Collaborative advantage will thus be limited.

If collaborative advantage is of interest, the structure should incorporate functions that could not be performed by the collaborators acting independently. This generally involves collaboration among partners with complementary capacities. The simplest case of this would still be two individuals, perhaps from differing disciplines, as shown in Case 2 of Figure 5.2. Now A is from one discipline, represented by a black box and line, and B is from another, indicated by a gray box and line. Each brings something different and thus complementary to the collaborative relationship. In the process, they are crossing a disciplinary boundary, represented by the dashed line in be-

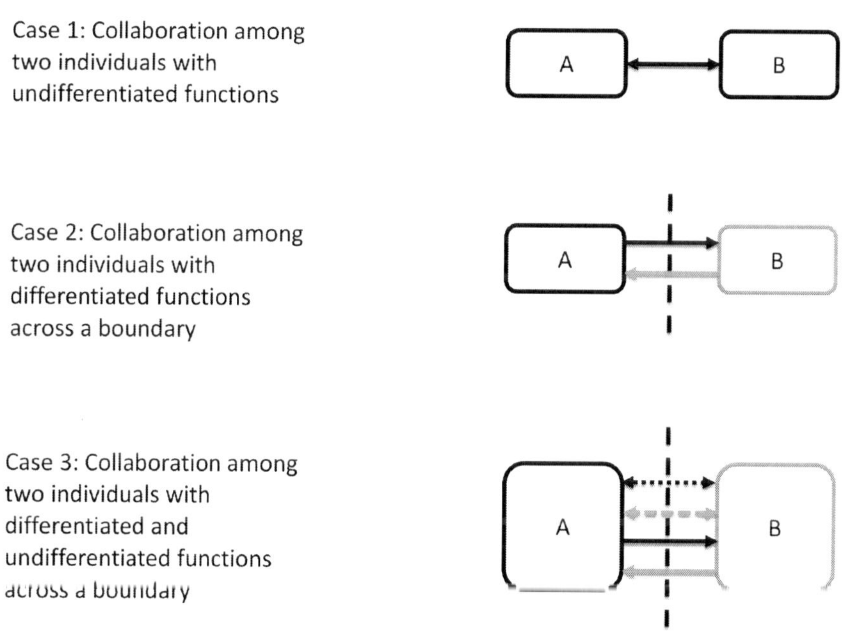

Figure 5.2. Very Simple Collaborative Structures.
Source: Created by author.

tween A and B. Now the design highlights complementarity rather than redundancy.

Of course, both figures are simplifications. One could easily imagine a suite of functions performed by A and B, and thus numerous arrows running between them. Redundant functions would be represented by one type of line with arrows at both ends, and complementary functions by another type of line with an arrow at only one end. Case 3 in Figure 5.2 therefore redraws the relationship between A and B by breaking out distinct types of functions, with dual-arrow lines indicating redundant functions, and single-arrow lines indicating complementary functions. For tractability in the ensuing exposition, I will visualize relationships among partners with a single line, though I will employ dual-arrow lines to indicate redundancy and single-arrow lines for complementarity.

Table 5.1 summarizes the characteristics of the collaborative relationships in the designs presented thus far. The table is organized around the components of the framework presented in the first half of this chapter (shown across the columns) and the cases discussed (listed down the rows). All three cases mentioned thus far are small in scale and have simple structures with a collaborative relationship between two agents. They vary in terms of their functional differentiation, such that Case 1 displayed undifferentiated functions while Case 2 presented differentiated functions and Case 3 included both. Consequently, the cases vary in their modes of collaboration. Whereas Case 1's undifferentiated functions mean the mode highlights redundancy, the differentiated functions in Case 2 imply complementarity, and Case 3's various functions imply instances of both. Because I presented Cases 2 and 3 as an instance of functional complementarity due to differing disciplinary backgrounds among A and B, they involve disciplinary boundary crossing.

Here it is important to further clarify the relationships of functions, modes, and boundaries. Functional differentiation implies that the mode of collaboration will be complementarity; similarly, undifferentiated functions imply a redundant mode of collaboration. Further, crossing boundaries generally means that there will be some degree of functional differentiation and thus complementarity, though this does not necessarily prohibit the possibility of some redundancy as well. Crucial in these cases is the identification of the key functions to be upheld among the collaborators, followed by determination of whether one or both can provide them.

Figure 5.2 and Table 5.1 present the simplest cases. We can now expand the scale of collaboration by adding more partners, as shown in Figure 5.3. Case 4 takes Case 1 and adds collaborator C. The structure is still small, the functions undifferentiated, and the mode redundant. But the number of relationships has jumped from one to three. This is not an insignificant change. As social network analysts know, increasing a network by one agent has a larger than proportional effect on the number of possible ties (Scott 2000). If all collaborators indeed exhibit direct relationships with each other, the complexity of the collaborative effort can rapidly become prohibitive. Case 5 adds a fourth collaborator, without modifying anything else. Now the number of possible relationships leaps up to six, as each collaborator has relationships with three others. Case 6 adds a fifth collaborator, again without other modifications. There are suddenly ten relationships, and each collaborator works with four others. At the risk of belaboring the point, network analysis has shown that for n collaborators, the maximum possible number of relationships equals $n(n-1)/2$ (Scott 2000: 75). Adding a sixth collaborator would raise the number of possible relationships to 15; adding a tenth would yield 45. Complexity thus rises rapidly.

Collaborative networks are rarely so dense. Instead, there are usually keystone collaborators who bring many others together. This manages the complexity for most collaborators, but it also means that the keystone collaborators spend considerable time playing the role of network brokers to establish strategically important ties (cf. Manring 2007). Figure 5.4 presents an extreme instance in Case 7, which has seven collaborators organized in a

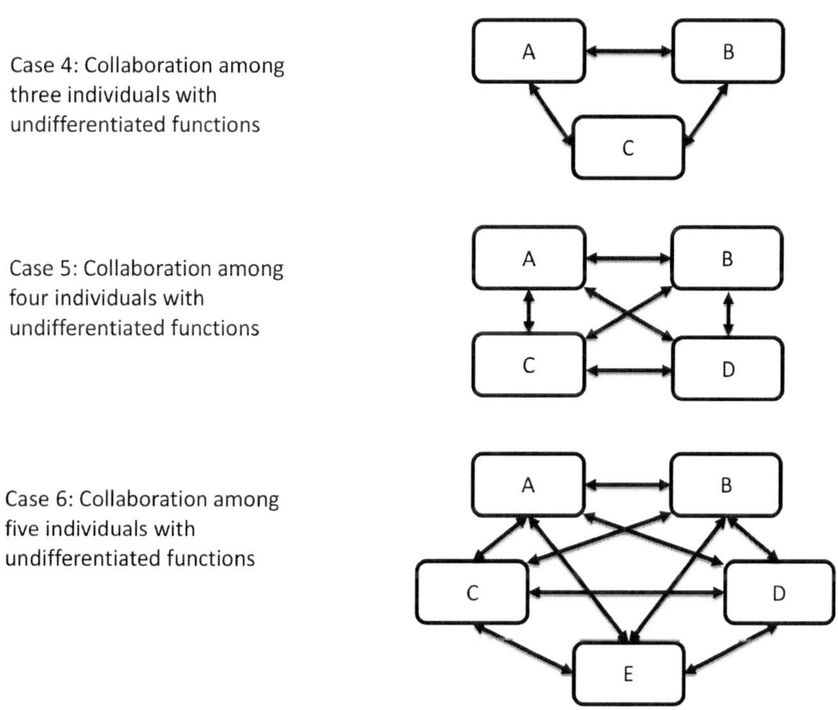

Figure 5.3. Simple Collaborative Structures.
Source: Created by author.

highly centralized structure where collaborator X has relationships with all six of the others (A-F), but none of them have any relationships with each other. Hence, X has six relationships to manage and the others only one. This highly centralized structure places substantial responsibility on X and arguably too little on the others; it also limits information flow. In this case, the network begins to take on characteristics of a hierarchy. Here it is important to note that some functional differentiation necessarily emerges, as X plays a coordination role, so X's box is gray rather than black for A-F. Further, because X has a coordinating function but the others do not, gray lines from X have one arrow each pointing toward A-F, whereas the black lines from A-F each have one arrow pointing toward X. This is because the function associated with the information running in the two directions differs. From X to A-F, there is information pertaining to coordination; from A-F to X, there is technical information that X is coordinating.

Something approaching a happy medium between Cases 6 and 7 might be Case 8, which has the same scale and the same differentiated relationships between X and A-F, but also has selected collaborative relationships among A-F. Hence, A and B now collaborate, as do C and D as well as E and F. In this structure, A and B share the same function, thus their dual-arrow relationship matches their box motifs with solid black lines. The same applies to

Case 7: Highly centralized collaboration among seven individuals with somewhat differentiated functions

Case 8: Moderately centralized collaboration among seven individuals with highly differentiated functions

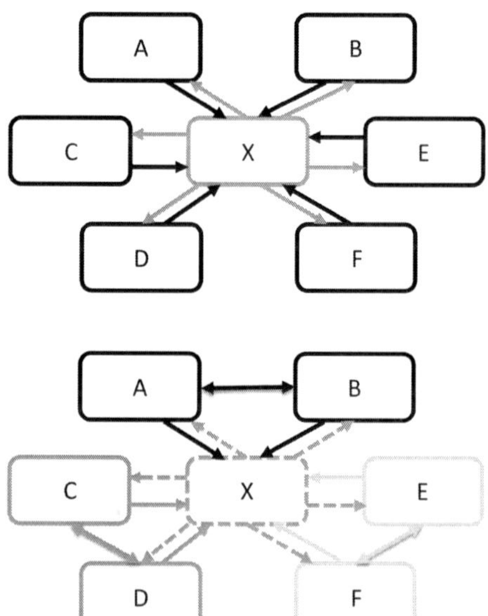

Figure 5.4. Centralized Networks.
Source: Created by author.

C and D, which share a different function and thus have a distinct relationship, as do E and F. In turn, X has relationships with A-F, shown with two single arrows in each case, to reflect their functional differentiation. The key point in Case 8, however, is that there is now communication between the dyads A-B, C-D, and E-F, and not merely between A-F and X. In this regard, the collaborative structure in Case 8 is more centralized than Case 6 but less so than Case 7.

Figure 5.5 shows an alternative in Case 9. There, A and B still collaborate on one function, as do C and D on another, and E and F on a third. But now, A coordinates with X for A and B, C communicates with X for C and D, and E does the same for E and F. This further reduces the number of ties, making things simpler for X. This of course comes at the risk of leaving out B, D, and F from direct communication with X.

Here I want to recognize that in Cases 8 and 9, I differentiated the functions of A-F on the justification of reducing the coordination burden on X. Two issues arise here. The first is the observation that X has the same number of relationships in the "highly centralized" structure of Case 7 and the "moderately centralized" structure of Case 8. There is a relative reduction in centralization because the other collaborators now communicate with each other. The presumption is that if A and B are collaborating directly on the same function, X will not have to spend as much time talking to both A and

Case 9: Working groups: Moderately centralized collaboration among seven individuals with differentiated functions

Case 10: Working groups: Moderately decentralized collaboration among nine individuals with differentiated functions

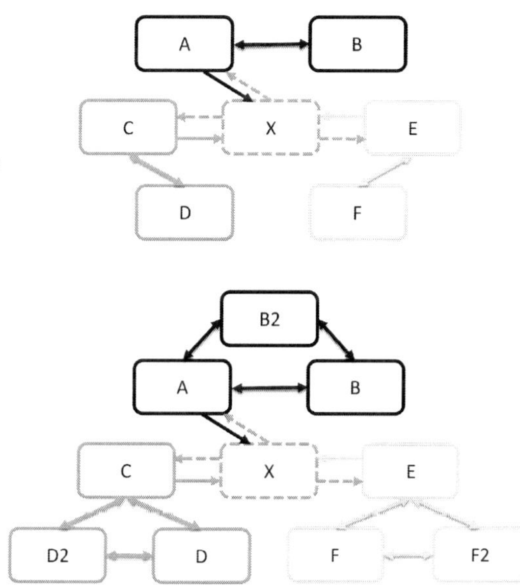

Figure 5.5. Working Groups.
Source: Created by author.

B, because repetition becomes unnecessary. The second issue concerns my differentiating functions in Cases 8 and 9. In both cases, there could be undifferentiated working groups doing the same thing. However, when there are ties among some collaborators but not others in moderately centralized networks, the implication is that those linked are in working groups doing different things. This is a tendency, not a requirement or iron law. When networks grow in scale, and direct communication among all members becomes more time-consuming, working groups tend to form which take on distinct functions.

Of course, working groups may well have numerous members and their own coordinators. Figure 5.5 therefore presents Case 10, which adds one more collaborator in each functional working group. To preserve the letter designations used thus far, to A and B I add B2, and D2 to C and D along with F2 to E and F. The same basic structure applies: X coordinates via collaboration with A, C, and E. But now with the addition of B2, D2, and F2, the structure is "moderately decentralized" because most of the relationships no longer involve X, but instead reside within the three working groups (A, B, and B2; C, D, and D2; and E, F, and F2). Similarly, A, C, and E now have the same number of relationships as X. Here the symbology of the lines

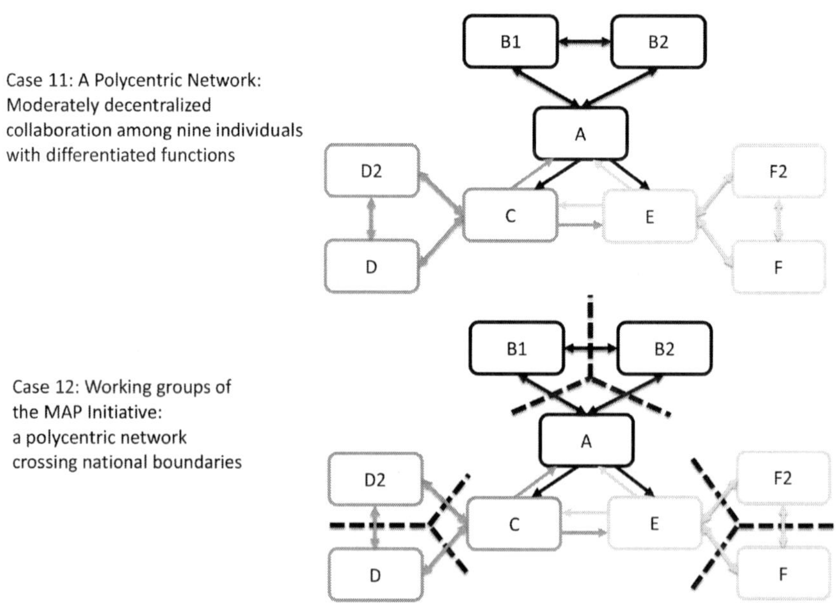

Figure 5.6. Polycentric Networks.
Source: Created by author.

begins to break down, as for example A is serving a coordination function within A's working group but not B or B2.

If there are doubts that Case 10 is moderately decentralized, Figure 5.6 presents Case 11, which retains the same three working groups and group coordinators but removes X. A, C, and E now interact directly to perform coordinating functions, indicated with distinct line motifs that denote their functional complementarities. There is the same number of relationships; only the central coordinator X is now missing. However, the three coordinators (A, C, and E) have more relationships in Case 11 than in Case 10. This design constitutes a "polycentric network" configuration since there is no central coordinating figure, but rather three coordinators who work together.

This design is close to that employed by the MAP Initiative. The main difference is that the MAP Initiative had working groups and coordinators crossing national boundaries between Bolivia, Brazil, and Peru, particularly at the level of the working groups. I have attempted a representation in Case 12 of Figure 5.6. The thick dashed lines between the collaborators in each working group represent national boundaries among the partners. This representation simplifies the MAP Initiative, in that the coordinators may or may not have national boundaries between them.

Collaborative Structures among Organizations, Part 1: Environmental Science

The exposition thus far is guilty of the gross simplification of assuming that individual collaborators are independent agents who have no institutional homes. The remainder of this discussion addresses that simplification by situating agents within organizations. This is much more accurate to the infrastructure required for collaboration. However, it adds considerably to the complexity since the cast of characters becomes larger, the number of functions increases, and the number of relationships grows.

This section focuses on collaboration among organizations for environmental science as it contributes to conservation and development. Key collaborative relationships therefore focus on research. The organizations at play will be universities. The emphasis will thus be on crossing international and interdisciplinary boundaries.

Table 5.2 provides a summary of the collaborative structures involving organizations. As in Table 5.1, for each of the following cases of organizational collaboration, Table 5.2 notes the scale, structural complexity, functional differentiation, collaborative modes, and boundaries of the structures presented.

Figure 5.7 presents Case 13, which involves two simple organizational charts, for universities A and B. Both are hierarchical. Within each hierarchy,

Table 5.2. Key Characteristics of Structures of Collaborating Organizations for Environmental Science.

Case	*Scale*	*Structure*	*Functions*	*Modes*	*Boundaries Crossed*
13. Two organizations	Small	Simple (1 organizational tie)	Differentiated (administrative, technical, staff support)	Redundant (but limited; one person per function per organization)	None
14. Two organizations	Small	Moderate complexity (multiple organizational tiers, 2 inter-organizational ties)	Differentiated (administrative, technical, staff support)	Redundant (but limited; one person per function per organization for most functions)	None
15. Two organizations	Small	Moderate complexity (multiple organizational tiers, 3 inter-organizational ties)	Differentiated (administrative, technical, staff support)	Redundant (but limited; one person per function per organization for most functions)	National
16. Two organizations	Small	Moderate complexity (multiple organizational tiers, 3 inter-organizational ties)	Highly differentiated (administrative, two technical disciplines, staff support)	Redundant (one or more persons per function per organization) and complementary (two technical disciplines in one organization)	National, disciplinary
17. Four organizations (NSF Roads project)	Medium	Complex (multiple organizational tiers, 6 inter-organizational ties)	Highly differentiated (administrative, three technical disciplines, staff support)	Redundant (one or more persons per function per organization) and complementary (three technical disciplines in each organization)	National, disciplinary

I employ a letter-number system to indicate the "positional address" of individuals. The letter indicates which organization; the first number the tier in an organization's hierarchy, and second and subsequent numbers indicate divisions in the hierarchy as well as individuals where there are more than one in the same division. Hence, the hierarchical organizational charts in Case 13 have at the top an administrator (A1 and B1) who is situated above staff (A2.# and B2.#). Staff include environmental researchers (A2.2 and B2.1) as well as administrative support staff (A2.1 and B2.2). Figure 5.7 uses gray shades to differentiate the functions, with administrative oversight in black, research in dark gray, and administrative support in light gray. The relationships portrayed with the arrows thus reflect the functions of personnel collaborating. One key point is that where relationships are hierarchical, I generally represent them using one-way arrows to denote the direction of authority. Hence, A1 has authority over A2.1 and A2.2, since the arrows point toward the latter two agents. The key inter-organizational relationship involves the two-way arrow representing collaborative research between A2.2 and B2.1, shown in dark gray. In addition to their research collaboration, the researchers also answer to their administrators (black one-way arrows) and work with support staff (light gray two-way arrows). Hence, the research collaboration across organizations also involves work with other personnel within one's home institution.

Case 13 is the simplest case. It ignores the fact that many other people work in each organization, and thus that there are numerous other relationships in those hierarchies. It only contemplates two collaborating researchers; we could add many more. Case 14 therefore expands university A, which now has three tiers, the top two focusing on administration, with the second tier indicating a divisional difference among two departments. Hence, A1 is

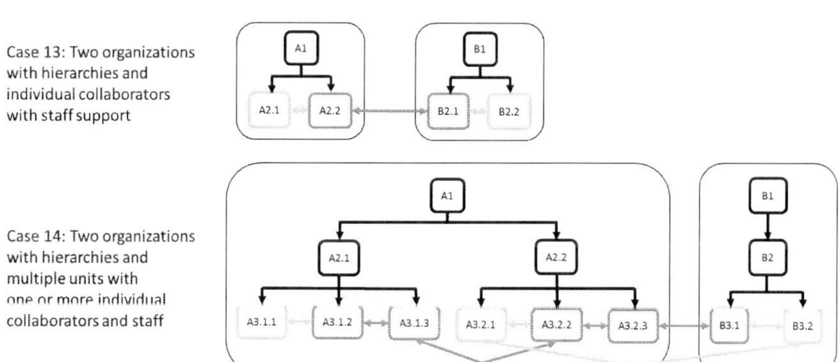

Figure 5.7. Inter-organizational Collaboration.
Source: Created by author.

now a top-level administrator over two or more departments, whereas A2.1 and A2.2 are mid-level administrators at the departmental level. The third tier of university A now has multiple researchers in each unit, and shows the collaborating researchers and their ties. Researcher A3.1.3 does research with A3.1.2 as well as A3.2.2 over in the other department, who is similarly working with A3.2.3. A3.2.3 in turn serves as the point of contact with university B via researcher B3.1. These linkages are represented on the same tier in Figure 5.7, and are thus horizontal. There are few direct ties between the researchers and support staff, and support staff in different departments of university A do not have direct ties for purposes of the collaboration. Much depends on the hierarchical organization in universities A and B, so everyone in the bottom tier knows their role. There is an administrative link from A3.2.1 to B3.2, which permits an individual contractual agreement, such as a consultancy, for B3.1 to collaborate as an individual.

Figure 5.8 presents two more cases which add further complexity. Case 15 modifies Case 14 in two significant and interrelated ways. First, Case 15 adds a higher level linkage, a one-way arrow from A1 to B1, which represents an organizational agreement such as a subcontract. This is likely if the collaboration is of strategic importance to both institutions and/or will be of long duration (i.e., more than a year). Case 15 also adds a boundary, between the countries of universities A and B. Because crossing a boundary can be difficult, it is important to have multiple collaborative relationships spanning said boundary, ideally at differing organizational levels and for differing functions. The boundary crossing at the top tier between administrators A1 and B1 occurs via the mechanism of the organizational agreement, which is important for spelling out the terms of reference and specific obligations of each partner. Contracts are always important for collaboration between organizations, but especially so here since A and B are in different countries with distinct legal frameworks, histories, and cultures. Similarly, the boundary crossing between researchers A3.2.3 and B3.1 is important, for they may share the same discipline and thus technical functions even though they operate in different countries. While there is an inter-organizational link between support staff A3.2.1 and B2.2, staff person A3.2.1 does not have a direct link with researcher A3.2.3, which could be significant. This is an important potential weakness in Case 15. This case thus begins to reveal the importance of balancing complementarities in functions against redundancies, especially where boundaries are being crossed.

Case 16 adds a second boundary, this time between disciplines. University A has two departments with differing missions, so there is a disciplinary boundary between the units administered by A2.1 and A2.2. I have therefore represented the distinction between the two disciplines with solid or dashed lines around the researchers, on the argument that they bring differing technical backgrounds, experience, and perspectives to the collaboration. Re-

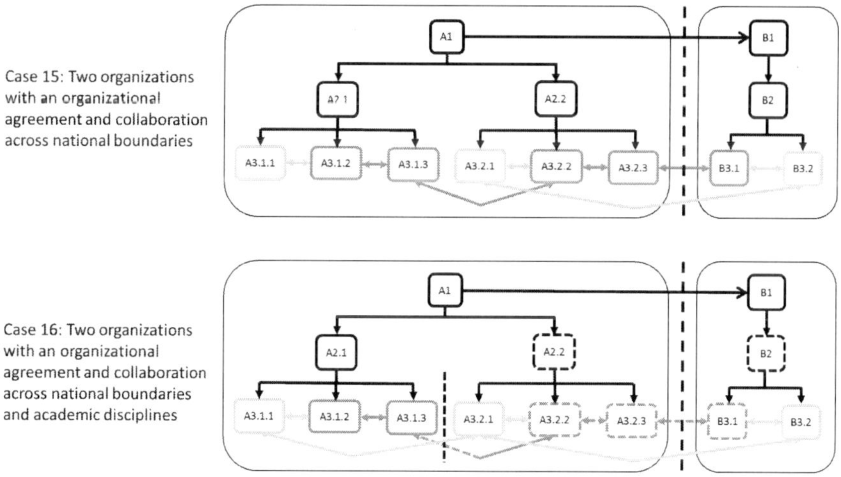

Figure 5.8. Environmental Science Collaboration among Universities.
Source: Created by author.

searchers A3.2.2 and A3.2.3 are in one discipline, the same one as collaborating researcher B2.1. However, researchers A3.1.2 and A3.1.3 are in another discipline. Now a key link is between researchers A3.1.3 and A3.2.2. This is a cross-disciplinary relationship. I therefore indicate the relationship using a dual-line motif, such that the information from A3.1.3 to A3.2.2 is represented by a solid line (per the motif of A3.1.3's department), whereas the information from A3.2.2 to A3.1.3 is indicated with a dashed line (per A3.2.2's department). Because this is a boundary-crossing relationship, it likely requires additional supporting links, such as among support staff in the two units. This depends on how complementary the functions in organization A's two departments are. Case 16 thus shows an inter-departmental link between support staff A3.1.1 and A3.2.1.

Given the discipline-crossing in Case 16, it is likely that the functions of participating faculty in different disciplines will be distinct. This, however, begs questions about how exactly they will be collaborating, and whether researchers in different disciplines will actually be serving distinct functions. Here the nature of the discipline-crossing becomes crucial. In multidisciplinarity, there will be specialization among the researchers and thus functional differentiation. But if there is interdisciplinarity, then the focus is on integration of perspectives. Consequently, the functions of the participating researchers are similar, even if specific perspectives differ. Case 16 thus implies multidisciplinarity, though it could also be representing interdisciplinarity insofar as the disciplinary perspectives being integrated differ substantively with regard to the research problem.

These observations beg questions about crossing boundaries and functional complementarity and redundancy. If there is interdisciplinarity and researchers are performing similar functions but offering different perspectives, does this amount to complementarity or redundancy? The answer of course is both, and that reveals limitations in the symbology at hand. But this is not the crucial question; there are several which are more important. For example, it is more important to ask whether the perspectives represented are necessary and sufficient for the collaborative problem at hand. Beyond that, there are equally important questions about the complementarity of collaborators with regard to the overall task set, both those specific to the technical work at hand (i.e., who conducts the fieldwork) and those specific to the collaboration itself (i.e., how goals are set).

Table 5.3 and Figure 5.9 go beyond stylized cases to present a concrete example, the "NSF Roads" project. As noted in Chapter 2, this project involved four universities in four countries (UF/USA, UAP/Bolivia, UFAC/Brazil, and UNAMAD/Peru), in which we organized three components of data collection: social science, remote sensing, and botany/ecology. This project thus crossed both national and disciplinary boundaries. Table 5.3 presents the basic structure we designed to coordinate the project. Each university partner had one or more coordinators serving each of four distinct functions: general administration, and technical coordination in one of the three science components. Because of the multiple types of boundaries being crossed, there was particular need to have multiple linkages among coordinators. The structure in Table 5.3 provides for both administrative as well as technical coordination across universities, as well as within universities among components. There were multiple individuals working in each country and in each component, so each coordinator had multiple peers in other countries and project components. This strengthens institutional ties across both national boundaries and disciplines. In particular, the administrative coordinators played key roles in bridging boundaries by supporting the other coordinators.

For NSF Roads, Case 17 in Figure 5.9 presents a partial outline of the structure, with some of the functions differentiated, and a simplified representation of the participating agents. I draw on the symbology and positional address system from previous cases to denote the agents, and the arrow motifs to represent hierarchical and horizontal relationships. Each university has its own box with individuals linked by hierarchical relationships, from managers in the higher tiers in black boxes (A1 and B1) down to support staff in light gray boxes (A2 and B2) and research faculty with dark gray boxes (A2 and B2). Below the researchers are students, also in dark gray boxes but lower in the hierarchies (A3 and B3). Because the project crossed

Table 5.3. Coordination Structure of the NSF Roads Project.

	General Adminis-tration	*Social Science Component*	*Remote Sensing Component*	*Botanical Component*	*Total*
University of Florida, USA	1	1	1	2	5
Amazonian University of Pando, Bolivia	2	2	1	2	7
Federal University of Acre, Brazil	1	1	1	1	4
National Amazonian University of Madre de Dios, Peru	2	1	1	1	5
Total	6	5	4	6	21

Source: Perz (2010).

national boundaries, Case 17 indicates the cross-national dimension via dashed lines between the universities.

Within each university partner, the project had social science, remote sensing, and botanical/ecological components. The numbering system in Case 17 designates the social science component as Component 1, remote sensing as Component 2, and botany/ecology as Component 3. Because the project components fall within the purviews of differing departments, Figure 5.9 shows disciplinary boundary lines between departments associated with different components. Following the positional address system, research faculty at UF in the social science component are designated A2.1; similarly, the social science researchers in the partner universities are designated B2.1, C2.1, and D2.1. Further, the students are indicated as A3.1, B3.1, and so forth. Because there were multiple students participating in each project component in each university, Case 17 notes 1 . . . n students.

Case 17 has several characteristics that merit comment. First, the positional addresses of some collaborating agents are not the same across universities. For example, the administrative coordinator at UF, agent A2.1.2, had hierarchical relationships with administrators in higher hierarchical tiers, like B1.1 in the partner universities. This reflects the hierarchical relationship among the participating universities: UF served as the project lead; hence, its Principal Investigators served as Project Coordinators and thus took on managerial roles on the project. In particular, the project PI, agent A2.1.2, per-

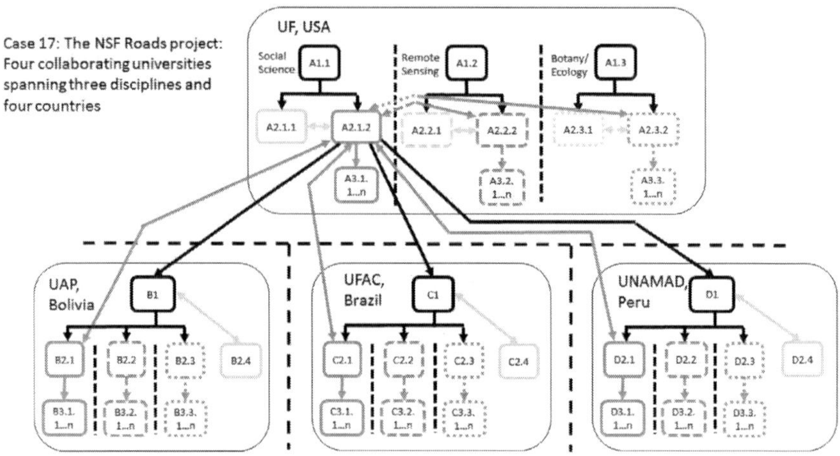

Figure 5.9. Environmental Science Collaboration among Universities: The NSF Roads Project.
Source: Created by author.

formed managerial functions in a hierarchical fashion with the in-country coordinators, B1, C1, and D1. This reflects the centralized organization of the project at UF, as required by the funder. The hierarchy also constituted a clear chain of command and permitted managerial efficiencies, an important consideration in crossing boundaries.

At the same time, agent A2.1.2 was collaborating in network fashion with social science coordinators B2.1, C2.1, and D2.1 on technical matters tied to data collection. These collaborative relationships are denoted by the bi-directional dark gray lines. This complicates the relationships at play, since A2.1.2 has a hierarchical relationship with B1.1, who in turn has a hierarchical relationship with B2.1, while A2.1.2 also has a network relationship with B2.1. This could cause significant confusion. The problem was avoided insofar as the hierarchical relationships involve one function, project management (denoted by black lines), whereas the network relationships pertain to another function, data collection. In this component as the others, we combined hierarchical and network relationships to permit efficient project management as well as collaborative science.

Horizontal relationships predominated in the NSF Roads project. Project co-PIs at UF worked in network fashion with departmental staff and with each other. Of particular note are the horizontal relations from A2.1.2, A2.2.2, and A2.3.2. I have denoted those relationships using the line motifs of the co-PIs. Hence, the relationship with the remote sensing coordinator A2.2.2 is denoted with a dashed line; the same is true of the relationship with the botanical/ecological coordinator A2.3.2. This reflects the interdisciplinary integration at hand, from the perspective of A2.1.2. There were also non-hierarchical relationships between the component coordinators at UF and their counterparts at the collaborating universities. Finally, there were horizontal ties among students from different universities. Those relationships are not shown, as there were many and the large number of lines would clutter up the figure.

Table 5.4 therefore presents a network matrix that summarizes the ties among the collaborators on the NSF HSD project, as contemplated by the project design. In the parlance of social network analysis, Table 5.4 is a "square case-by-case" matrix because it lists the collaborators across the columns as well as down the rows (Scott 2000). Table 5.4 focuses on the administrators and researchers; if we added the staff and students, the table would be rather larger and harder to interpret. In each cell, I indicate whether a pair of agents were directly collaborating, in two respects. First, I indicate the presence of collaborative relationships with ones and the absence with zeros. I also denote in parentheses whether the relationship crosses a boundary, with "ID" for crossing a disciplinary boundary and "IN" for crossing an international boundary. Relationships can simultaneously cross two boundaries. Along the diagonal, I indicated "N/A" since an agent cannot have a collaborative relationship with him- or herself. I indicate relationships above and below the diagonal, as this permits sums of all collaborative relationships for all of the agents in the table in each column. Table 5.4 thus indicates more relationships for Case 17 than Figure 5.9. The sums appear at the bottom of Table 5.4, and include the grand total of collaborative relationships, along with those which are international and interdisciplinary.

Table 5.4 permits a numerical analysis of the collaborative relationships on the NSF Roads project. This reveals much about the design of the project, not only in terms of general network characteristics, but specifically in terms of ties across boundaries. One key point concerning project design is that all agents have two or more ties. All agents had ties with multiple collaborators. A second key point is that all agents cross more than one type of boundary. Conversely, crossing of a given boundary is not dependent entirely on any one agent. A third key point is that some agents are much more connected than others, which means that the network depends more on some collaborators than others. The project PI, agent A2.1.2, has a total of eight ties, all of which are interdisciplinary or international. Loss of the PI means the loss of key international ties to the in-country administrators, indicating vulnerability due to limited redundancy. That said, multiple collaborators have international ties. The key design feature that emerges out of Table 5.4 is that there is a modicum of tie density insofar as all collaborators have multiple ties, and in particular all collaborators have at least one interdisciplinary and one international tie. As a result, if one sums up the ties among institutions, there

Table 5.4. Network Structure among Administrators and Researchers on the NSF Roads Project.

Collaborator	*A2.1.2*	*A2.2.2*	*A2.3.2*	*B1*	*B2.1*	*B2.2*	*B2.3*	*C1*	*C2.1*	*C2.2*	*C2.3*	*D1*	*D2.1*	*D2.2*	*D2.3*
A2.1.2	N/A	1 (ID)	1 (ID)	1 (IN)	1 (IN, ID)	0	0	1 (IN)	1 (IN, ID)	0	0	1 (IN)	1 (IN, ID)	0	0
A2.2.2	1 (ID)	N/A	1 (ID)	0	0	1 (IN)	0	0	0	1 (IN)	0	0	0	1 (IN)	0
A2.3.2	1 (ID)	1 (ID)	N/A	0	0	0	1 (IN)	0	0	0	1 (IN)	0	0	0	1 (IN)
B1	1 (IN)	0	0	N/A	1 (ID)	1 (ID)	1 (ID)	0	0	0	0	0	0	0	0
B2.1	1 (IN, ID)	0	0	1 (ID)	N/A	0	0	0	0	0	0	0	0	0	0
B2.2	0	1 (IN)	0	1 (ID)	0	N/A	0	0	0	0	0	0	0	0	0
B2.3	0	0	1 (IN)	1 (ID)	0	0	N/A	0	0	0	0	0	0	0	0
C1	1 (IN)	0	0	0	0	0	0	N/A	1 (ID)	1 (ID)	1 (ID)	0	0	0	0
C2.1	1 (IN, ID)	0	0	0	0	0	0	1 (ID)	N/A	0	0	0	0	0	0
C2.2	0	1 (IN)	0	0	0	0	0	1 (ID)	0	N/A	0	0	0	0	0
C2.3	0	0	1 (IN)	0	0	0	0	1 (ID)	0	0	N/A	0	0	0	0
D1	1 (IN)	0	0	0	0	0	0	0	0	0	0	N/A	1 (ID)	1 (ID)	1 (ID)
D2.1	1 (IN, ID)	0	0	0	0	0	0	0	0	0	0	1 (ID)	N/A	0	0
D2.2	0	1 (IN)	0	0	0	0	0	0	0	0	0	1 (ID)	0	N/A	0
D2.3	0	0	1 (IN)	0	0	0	0	0	0	0	0	1 (ID)	0	0	N/A
Total Ties	8 (5 ID, 6 IN)	5 (2 ID, 3 IN)	5 (2 ID, 3 IN)	4 (3 ID, 1 IN)	2 (2 ID, 1 IN)	2 (2 ID, 1 IN)	2 (2 ID, 1 IN)	4 (3 ID, 1 IN)	2 (2 ID, 1 IN)	2 (2 ID, 1 IN)	2 (2 ID, 1 IN)	4 (3 ID, 1 IN)	2 (2 ID, 1 IN)	2 (2 ID, 1 IN)	2 (2 ID, 1 IN)

Notes. Numbers in the cells indicate whether there was a direct collaborative tie, where 1=yes and 0=no. IN=international collaborative tie. ID=interdisciplinary collaborative ties. Columns provide sums of ties.
Source: Perz (2010).

are several, at least between the lead (UF) and the in-country partners (UAP, UFAC, and UNAMAD).

Having multiple ties among collaborating institutions is crucial, especially when crossing boundaries. As noted in the discussion of Table 5.3, there were administrative coordinators and technical coordinators with counterparts in each university partner, for a total of up to four ties (the administrator, the social science coordinator, the remote sensing coordinator, and the botanical coordinator). Further, during fieldwork, students collaborated actively while gathering data, forming additional ties. Case 17 makes clear that agent A2.1.2, the PI, had many ties, but the same was also true of some of the other coordinators, particularly those in components with field campaigns. With this structure, the NSF Roads project incorporated functional complementarity among counterparts as well as redundancy in ties among institutions. If two technical coordinators did not know each other or could not communicate in the same language, the administrative coordinators provided backstopping.

That last observation raises a vital point: when crossing boundaries, not everyone goes all the way, as it were, because not everyone can. Table 5.4 represents the design rather than implementation of the NSF Roads project; the actual structure of the network on the project differed somewhat. For example, due to language issues, the remote sensing coordinators had virtually no direct contact. Rather, they relied on bilingual students and the administrative coordinators to fortify the bridge across the gap. Hence, the ties indicated in Table 5.4 did exist, but via other agents in the collaborative network than those indicated. This reveals additional functional redundancy in the NSF Roads project as regards boundary crossing. A given agent can perform multiple functions, including bridging, to strengthen ties across boundaries and thereby make a collaborative project work.

Collaborative Structures among Organizations, Part 2: Environmental Management

The cases discussed so far permitted analysis of the structural exigencies of crossing boundaries for collaboration. The concrete cases offered specific examples of crossing boundaries for objectives tied to conservation and development. The example of a MAP Initiative working group (Case 12) highlighted a polycentric network crossing national boundaries for the purpose of environmental planning. The NSF Roads project (Case 17) paid attention to crossing disciplinary and national boundaries for environmental science. These concrete examples, however, beg questions about other collaborative structures, particularly those involving different types of organizations and those focusing on environmental management.

Table 5.5. Key Characteristics of Structures of Collaborating Organizations for Environmental Management.

Case	*Scale*	*Structure*	*Functions*	*Modes*	*Boundaries Crossed*
18. Two organizations	Small	Medium complexity (multiple organizational tiers, 3 inter-organizational ties)	Highly differentiated (administrative, three technical disciplines, staff support)	Limited redundancy (usually one person per function per organization), highly complementary (different technical disciplines in different organizations, two organizations with extension personnel)	National, disciplinary, organizational
19. Four organizations (MABE Consortium)	Medium	Complex (multiple organizational tiers, 8 inter-organizational ties)	Highly differentiated (administrative, multiple technical disciplines (not shown), extension personnel, staff support)	Limited redundancy (usually one person per function per organization), highly complementary (different technical disciplines in different organizations, two organizations with extension personnel)	National, disciplinary (not shown), organizational
20. Five organizations (MDD Consortium)	Medium	Very complex (multiple organizational tiers, 18 inter-organizational ties)	Highly differentiated (administrative, multiple technical disciplines (not shown), extension personnel, staff support)	Limited redundancy (usually one person per function per organization), highly complementary (different technical disciplines in different organizations, two organizations with extension personnel)	National, disciplinary (not shown), organizational
21. Five organizations (BACC Consortium)	Medium	Very complex (multiple organizational tiers, 15 inter-organizational ties)	Highly differentiated (administrative, two key technical disciplines, extension personnel, staff support)	Redundant (usually one person per function per organization; technical disciplines and extension personnel in most organizations); limited complementarity	National, disciplinary, organizational

Case	*Scale*	*Structure*	*Functions*	*Modes*	*Boundaries Crossed*
22. Six organizations (M-P Consortium)	Medium/ large	Very complex (multiple organizational tiers, 14 inter-organizational ties)	Highly differentiated (administrative, multiple technical disciplines (not shown), extension personnel, staff support)	Limited redundancy (usually one person per function per organization), highly complementary (different technical disciplines in different organizations, two organizations with extension personnel)	National, disciplinary, organizational
23. Nine organizations (G-MAP Consortium)	Large	Extremely complex (multiple organizational tiers, 16 inter-organizational administrative ties, 8 financial coordination ties, 15 technical ties)	Highly differentiated (administrative, multiple technical disciplines (not shown), extension personnel, staff support)	Limited redundancy (usually one person per function per organization), highly complementary (different technical disciplines in different organizations, some organizations with extension personnel, highly connected personnel in different organizations)	National, disciplinary, organizational

I therefore turn to more elaborate and thus more formidable cases of consortia of organizations that crossed various boundaries for environmental management. As elsewhere, I draw on my own collaborative experiences, stemming from the model of environmental management discussed in Chapter 2. There are several key hallmarks of that model that will be important for the exposition that follows. First, the model presumes that the science being done is interdisciplinary and has a strong applied orientation. Second, the model marries environmental science to local knowledge as joint inputs into management planning. Third, the model highlights the crossing of boundaries among organizational types in consortia, as well as the inclusion of other stakeholders in planning. And fourth, it contemplates replication of local experiences across national borders where there are shared conservation and development problems. Hence, there are more individuals and organizations involved, there are more boundaries being crossed, and there are more functions being performed. Consequently, the complexity is greater. Table 5.5 provides a summary of the key design elements of these cases. As before, I include the structural design elements that permit direct comparisons of the cases.

Figure 5.10 presents a relatively simple example, Case 18, which has organizations in two different countries. At first glance, Case 18 looks like Case 16: both present two organizations, where the one on the left (A) is a university with two departments containing researchers doing interdisciplinary environmental science. In Case 18, the organization on the right (B) is, however, not a university but instead an NGO. Hence, A is a university that does environmental science, whereas B is an NGO with manifold ties to local governments, communities, and other stakeholders. B has a manager (B1) with signature authority for an organizational agreement, as well as a project director (B2.1) along with support staff (B3.2). The project director oversees a group of extension agents (B3.1, from 1 . . . n).

The key issue in Case 18 is the fact that the two organizations are simultaneously crossing multiple boundaries. A and B are different organizational types, and they are situated in different countries. Case 18 therefore shows two distinct boundaries between A and B. Further, there are multiple departments inside university A participating, so there is also a disciplinary boundary being crossed, and thus a boundary line between the departments. These boundaries present a series of potential obstacles for the collaborating organizations. For one thing, faculty A3.1.3 and A3.2.2 must work across a disciplinary boundary. This is occurring without supporting collaboration among staff. Hence, a linkage from A3.1.1 to A3.2.1 might become important. At the same time, A3.2.3 is working with B2.1. This time there is staff support from A3.2.1 to B3.2, which is key. But now another issue arises: indirect collaborative ties. A3.2.3 is a key point of contact from university A, and has as a counterpart B2.1 in NGO B. A3.2.3 is linked with A3.2.2, who

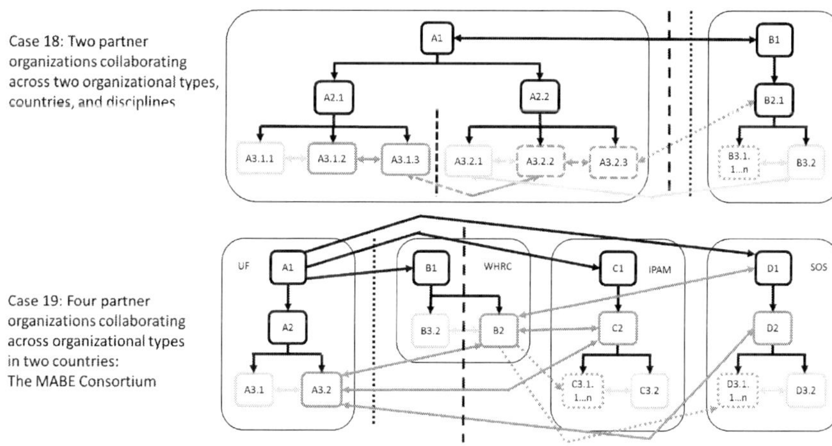

Figure 5.10. Consortia for Environmental Management: The MABE Consortium.
Source: Created by author.

works with A3.1.2. On the other side, B2.1 directs B3.1.1 . . . n. Hence, from the interdisciplinary environmental science between A3.1.3 and A3.2.2, there are up to four steps in the network to the extension agents B3.1.1 . . . n. Interlocution is thus substantial; perhaps it is necessary due to language or cultural issues. Adding direct linkages might improve the collaboration by broadening the channels of communication, though adding numerous horizontal linkages might undermine the efficiency with which the work gets done. Here it is worth noting that periodic meetings of the entire group will likely not be sufficient to ensure effective collaboration. The issue then is to identify the key players, notably those with direct ties across boundaries, as individuals who should participate in regular meetings.

The partnership among the organizations in Case 18 seeks to link environmental science and management via functional complementarity across multiple boundaries. Whereas university A engages in cross-disciplinary environmental science among two participating departments, key players in one of the departments also directly engage NGO B, which uses scientific findings to support management planning activities with local communities. At the same time, in each case of boundary crossing, there are multiple collaborators involved. Due to the differing functions performed by the collaborators engaged in bridging, having "multiple planks" in the bridge strengthens the ties across boundaries. This partnership relies heavily on in-house administration and support staff. This is not always possible. But since there are only two collaborating organizations, in-house administration should not be too onerous, even when crossing boundaries.

The remaining cases modify these parameters in specific ways. For one thing, I will increase the structural complexity, in terms of the number of organizations, the number of ties among collaborators, the number of functions performed by collaborators, and other factors. For another, I will employ concrete cases, which will present additional complications stemming from particular partners or specific structural difficulties. In particular, I will highlight complications that seem obvious in retrospect but were unanticipated at the time. I have sought to order the cases in a sequence where one is instructive for understanding the next, though this is not always possible due to multiple differences among some cases. I also note that the cases do not follow a chronological order; they are organized for purposes of exposition about collaborative structures.

Figure 5.10 presents Case 19, the MABE Consortium, one of the projects noted in Chapter 2. MABE was relatively small, with four partner organizations, but is nonetheless more complicated than Case 18, which had only two. MABE was funded by USAID and led by UF, which wrote sub-agreements to the NGOs WHRC, IPAM, and SOS Amazonia. UF administration thus performed a coordination role under A1, indicated by the black unidirectional arrows leading to directors of the partner organizations. In addition, UF faculty (A3.2) engaged in technical collaboration with personnel including project directors at the partner organizations (B2, C2, and D2), indicated by dark gray two-way arrows. In turn, technical personnel managed fieldwork conducted by extension teams (C3.1.1 . . . n, D3.1.1 . . . n), who supported environmental management planning by local governments and rural communities.

At first glance, MABE appears a relatively more straightforward case than Case 18 in that only two boundaries are explicitly being crossed. Case 19 shows an organizational boundary between UF (a university) and the three partners (all NGOs). Work done under MABE occurred in Acre, Brazil, so there is also an international boundary. Here the case-specific vagaries begin to appear: whereas WHRC (B) is based in the United States, the participating WHRC technical person (B2) was based in Brazil. Hence, the national boundary runs right through WHRC. This is an effective way of leveraging collaborative advantages when crossing a national boundary: one gains the administrative capacity of an established organization like WHRC, while also affording committed personnel in country. Unlike Case 18, Case 19 indicates no disciplinary boundary crossing among personnel. This is not to say there is no interdisciplinarity. Whereas Case 18 had cross-disciplinary research among researchers in different university departments, Case 19 also has interdisciplinarity, but as practiced by individuals in MABE. All MABE organizations included technical personnel with substantial interdisciplinary experience. Indeed, technical personnel at every partner organization led their own interdisciplinary research and outreach activities: B2 on climate change and civil defense planning; C2 on social-ecological impacts of highways and mitigation planning; and D2 on watershed management planning. Case 19 therefore presents all technical personnel as providing the same functions. However, the functions are themselves boundary-spanning, in the form of research combined with outreach, that is, transdisciplinary science. Further, the technical personnel contributed actively to each other's activities, so the arrows look the same and they are all bidirectional. For these reasons, Case 19 emphasizes MABE's considerable functional redundancy among its technical staff.

Case 19 also indicates substantial bridging of boundaries among the partners. UF administration and partner administrators had clear links via subagreements, and UF faculty had ties to technical personnel in every partner organization. Both provided important supporting functions, both administratively and technically, to the MABE partners. Further, WHRC personnel provided complementary support from within country, specifically to field extension staff. This complements the UF support via boundary crossing, as well as connections within each partner organization to field staff.

One final comment is in order for Case 19. Absent in Case 19 are direct ties among support staff. Also absent are individuals with specific management roles. Both result from the fact that all of the participating organizations had established administrative hierarchies that functioned efficiently. Subagreements with signatures by high-level administrators made it abundantly clear how financial procedures were to be accomplished by staff, how technical personnel were to orient field staff for activity implementation, and how field staff were to document their work and their expenses for purposes of reporting. MABE is a relatively simple case in part because the partners had sufficient administrative capacity. This, however, is not always the case, and should be carefully evaluated beforehand.

Figure 5.11 presents another consortium, Case 20, the Madre de Dios (MDD) Consortium, which exhibits several similarities to MABE but also some significant differences. The similarities are evident in terms of the number of participating organizations (four in MABE, five in MDD), the organizations participating (UF, WHRC were in both consortia), the boundaries being crossed (national and organizational), and the organizational structures among the partners (with authorities, project directors, and support staff and field teams). MABE and MDD were also based on the same underlying conservation and development model that combines environmental science and management. However, several key structural differences also appear: MDD exhibits greater variation in organizational types by including a GA, there is a much greater density of administrative ties among organizations (black lines), and there is a consortium-level management team (A4.1 and A4.2). MDD was also operating in a much more rapidly changing political and environmental context in southeastern Peru, with a gold mining boom, multiple floods, political contention over corruption accusations, and social unrest over legislation tied to unsustainable extractive activities.

MDD faced significant challenges from within and without. For present purposes, I focus on internal challenges but note instances where external issues exacerbated internal difficulties. One internal challenge concerned greater organizational diversity, which meant greater challenges for collaborating. There are numerous dark gray arrows among technical personnel in different organizations in MABE, but relatively few in MDD. MDD sought to manage the challenges of crossing boundaries among organizational types by incorporating "consortium scientists" who amped up MDD's technical staff. Those people (A4.3 hired by UF and A2.2 from WHRC) exhibit numerous dark gray arrows to technical personnel in each partner organization. Adding technical personnel proved a useful mechanism to offset limited partner technical capacity.

Another internal challenge concerned limited administrative capacity. This can mean either of two things. First is a young and/or small entity with limited staff and experience. The other involves entities that face large administrative burdens, as often is the case in public entities. MDD had examples of both: a small NGO and a public GA and university.

These considerations made it especially important to have a consolidated management team, able to operate autonomously from the partner organizations for purposes of efficiency. Without investment in managerial capacity, any collaborative advantage afforded by organizational complementarities could be threatened by administrative delays. UF therefore hired a program manager, A4.1, and a financial manager, A4.2, who also incorporated the

Case 20: Five partner organizations collaborating across organizational types in two countries: The MDD Consortium

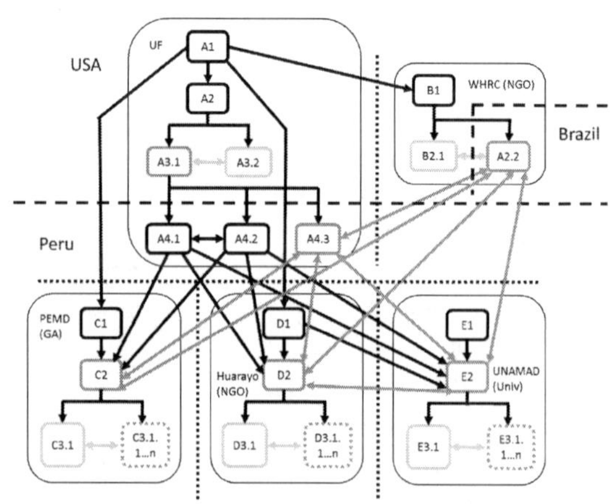

Figure 5.11. Consortia for Environmental Management: The MDD Consortium. ***Source: Created by author.***

role of an operational manager. I note that this is the most recent structure. Previously, the role of the operational manager resided with E2 in UNAMAD, which proved unworkable. This became in effect a "functional hole" left in the middle of the MDD Consortium, and resulted in a series of administrative impediments. Centralization of administration at the consortium level became necessary to improve administrative coordination between UF and the partners, and to seek to offset the limited administrative capacity of the partners. The consequence is a much more complicated consortium diagram, which looks like a tangled spider's web of black arrows, running in one direction to reflect the chain of command from above to below. The number of such arrows in turn implies that considerable effort is required by the management team. This is an eventuality of limited partner capacity.

Hence, whereas the presentation of MABE was rather horizontal, MDD is rather more vertical. To repeat, both consortia had numerous similarities, including their sizes and goals. But institutional differences among consortia, as in terms of organizational diversity and administrative capacity, can modify collaborative structures. MABE and MDD are distinct puzzles with different solutions. As Cases 19 and 20 make evident, some solutions are more complicated than others. Complicated solutions may be painful, but they are sometimes necessary. I conclude my comments on MDD with the reminder that the environmental threats in Madre de Dios are unfolding rapidly.

Figure 5.12 presents another case, which falls somewhere in between Cases 19 and 20. Case 21 is the BACC, a consortium of universities and NGOs that proposed to collaborate in Bolivia. The underlying conservation and development model is the same: do environmental science in the service of environmental management. This time, however, the focus was on building the capacity of an in-country university, UAP, as supported by other universities for the science and NGOs for the management applications. The BACC is similar to the two previous cases in several ways: the number of organizations (five), the participation of universities and NGOs, and the crossing of national and organizational boundaries. The BACC falls in between the MABE and MDD Consortia in terms of organizational capacity: while there were two public Bolivian universities participating (UAP and UAGRM), one (UAGRM) was larger and older, and the BACC also had two NGOs, specifically participating to support the focal partner (UAP). The focus on capacity building and the explicit design of a support structure helped address the question of institutional capacity.

The BACC incorporated several design features to ensure efficiency. UF served as leader and established administrative ties in multiple ways. First, UF signed sub-agreements with UAP and an NGO partner, IBIF, thereby linking authorities in these organizations (A1, C1, and D1). This established a hierarchy with clearly differentiated responsibilities for efficient implementation. Further, the project PI (A3.1.1) established bidirectional administra-

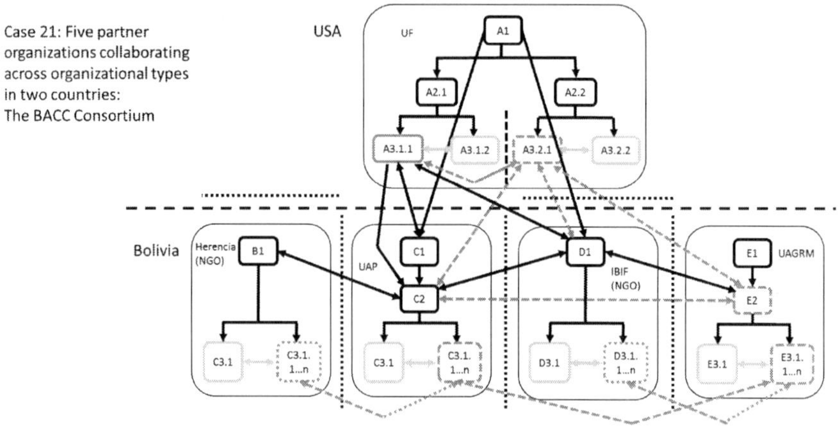

Figure 5.12. Consortia for Environmental Management: The BACC Consortium. ***Source: Created by author.***

tive ties with authorities in the institutions with sub-agreements for administrative oversight. Finally, in consultations with UF, the focal institution, UAP, hired a project manager (C2), who responded to A3.1.1 and oversaw participating faculty (C3.1.1 . . . n). The last two measures provided a modicum of redundancy via integrated managerial oversight among organizations, all focused on supporting UAP.

The BACC sought to maximize collaborative advantage. The consortium partners spanned disciplinary, organizational, and national boundaries. Participating UF faculty included personnel from the social sciences (A3.1.1) and the biological sciences (A3.2.1). Hence, there is discipline-crossing between A3.1.1 and A3.2.1. Whereas A3.1.1 primarily served in an administrative role, A3.2.1 led the science agenda, focusing on plant functional ecology and its implications for forest management in the context of climate change. A3.2.1 thus gets dashed lines and exhibits ties to a suite of in-country collaborators. The national boundary crossing between the United States and Bolivia is similarly indicated by the dashed line between UF and the other partners. Finally, there is organizational boundary crossing via the dotted lines between the universities and the NGOs. I underscore the strategic importance of the design, which paired Bolivian universities with NGOs. UAP was paired with Herencia for work in Pando, in northern Bolivia, and UAGRM was paired with IBIF in Santa Cruz, in eastern Bolivia. In turn, the universities were linked via the scientific coordinators (A3.2.1, C2, and E2). This structure thus highlights organizational complementarities in each of the participating regions in Bolivia.

The BACC thus exhibits design elements for vertical as well as horizontal coordination. In particular, the administrative structure includes hierarchical elements (the sub-agreements, concentrated management responsibilities with A3.2.1 and C2) but also horizontal network features (numerous bidirectional arrows among Bolivian partners). The goal was to incorporate elements of both hierarchical and network organization as a means of capturing the efficiencies of the first and the manifold information flows in the second. Under this design, the BACC was functioning well by those criteria. For reasons that will be explained in Chapter 6, however, its life was cut short.

To this point, the concrete cases of collaboration for environmental management have featured crossing of organizational boundaries and limited the number of countries to two. Given the complexities of crossing national boundaries, the remaining examples add countries. I therefore feature discussion of the complications that may arise from crossing multiple national boundaries, along with design strategies for managing the challenges. Figure 5.13 presents Case 22, the M-P Consortium, which involved partners in the United States, Bolivia, and Peru. By now the cast of characters should be familiar: UF served as lead organization, WHRC provided technical assistance, and in Peru, PEMD and UNAMAD were partners whereas in Bolivia, Herencia and UAP were partners. The case of M-P is thus more complicated as regards the number of countries; it also had the organizational diversity seen earlier in the MDD Consortium, with multiple universities and NGOs as well as a GA.

The administrative structure of Case 22 was multi-leveled. M-P was established via sub-agreements from UF to WHRC, PEMD, and Herencia. In turn, PEMD managed funds for UNAMAD, while Herencia did the same for UAP. This reduced the number of sub-agreements and thus monitoring effort for UF, and distributed administrative responsibilities to PEMD (for UNAMAD) and Herencia (for UAP). In addition to sub-agreements, UF hired a consortium program manager (A4.1) to provide in-country administrative oversight of the partners. This reinforced the chain of command from UF to partners with sub-agreements and then to other partners. The program manager worked directly with key personnel in partner organizations with sub-agreements, who in turn oversaw the other partners for which they were managing funds. The program manager was based in Peru, which facilitated oversight of the Peruvian partners. This, however, had the effect of hampering oversight and relations with the Bolivian partners, who were operating in a distinct legal and administrative context. This in turn required greater time and attention from A3.1 in oversight of the Bolivian partners.

Technical support for the M-P partners in Peru and Bolivia came from UF and WHRC. UF hired post-docs who were placed in each of the partner universities. The goal was to add technical capacity in the form of researchers without the teaching responsibilities of faculty. Both post-docs developed

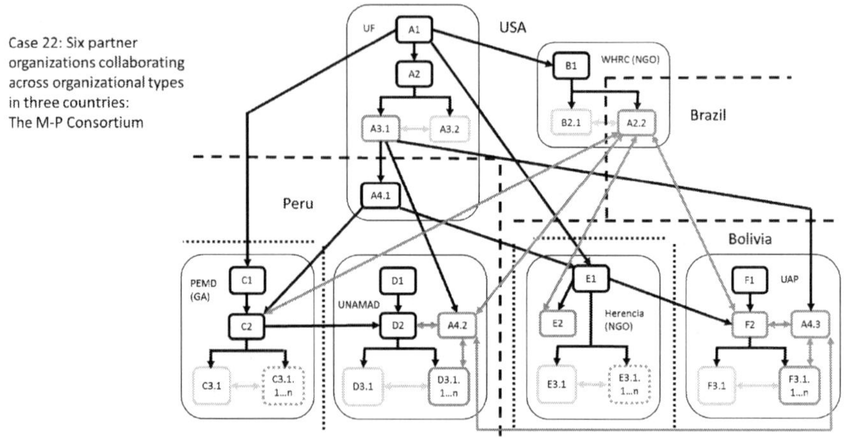

Figure 5.13. Consortia for Environmental Management: The M-P Consortium. ***Source: Created by author.***

new training programs which featured the use of newly established laboratory facilities, which in turn supported data production for management planning by field teams. WHRC supported these efforts, as did Herencia via hiring of additional technical staff. The added technical capacity was thus focused on universities but complemented by other partners. Most of these relationships were horizontal, which facilitated lateral communication among partners and thus collaborative advantages from production of scientific data to applications in management planning and disaster risk reduction.

Importantly, this included cross-fertilization among the post-docs in Peru and Bolivia, one of the few direct relationships among personnel working in the two sides. A major challenge in the M-P Consortium was the establishment and maintenance of ties between the two countries. The administrative and technical demands were sufficient to preclude regular meetings, much less directly collaborative work. UF and WHRC provided the main integrative mechanisms of the M-P Consortium, via the program manager or technical support via the post-docs.

The M-P Consortium (Case 22) thus falls somewhere near the BACC (Case 21) by combining vertical and horizontal design elements. The hierarchical design elements for administration (sub-agreements and program manager) were mutually reinforcing, a key feature for a larger consortium crossing multiple boundaries. The sub-agreements were important, as the chain of command when crossing national boundaries encountered difficulties since the program manager was based in one country and less familiar with the other. The other key challenge in the M-P also concerned the difficulties of spanning national boundaries, this time in scientific and technical work. Successful initiatives in Bolivia, such as production of early warning flood bulletins, were slow to catch on in Peru, and this only occurred due to participation by WHRC personnel. A key lesson from the M-P Consortium is to have personnel who are willing to "float" among partners and across boundaries in order to facilitate bridging. WHRC personnel and UF post-docs provided this key function in the M-P; the consortium scientists in the MDD Consortium performed a similar function.

Figure 5.14 presents G-MAP, aka Case 23, a rather more complicated consortium. G-MAP had more organizations (nine) and spanned more countries (four). Consequently, the complications and difficulties noted already all apply here. But G-MAP was even larger, so some additional challenges arose due to scale effects. Case 23 includes four universities (UF, UAP, UFAC, and UNAMAD), four NGOs (WHRC, IPAM, SOS, and Herencia), and one GA (PEMD). Together they spanned the United States and the three countries of the MAP frontier.

Recall that the number of possible relationships in a network expands faster than the number of network members. Since the G-MAP Consortium had more organizational partners, it is impossible to tractably present the various functional relationships at the same time. Figure 5.14 therefore focuses only on the administrative coordination relationships, indicated as before by black arrows. As in other consortia, UF served as the lead organiza-

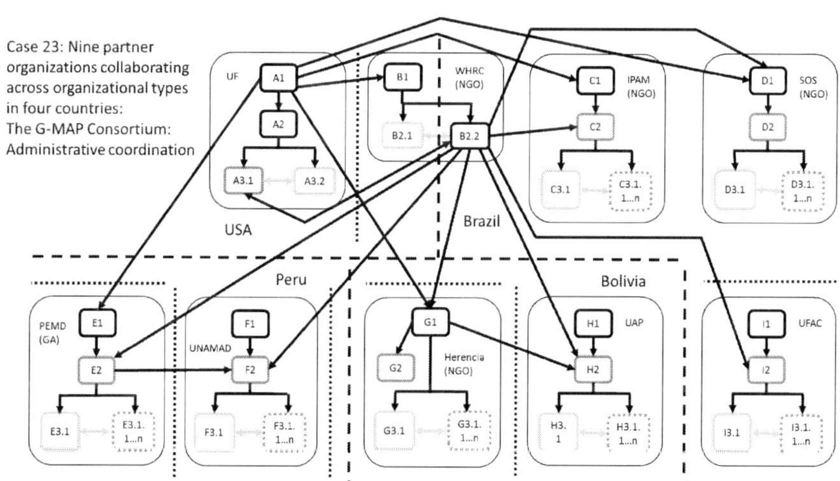

Figure 5.14. Consortia for Environmental Management: The G-MAP Consortium, Part 1.
Source: Created by author.

tion that received an agreement from the funder (USAID) and in turn wrote sub-agreements with selected partners. For G-MAP, UF wrote five: one for each of the NGOs plus the GA. We opted against sub-agreements with universities out of the consensus that administration would be more agile via NGOs. In turn, some partners with sub-agreements thus included budgets for partner universities. Hence, in Peru, PEMD had a budget for UNAMAD; in Bolivia, Herencia had a budget for UAP. This led to difficulties in Brazil, where the partners faced administrative complications for management of funds for university personnel. I elaborate on the travails of this question in Chapter 6. For now, it merits mention that we considered the "Plan B" of asking Herencia, the Bolivian NGO, to manage funds for UFAC, the Brazilian university. Plan B also had complications tied to international fund transfers. Establishing administrative protocols requires substantial creativity when crossing boundaries: what is feasible between an NGO and a university in one country may not be in another. Further, Plan B may not work, necessitating Plans C and beyond.

Figure 5.14 shows a suite of administrative relationships between agent B2.2, the program manager, and other key players of the G-MAP Consortium. Whereas sub-agreements established administrative relationships among organizations, the program manager was responsible for overseeing staffing and monitoring procedures implemented by personnel in each partner organization. As in previous cases, coordination by a program manager backed up the agreements and thereby strengthened consortium integration. G-MAP's manager operated in a largely horizontal fashion since G-MAP was a formalization of the MAP Initiative. This reflects the fact that the key personnel already knew each other. Figure 5.14, however, indicates hierarchical relationships via unidirectional arrows. This reflects the importance of building clear administrative hierarchies upon pre-existing relationships. In effect, we supplemented those informal horizontal relationships with formal vertical relationships, betting that the former would permit the latter. Doing this entirely through the medium of one person, the program manager, was, however, insufficient. Aside from risks of conflicts due to personalizing disagreements centered on the program manager, there were other administrative relationships in need of formalization to sustain the administrative structure of the consortium.

Figure 5.15 therefore replaces the administrative relationships based on sub-agreements and the work of the program manager with the financial relationships (light gray lines) established by the financial manager, D3.1. G-MAP partners collectively agreed that D3.1 had the highest qualifications in accounting, auditing, and other aspects of financial management. This placed D3.1 in a position of authority on financial matters with respect to the partners. The financial manager requested financial data from project directors in each partner organization, who in turn worked with support staff to compile the figures. The financial manager also worked in a non-hierarchical mode with A3.1 at UF as the lead organization, who in turn worked jointly with UF staff to arrange financial reports to the funder. Hence, there were primarily hierarchical elements to the financial management system, along with horizontal collaboration in the final steps of financial reporting.

Figure 5.16 again replaces relationships in order to present the technical and scientific relationships among the G-MAP partners. The program manager B2.2 again emerges as a key player, now with numerous bidirectional relationships with project directors at the partner organizations. Aside from playing the role of administrator, the program manager of G-MAP supported partners in their technical activities, as by participating in training events held by universities or management planning workshops run by NGOs and the GA. Figure 5.16 is perhaps the most complicated because several other personnel also played key roles for specific technical programs. C2 at IPAM led activities on road impact mitigation planning; D2 at SOS similarly led activities on watershed management planning. Both had numerous relationships on a technical level with project directors at partner organizations in other countries working on the same topic. In each case, project directors had ties with partner universities in the same countries. This horizontal network constituted the structure by which universities coordinated training activities with NGOs and the GA on topics on which NGOs and the GA then conducted management planning. UF supported these activities via the trainings

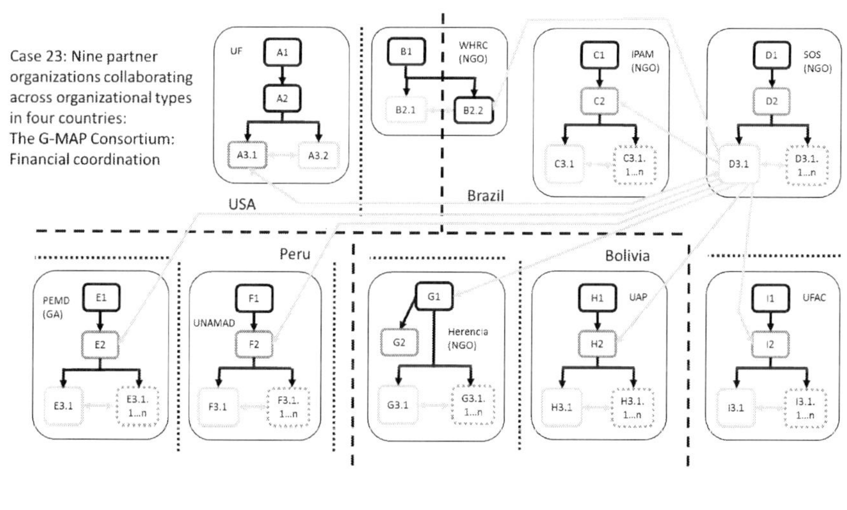

Figure 5.15. Consortia for Environmental Management: The G-MAP Consortium, Part 2.
Source: Created by author.

but primarily via the interest of A3.1 in road impacts, reflected in the relationship with C2.

G-MAP operated on a scale sufficient to require multiple figures in order to capture even a portion of the collaborative relationships at hand. With this number of organizations involved and boundaries being crossed, complexity is unavoidable if there is to be sufficiently strong management of personnel, time, and money, adequate support of bridging boundaries, and clear roles in organizational hierarchies to get work done. Despite the larger scale, G-MAP does not differ markedly from previous cases in that it incorporated a combination of hierarchical and horizontal elements in its organization. As in the other consortia, G-MAP employed primarily hierarchical relationships for administration, with mutually reinforcing mechanisms when crossing boundaries, and largely horizontal relationships for technical work, including across boundaries. The large number of relationships for each of several key functions ensured multi-planked bridges across boundaries.

Like the previous consortia discussed, G-MAP relied heavily on complementarities rather than redundancy. The emphasis on complementarity stemmed in part from the limited organizational capacity of the partner organizations, based as they are in a developing region. It also resulted from a desire to respond to the huge opportunities of working across boundaries on shared problems in the MAP frontier. Complementarity thus pervaded the design of G-MAP due to its agenda to achieve collaborative advantage. There was substantial learning occurring under G-MAP. One example con-

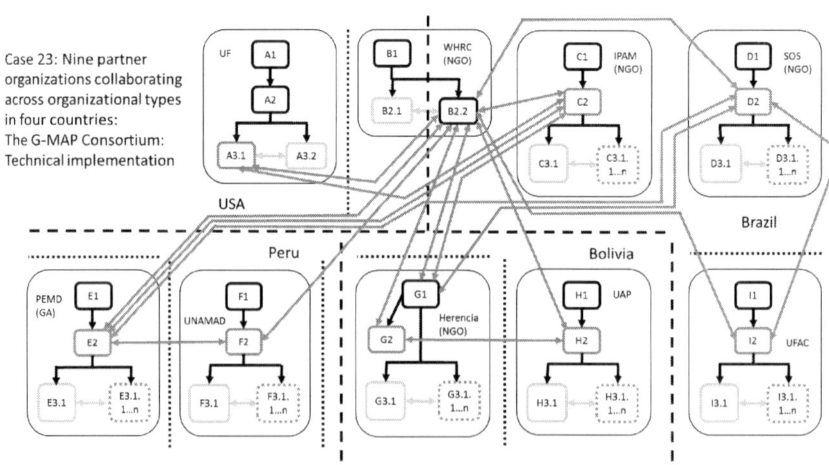

Figure 5.16. Consortia for Environmental Management: The G-MAP Consortium, Part 3.
Source: Created by author.

cerns the collaboration among organizational types to link training by universities with management planning by NGOs and the GA. Another involved crossing national boundaries, as the Brazilian NGOs led the road impact mitigation and watershed management planning by developing planning models for implementation on all three sides of the MAP frontier.

With the emphasis on complementarity came the risks stemming from a lack of redundancy. Here it is worth noting that G-MAP was broken up into two separate consortia, MABE and M-P. I will explain the reasons for the division in the next chapter. For now, the key point concerns the structural implications. When G-MAP was divided, the Brazilian partners constituted MABE, while the Peruvian and Bolivian partners constituted M-P. This had the effect of separating key players providing vital functions from some partners. One example is the financial manager, who ended up in the MABE Consortium. This contributed to MABE's administrative capacity. M-P, however, was left without a financial manager, something none of the M-P partners were in a position to readily provide. M-P thus struggled with financial management for a time thereafter. The other major example was that the division similarly affected coordinated implementation of the environmental management planning model for road impact mitigation and watersheds. Whereas MABE partners included coordinators for both of those programs, MABE was able to continue with minimal disruption beyond a reduced geographic scope for impact. Meanwhile, M-P took roughly a year to reorganize in order to proceed with its version of road impact mitigation and watershed management planning. For the cases of the environmental management planning programs, G-MAP did build in some measure of redundancy against the division, since each program had counterparts in each country. Nonetheless, the loss of program leadership mattered; the lack of distributed leadership across countries made the division a more severe test due to the loss of collaborative advantages, primarily among the Andean partners.

In the foregoing discussion, I periodically made reference to the technical work of conservation and development in the Amazon projects. But my comments on the stakeholder training, road impact mitigation planning, watershed management planning, and other such activities were few in number and limited in depth. For present purposes, describing that technical work was not the focus. Rather, the emphasis here resided with the design of collaborative structures to support technical implementation. This is because those collaborative relationships constitute the infrastructure which makes technical implementation possible in the first place. The infrastructure is too often overlooked, and doing so is perilous. The design of collaborative structures must be done with care if they are to be sufficient not only to support efficient performance of the technical work but also to manage the challenges to collaboration in general and the crossing of boundaries in particular.

More could be said about varieties of organizational hierarchies as well as network structures, and more could be done with the concrete cases reported here. This reflects the complexity of collaborative structures, particularly when crossing multiple types of boundaries simultaneously. Even with experienced project managers at hand, we have encountered many challenges to collaboration when attempting to cross boundaries for environmental science and management. I therefore spend the remainder of this chapter highlighting a limited suite of lessons learned out of the experiences of the Amazon projects. Readers working in other contexts can likely add to and amend these lessons.

1. Complementarity is golden, but so is redundancy. Complementarities are virtually unavoidable when crossing boundaries, so much so that crossing boundaries can yield vulnerabilities due to a lack of redundancy. This is especially likely to be the case where some partner organizations are young or small, and consequently have few personnel. Such instances are likely to arise in developing regions where conservation and development endeavors can benefit more from collaboration. Investing in redundancy is thus a priority when crossing boundaries for collaboration.

Ensuring redundancy can be seen as a luxury only affordable by large organizations in developed countries. I have evaluated redundancy in the organizational structures in the foregoing cases primarily by considering whether multiple individuals could or did perform a given function within an organization. While this is an obvious indicator of redundancy, it is an expensive one, and it undermines the efficiency of investments in collaborative enterprises. One means of sidestepping the problem of redundancy as a luxury is to foment cross-training among personnel with regard to functions, as was noted in Figure 5.1. In the empirical cases from the southwestern Amazon, this was a lesson learned: it is never too early to institute various forms of functional cross-training on the assumption that individuals will move around or leave. Cross-training reduces but does not eliminate the need to find new personnel, which can be difficult in small markets if the functional requirements are highly specialized.

Here the crossing of boundaries can be useful insofar as it expands networks in which partner organizations are situated, which in turn broadens the potential markets in which personnel can be sought. This, however, has the effect of scaling up (even globalizing) labor markets, which may substantially affect the asking price for particular specialists. In the empirical cases, when we sought to expand technical capacity and concluded that we needed PhDs and not just MAs (or MScs), searches became international and costs for technical personnel came out much higher.

2. Hierarchies work best for management, and networks for learning. If this seems a truism in prior work, the present point is that it also applies when crossing boundaries. The concrete cases in this chapter highlighted

collaborative structures based on both hierarchical and non-hierarchical organizational principles. This is to incorporate the contrasting advantages they present, while also managing their liabilities. Hierarchies permit efficiencies via a clear chain of command, but they also tend to restrict flows of information and the openness of people to learn. But information and learning are crucial in rapidly developing regions where threats to conservation and sustainability may emerge quickly. Hence, horizontal ties are also necessary in collaborative endeavors.

The question then is how best to design hierarchical and network relationships to maximize their distinct benefits without one undercutting the other. One approach that arises from the empirical cases is to design administrative hierarchies and technical networks. Let the technical thinkers think and talk as means of evaluating problems and threats, which then serve as the basis for developing ideas for technical proposals. They then bring said proposals to the administrators to design the organizational structures for efficient implementation via a hierarchical chain of command. If it is a mistake to impose hierarchies on the development of ideas, it is also problematic to impose networks on the implementation of ideas.

The complementarities of hierarchies and networks can be put to good use in a cycle of ideation, implementation, and reflection, as in the tradition of adaptive management (cf. Holling 1978; Lee 1993). The key is to know when to switch gears in the organizational structures employed at different points in the adaptive cycle. If networks permit the generation and development of ideas into proposals for action, hierarchies serve best to put plans into action; conversely, upon implementation, it is useful to switch back to network mode for open evaluation and other means of reflection. The complexity that arises concerns the ambiguities in roles when switching among hierarchical and network forms of organization. This can be threatening to those with authority in a hierarchy and catalyze the politicization of the work at hand and the relationships supporting it, a theme of the next chapter.

3. Define the key functions with care, and subdivide them when necessary. I have played fast and loose in this chapter in how I have defined functions. In particular, I admit my guilt as an academic in differentiating much more among technical disciplines than other functions in collaborative endeavors, such as administration and extension. I can well imagine that the administrators and staff at my home institution would have preferred more elaborate organizational diagrams as regards their functional roles, whether by differentiating proposal processing from award administration, contract management from financial accounting, or handling of sub-agreements from reporting to funders. Similarly, I suspect key players in partner NGOs would probably have liked to see some differentiation of functions among coordinators and technical personnel on field extension teams. I could go on.

This, however, motivates the larger point: functions can be differentiated in various ways. The key is to collaboratively identify the crucial functions of a collaborative enterprise with regard to the key objectives of said enterprise. I want to emphasize the word "crucial" here. In my experience, the tendency in group processes when designing collaborative structures is to be inclusive. The upshot is that numerous functions get enumerated so all those present are assured a role and thus able to justify whatever resources they plan to request. Differentiating a litany of functions is no doubt important, especially in large collaborative structures where a single oversight can threaten the whole enterprise. But this is not the same as identifying the key functions which are crucial to selling the enterprise to a funder, or to a new collaborator who brings something new to the game.

In turn, where there are key functions, they usually require the most attention with regard to subdivisions that recognize specialties, particularly if they are hard to come by. If a collaborative group has a long list of functions and the key functions are given equal time with more mundane, standardized tasks, there is great risk of operational problems later. On the NSF Roads project, the lack of attention to the demands of the botanical component became a liability. Whereas the socioeconomic component had a significant field activity and more limited demands for data processing, and the remote sensing component required limited fieldwork but considerable lab work, the botanical component required heavy investments in both, and that was not adequately recognized at the outset, which led to delays later. Similarly, the many tasks under the broad rubric of "financial coordination" were unrecognized at the outset of G-MAP. The importance of adequately vetting budgets, reviewing expenses, reconciling advances and liquidations by program activity, and preparing for external audits was not initially appreciated by anyone involved. Consequently, we relied heavily on the beleaguered financial manager and found it necessary to bring in additional support staff to perform very specific functions in order to keep up with the demands of financial management.

4. Crossing boundaries requires multiple planks in the bridge. By now it should be evident that the design of collaborative structures requires particular attention to the construction of strong bridges to span divides. Not only do individual relationships have to be strong, there need to be multiple ties that encompass a suite of distinct functions. Effective bridging thus requires "multifunctional" boundary crossing. Technical implementation across a boundary is likely to encounter problems if there is not adequate provision for administrative support. Every example beginning with Case 15 portrayed multiple ties across boundaries. The empirical cases therefore looked like circuit diagrams with bundles of wires representing various types of ties across boundaries.

A key structural lesson when crossing boundaries is to ensure that the bridging involves multiple people performing different functions so that the divisions being spanned are "fully served" and thus adequate to suffice the objectives of the collaborative structure. By "fully served" multifunctional boundary crossing, I mean to say that for every key function in a collaborative structure, there needs to be a separate linkage across that boundary. This ensures that key functions are supported where they are hardest to maintain.

Multifunctionality in boundary crossing also implies the existence of a peer group of boundary crossers in a given collaborating organization. Individuals crossing boundaries in the name of a given function also perform the additional function of building and maintaining bridges. One key way for an organization to support its boundary crossers is to have them regularly talking to each other. Even if they putatively serve distinct functions, they still share the bridging function. Difficulties in crossing a boundary for one function permit the responsible parties to consult with others in their home institution crossing the same boundary. That way, nobody is "on their own." This does not guarantee resilience via redundancy since multifunctional boundary crossing may not involve multiple individuals in an organization performing the same functions. But having multiple individuals in an organization crossing a given boundary still provides a reference group to which one can turn, at least with regard to bridging.

I note that few cases illustrated those in-house relationships. In practice, they often emerged over time in multi-year projects, especially when some boundary crossers had better language skills or more experience. The figures presented could thus add more arrows among individuals within an organization to represent either relationships involving advising about boundary crossing.

5. In-house support for crossing boundaries is especially important in more complicated collaborative structures. Investments in servicing the crossing of boundaries rapidly become greater in larger, more complex collaborative structures. The more organizations involved and the more boundaries crossed, the greater the need for investments to fully service multifunctional bridging. In the empirical cases considered, more complex collaborative structures unavoidably put pressure on key individuals, who may not only work with several partner organizations across various boundaries, but also with multiple departments and individuals performing different functions within their home institution. In other words, the need to invest heavily in multifunctionality for fully serving bridges across boundaries hampers functional redundancy within organizations. As noted earlier, boundary crossing pushes collaborative structures to emphasize complementarity. The key implication is that fully servicing multifunctional boundary crossing makes redundancy hard, especially in large collaborative structures.

Redundancy can be improved either via investments in more personnel, though there are alternatives such as cross-training. In larger organizations, a third way to manage this problem is to indicate other personnel with comparable skills as backup for personnel managing a given function in crossing a boundary. This supports boundary crossing while adding both complementarity (as only one individual is bridging boundaries) and redundancy (since both individuals perform that type of function). Backing up boundary crossers is easier for some functions (such as staff support) than others (such as technical specialists). Thus, bridges across boundaries should not only be built with planks for different functions, but in addition, each end of the proverbial bridge requires structural reinforcement in the form of redundancy, whether via additional dedicated personnel, cross-trained personnel, or backups.

The design of collaborative structures thus encompasses various devils in the details. Individuals who seek collaboration across boundaries for conservation and development often do not have the savvy or training to foresee the consequences of designing robust structures that will hold up in the face of the many rigors to come. To summarize, the problematic tendencies in structural design are 1) to emphasize complementarity and ignore or underestimate the importance of redundancy, 2) to fail to understand the very different assets and liabilities of hierarchical and network structures for particular tasks, 3) to underestimate the number of key functions or to misidentify the key functions, 4) to build weak bridges across boundaries, and 5) to fail to reinforce bridges across boundaries at both ends. Collaborative structures that avoid these tendencies are much more likely to survive.

This is not to guarantee their survival, much less their success. Sound structural design in light of collaborative objectives and key functions simply avoids one set of potential breakdowns. There is, however, a suite of other issues that pose additional threats to the viability of collaboration across boundaries. The next three chapters take up these issues: politics and inequality, uncertainty and surprise, and collaboration and the self.

Chapter Six

Inequalities and Politics in Collaboration across Boundaries

Everybody is getting along because there isn't any money.

—Marianne Schmink, on the MAP Initiative

Question: What is the worst enemy of a Peruvian? Answer: Another Peruvian.

—Popular joke in Peru

The previous chapter focused on the design of collaborative structures with an emphasis on efficient performance of key functions. Some social scientists may groan at such a rendition of collective action because it overlooks the problems of inequalities and politics, and the conflicts they engender. I am aware of the long-standing debates in social sciences such as sociology over functionalism and conflict (e.g., Horton 1966; Ritzer 1990). My intent in focusing on structures and functions in collaboration was not to neglect inequalities and politics. On the contrary, the previous chapter motivates a detailed examination of the ramifications of crossing boundaries for inequalities, politics, and conflict.

In collaboration, there are power inequalities between many pairs of social actors: managers and staff, faculty and students, salaried personnel and consultants, funders and recipients, and so forth. These inequalities often reflect broader inequalities defined by various social differences such as class, race/ethnicity, gender, age, and nationality. Previous treatments of inequalities among collaborators have often highlighted the dominance of stronger partners (e.g., Phillips, et al. 2000; Edmunds and Wollenberg 2002; Stewart and Gray 2006). Collaborators with larger organizations or more experience tend to impose hegemonic narratives about the problem, the approach, and so on. In this chapter, I suggest that inequalities are often increased when crossing boundaries, as between strong and weak disciplines, large and small organizations, and developed and developing countries. This is the point of departure.

Further, collaboration across boundaries can be understood as a political act in the broader context beyond the partners themselves. As noted in Chapter 4, boundaries exist for various reasons, including political reasons. Crossing politically important boundaries may thus be taken as a form of political transgression. This chapter elaborates on the observation that there are nonzero risks of "external" political interventions when collaborating across boundaries. Tread with care, or be trodden upon.

But the politics of collaboration across boundaries do not stop there. Precisely due to the inequalities among collaborating partners, there is also a high probability that collaborators will also encounter difficulties that lead to disagreements and "internal" politics. This chapter therefore asserts that the potential for internal political disputes is heightened when collaborators span divides. My experience is that even among initially enthusiastic courting partners with manifold compatibilities, when crossing boundaries, the honeymoon can end quickly. The additional challenges posed by boundaries readily prompt confusion and disagreements. Tempers flare, and because of boundaries, conflicts cannot be solved but only managed. Indeed, internal politics can readily become a major focus in the management of collaborative relationships.

The design of collaborative structures thus becomes an exercise in selecting the inequalities with which one is going to have to live. Any design involves inequalities among partners and thus creates the conditions for the potential politicization of goals, tasks, and relationships. Networks are celebrated as spaces for communication, but when tensions arise, they become spaces of conflict. Hierarchies are useful for arranging a clear chain of command, but if there are disputes, authority figures must choose carefully how to respond.

This chapter thus recognizes that there is no means of designing collaborative structures so as to entirely avoid conflicts due to inequalities and politics. If structures and functions are key for thinking through the design of relationships in collaborative enterprises, recognition of inequality and politics calls attention to the management of possible conflicts in those relationships. Successful collaboration thus amounts to navigating external and internal conflicts caused by inequalities and politics in order to pursue shared goals. Of course, a goal that emerges during the process of collaboration can also be the reduction of conflicts via the application of strategic practices to mitigate inequalities and politics, where feasible.

Whereas previous work on collaborative advantage highlights the issue of power in general terms (cf. Huxham and Vangen 2005), there remains a need to more closely examine a suite of issues concerning inequalities and politics in collaboration across boundaries. Arguably more germane to what follows is work on conflict management (cf. Burton and Dukes 1990), notably in collaborative contexts like multi-stakeholder negotiations (e.g., Edmunds and Wollenberg 2002; Hemmati 2002; Warner 2007; Wondolleck and Yaffee 2000). Disputes irresolvable by established institutional channels require the design of collaborative processes that explicitly recognize inequalities and politics as a means of managing conflict (e.g., Buckles 1999; Carpenter and Kennedy 1988; Constantino and Merchant 1995). The relevance of conflict management stems from the fact that multi-stakeholder negotiations amount to a case of inter-organizational collaboration. Broadly similar issues apply when crossing disciplinary and national boundaries (e.g., Bercovitch and Rubin 1992; Buckles 1999; Morris, et al. 1998; Obuchi, et al. 1999). But the complications of crossing boundaries can be compounded when spanning multiple types of divides at the same time (Perz, et al. 2010). What is more, a key difference between multi-stakeholder negotiations and the cases that follow is that in the first, the players are often forced by circumstance to collaborate, but in the second, they choose to do so. Hence, inequalities and politics in collaboration are arguably riskier, since unhappy collaborators can walk away.

This chapter therefore discusses the complications of inequalities and politics when crossing boundaries for collaboration. I first discuss the significance of boundaries for inequalities and politics. Here I make the key assertion that crossing boundaries leads to the wider inequalities and more intense politics of collaboration. In addition, I argue that boundaries also hamper efforts at discussing inequalities and managing conflicts. To back up such arguments, I offer a suite of "vignettes," that is, short accounts based on firsthand experiences. For the vignettes, I draw on experiences from collaborative projects crossing boundaries in the southwestern Amazon. The vignettes serve as illustrations rather than universal or definitive accounts. Readers can judge whether similar situations in other contexts would entrain comparable processes and outcomes. The opening vignettes highlight upfront issues of inequalities among courting partners seeking to span boundaries. I then turn to a discussion of the travails of external politics as they may affect collaboration across boundaries. This section also comes with illustrative vignettes from collaborative projects in the southwestern Amazon. The crossing of national boundaries is particularly salient in these vignettes, though other boundaries are also important. The chapter then focuses on a discussion of inequalities among partners spanning boundaries in later stages of collaboration. When inequalities are large, even if they are addressed up front, downstream complications will arise and test the resolve of collaborators. More vignettes from the southwestern Amazon serve as illustrations of the consequences of inequalities among collaborators when spanning boundaries. The chapter then turns to the internal politics among collaborators. This is a distinct section since conflicts caused by politics may not result from inequalities per se. I again include vignettes from the southwestern Amazon, this time to illustrate how collaborators initially enthusiastic about partnering can become either frontal combatants or back-stabbing conspirators. The final section of the chapter offers a reflective discussion, noting the additional challenges of managing conflicts across boundaries, especially when collaboration crosses multiple boundaries simultaneously.

THE IMPORTANCE OF BOUNDARIES FOR INEQUALITIES AND POLITICS IN COLLABORATION

Collaborators crossing boundaries face two key challenges with regard to inequalities and politics. The first is that inequalities are greater and politics more complex when crossing boundaries. The second is that despite the greater necessity to openly discuss these issues, boundaries also make such discussion more sensitive and thus more difficult.

Some obvious reasons for these challenges present themselves. More powerful collaborators often adopt hegemonic narratives and thereby dominate proceedings by setting the terms of reference from introductions on through setting of goals and all the rest. These problems can be managed using the strategic practices outlined in Chapters 3 and 4. The more difficult issues that may arise when crossing boundaries are more subtle, and pertain to the adequate recognition of inequalities. Whereas it may be easy to call out a loud-mouthed collaborator in order to open more space for broader discussion, this does not guarantee that inequalities have been fully interrogated. When crossing boundaries, interrogation is likely to have to unfold over time rather than in a weekend workshop.

Inequalities among collaborators crossing boundaries are often complex and take time to dredge up. In large measure this is because inequalities are tied to cultural assumptions which are often unconsciously held and yet are deeply felt and thus carefully guarded. Make no mistake: disciplines, organizations and countries all have cultures with worldviews and assumptions. Pointing out those assumptions is necessary for a full accounting of inequalities, but at the same time may touch on cultural sensitivities. This is because learning about a partner in order to understand it generally requires getting beyond the front door propaganda that puts the partner in the best light. Digging down farther, identifying the partner's past foibles, failures, and limitations, can reveal issues that can be characterized in a range running from dirty laundry to illegal activity. This can be very awkward, as representatives of a given partner organization find themselves talking about their issues in front of prospective collaborators they do not yet know well, and moreover in front of their colleagues. This may involve exposure of stories heretofore unknown to others in the same institution. Hence, the pace has to be slow, and the conversations with the dirty laundry private. Meanwhile, other clocks are ticking, which puts a premium price on trust in the short term that initial joint actions will not be viewed in a negative light for reasons as yet unknown.

This leads to Vignette 6.1. This vignette stems from the relationship established by my home institution with a partner university in the southwestern Amazon. (For purposes of the vignettes, I withhold identifying names to preserve anonymity and focus on the issues at play rather than the individuals and organizations involved.) Initial meetings with the partner university focused on dialogues about shared interests in social-ecological research. I presented on UF and the proposed activities of the project, which became NSF Roads; the partners presented on their institution and relevant departments. We then discussed complementarities among the two organizations and how those complementarities could be put to positive use on the project. Since the project highlighted field data collection, UF would handle the funding and coordinate the methodological protocols, and the partner university would provide comments to improve the protocols, including pretesting to confirm their validity, and then collaborate with UF graduate students in the field data collection. All this was clear enough until we came to the issue of participation and funding. Specifically, there was a need to identify faculty coordinators at the partner institution, who would serve as points of contact for UF personnel and field teams and handle funds for local activities. Crucially, the partner coordinators would receive a consultancy fee for their contribution.

After a pronounced silence in the room, one of the collaborators assured me that there would be a "democratic" process for identification of the coordinators. There would be consultations with the partner university *rectorado* (president's office), which would oversee the proceedings, since the activities would span multiple departments. After a time, I received the results, evidently of a vote, which indicated some people who had been at the initial meeting and some who were new faces. Wanting to be a good collaborator, I accepted the results. Protocols were reviewed, discussed, and edited; fieldwork logistics got sorted out and data came in; people got paid for their efforts.

Or so I assumed. The last point became an issue of contention among faculty at the partner university. Eventually I was contacted because some people didn't get their money. The response from the main coordinator was that they hadn't turned in their final reports. They said they had; he repeated they hadn't. My hands were tied. A productive collaboration suddenly became politicized and I looked like a dupe.

Thus began the sleuthing. I made discrete inquiries, trying to find openings in professional networks surrounding the partner university to figure out why this had happened. The initial picture was very limited and fragmentary, resembling a dozen pieces out of a 100–piece puzzle. Over the next few years, however, as leadership turned over and I hung around, collaborating with some known and some new personnel, I began to receive more bits of information as to how things operated. The story that gradually emerged was that some people had struck deals with others in order to get "elected" to the project. Some people didn't like some others who had managed to get on the project. Some funds disappeared. The web of intrigue proved formidable. It did, however, make much more evident who the truly capable were, who worked for the money rather than the science, and who were politicians waiting for their next move up the ladder.

As my descent into cynicism got under way, a crucial piece of information came out: the faculty faced significant employment insecurity. Suddenly, the secretive behavior became understandable. The consultancies were supremely important not only as temporary salary bumps but as hedges against uncertain futures. This was so foundational to the calculus that nobody had bothered to explain it to me, an outsider. This reflected the big differences between the two institutions across national boundaries. The insecurities facing faculty at the partner institution stood in stark contrast with the handling of faculty contracts at UF, and thus reflected the substantial inequalities between the two institutions. Equally true, however, was the reticence of collaborators to elaborate on the implications of employment insecurity, precisely because they were sensitive, as they involved salaries. The insecurities at play thus helped to explain the internal politics, largely hidden from view, which necessitated the sleuthing in the first place.

In the case of Vignette 6.1, the partner faculty were well aware of the issues impinging on their decision making, though they did not initially think to explain it to a collaborator. This raises issues of the "subjectivities" of perspectives (e.g., Berger and Luckman 1967; Ferguson 1993; Holstein and Gubrium 2008; Searle 1995), notably those which stem from cultural differences when spanning boundaries. Subjectivities refer to the values and interests underlying the assumptions behind what one thinks, says, and does. Subjectivities thus greatly influence what one thinks is important, and how one responds to specific situations. Two people responding to a situation via hyperbole and a low-key discourse exhibit strong subjectivities with regard to the situation.

In the context of collaboration, subjectivities help define what one chooses to pay attention to or emphasize in relationships. Subjectivities are closely tied to how individuals construct their understandings, including of themselves and their relationships with collaborators. The stronger the subjectivities at play, the more difficult communication and understanding become. The stronger the subjectivities, the larger the differences in the constructions of a situation, and the higher probability of conflicts over said constructions.

Boundaries strengthen subjectivities by separating cultures with different assumptions that lead to contrasting constructions of situations. Points of view often differ profoundly from opposite sides of a boundary. Crossing boundaries therefore means navigating very different subjectivities about situations that may arise as part of the process of collaboration. Scholars from a discipline often presume to find the same cognitive tendencies and intellectual commitments in other disciplines. Personnel from one organizational type tend to expect counterparts in another type to have the same capacities and priorities. Citizens of one country frequently impose their cultural beliefs on those from another, and denigrate them if they don't measure up to values back home. Consequently, even with the same problem, the same language, and the same words, subjectivities across boundaries can yield fundamentally different constructions of a situation. By extension, subjectivities across two or more boundaries compound miscomprehension. When constructions of a situation differ, they often generate incommensurate proposals for what ought to be done. This greatly complicates problem solving because different constructions cause disagreements and politicize even mundane topics.

This leads to Vignette 6.2. At the very beginning of G-MAP, a question arose concerning how funds might be managed for one of the partner universities. Public universities in Latin America often face significant bureaucratic obstacles for handling external funds, so they routinely create private foundations with greater administrative flexibility. The complication that arose for one university partner in G-MAP is that the university's two foundations were both dealing with complications arising from previous projects. The consortium partners agreed that it would be better to find an alternative. We therefore considered a Plan B. The consortium had NGOs in the same country, which might have filled the same role. Both, however, had relatively small offices and large sub-agreements already, making it difficult for them to take on additional responsibilities. What is more, they had legal limitations in handling funds for personnel at a public university in their country of operation.

This necessitated Plan C. Another NGO partner, from a different country, stepped forward. Since it was from a different country, it lacked the legal restrictions noted above. But because it was from another country, it faced other complications in handling money, since the currencies would be different. This led to tensions due to the subjectivities at play, which stemmed from the organizational as well as national boundaries being crossed. For the NGO partner, the procedures were clear; for the university partner, they most certainly were not. An increasingly heated discussion ensued, in which the NGO partner representative explained what the procedures would have to be and why, and the university partner representatives indicated they didn't understand, and suggested an alternative that was simpler, but unworkable for the NGO partner. The same points were made in a series of repeated back-and-forths. People were indeed speaking different languages, which didn't help. It became evident that the same words did not have the same meanings in the two languages, notably the word *repase*. What is more, it became evident that such words had differing meanings in the distinct administrative contexts of a university and an NGO. Finally a network broker in the group began to ask questions about how certain words were being used and understood. It turned out that there were indeed differences in assumptions, leading to misunderstandings pertaining to bank fund transfers. By then, the university personnel were exasperated and began to express concern about whether there was confusion about other words. The discussion continued. More misunderstood words began to trickle out. We moved into the exercise of dredging up the implications of specific meanings of terms for the fund transfers. Slowly, they became clearer. I looked at my watch: two hours had passed.

But we weren't finished. Indeed, clarifying terms led to complications stemming from deeper, more implicit subjectivities. The NGO partner director began to get cold feet, expressing doubts that Plan C was a good idea. The university personnel complained that there seemed to be no workable solution, implying that nobody truly wanted to help. The NGO director observed that he had offered to help when others had not. Hard feelings were being vented. The network broker offered the sage interpretation by noting the irony of an NGO from a smaller country helping a university in a larger country. Everybody else kept their heads down. These tensions would arguably have consequences later on, as I will describe in Vignette 6.12.

Vignette 6.2 illustrates the implications of inequalities in capacities for expectations about what is workable. Crossing boundaries makes management of expectations among unequals especially tricky. Inequalities and boundaries beget differences in subjectivities about what is possible and thus what one partner can expect from another. Such expectations led to multiple disappointments as we moved from Plan A to Plan B to Plan C, which then also ran into complications. Especially important at that point were subjectivities that complicated discussion of the bureaucratic devils in the financial management details. There were further implications of inequalities and subjectivities in this vignette. Implicit in the conflict was the expectation that UF could resolve the problem. Whereas weaker partners may avoid discussing their limitations, putting their best feet forward, they similarly expect the same from stronger partners, resulting in unrealistically high expectations. As a large entity based in a powerful country, UF has been seen by some partners as capable of most anything in the context of a collaboration. At the same time, being cautious in making commitments to weaker partners can lead to suspicions that stronger partners are holding back on providing support. There is a careful balancing act at play.

Vignette 6.3 focuses more squarely on the problem of managing expectations among unequal partners. In the early days of G-MAP, it came time to arrange sub-agreements and begin transferring resources to the partners. The sub-agreements from UF to partners were large, and partners were excited to get their first infusions of money. Then the financial officer asked me when the money would arrive. In a classic case of dashed expectations, I had to reply that I hadn't a clue. The gasps from the partners were audible even in e-mail. Trying to save face, I hastily added that I would consult with UF administrative staff. UF administration replied that transfers could only be made as reimbursements in reply to invoices for expenses incurred and duly reported. The partners all but shouted that their in-house budgets were far too small for expenses on the large scale of the sub-agreements. G-MAP after all was supposed to permit scaling up of activities among MAPiense partners. I still vividly remember the testy one-word reply from the financial officer, who demanded an answer: *"VAMBORA!"* This is a contraction of two words in Portuguese, *vamos embora*. The words mean "let's go." The contraction (especially in all caps) carried a sentiment more along the lines of "HURRY UP!"

Further consultations with UF staff involving explanations about G-MAP then transpired, and led to deep dives into the rules manuals for sub-agreements using US government funds. It came out that advances were viable; standard forms existed for such procedures. We were off to the races with translations of instructions. I was off the hook, but the secret was out: UF was subject to regulations and strictures and couldn't do just anything it wanted. Thus began a campaign to manage partner expectations about UF administration of sub-agreements in light of the exigencies of G-MAP.

The high anxiety e-mails prompted additional consultations with UF administrative staff concerning advances, disbursements, cost sharing, liquidations, financial reporting, modifications to agreements, and numerous other topics that suddenly looked crucial. This led to another memorable quote, this time from UF administration: "Are you sure you want to do all this?" What they really meant was, "Are we really going to have to do all this?" UF staff thus made evident that the demands of G-MAP were greater than typical. But I'd committed, so the campaign expanded to find ways to support UF staff as they worked to support G-MAP. It was crucial to be proactive in consulting with staff, including with regard to their favorite types of chocolate, duly dispensed after trips abroad and around the holidays. Crossing boundaries among unequals requires careful management of expectations, on all sides, including at home. It became clear that at UF, I might have been faculty, but on administrative matters, I was a student. Fortunately the chocolates encouraged a modicum of shared understanding. When subsequent moments of panic ensued, I received maternal reassurances from administrative staff such as, "We're here to help" and "Everything is going to be okay."

Vignettes 6.1 to 6.3 offer illustrations of challenges to collaboration that arise due to partner inequalities when crossing boundaries. Whether due to the slow process of acknowledging limitations, the strong subjectivities at play, or the need to manage expectations, crossing boundaries for collaboration can involve large inequalities that readily lead to conflicts over seemingly innocuous issues. Acknowledging limitations, recognizing subjectivities, and managing expectations therefore become priorities for addressing challenges posed by boundaries to strategic practices for managing inequalities. Given these opening vignettes, I spend the remainder of this chapter discussing particularly salient difficulties stemming from external politics, consequences of inequalities among collaborators, and politicization of disagreements among collaborators.

EXTERNAL POLITICS AND COLLABORATION ACROSS BOUNDARIES

Collaboration for conservation and development can be politically sensitive due to the content of the technical work involved. As noted in Chapter 4, crossing boundaries may itself be perceived as a political act of trespass or infringement. Combining the crossing of boundaries for collaboration with conservation and development work complicates matters by expanding the sources of sensitivities and thereby multiplying the dimensions of the difficulties that might emerge. Further, the importance of external politics is likely to increase further when crossing multiple boundaries.

The fundamental problem of external politics with regard to collaboration across boundaries stems from prevailing disagreements about the value of crossing boundaries itself. One may find willing and engaged collaborators in another discipline, organization, or country, but the colleagues, coworkers or governments of those entities may not be so well disposed. The common denominator underlying such disputes within a discipline, organization, or country is the concern of vested interests to protect their claims against perceived threats from the other side of a boundary. As noted in Chapter 4, proactive engagement to introduce and inform is an obvious means of using diplomacy to alleviate such concerns. But diplomacy presumes that persuasion without threats is viable. And if one's discipline, organization, or country has a reputation for issuing threats, diplomacy is likely to fail. Crossing boundaries thus requires the navigation of political vagaries over which one usually has no control. This is especially important for political questions of long standing among countries. My vignettes in this section therefore focus primarily on external politics in terms of international relations as they affect collaboration across boundaries.

Vignette 6.4 highlights some of the international issues at play. In previous chapters, I mentioned that G-MAP was divided and became MABE and M-P. The reason for the division stems from external political considerations. G-MAP originated as one consortium among others under USAID's Amazon Basin Conservation Initiative (ABCI). ABCI was designed in part by drawing inspiration from the MAP Initiative, which crossed various types of boundaries, most notably national borders between Bolivia, Brazil, and Peru. USAID similarly proposed ABCI as a conservation and development program across the pan-Amazon (USAID 2006). Offices for ABCI institutional support staff were based in Brasília, Brazil's national capital. The government of Brazil, via the Ministry of Foreign Relations, expressed concerns that the US government was organizing an international program of activities, including in Bolivia, Colombia, Ecuador, and Peru, out of offices based in Brazil. A suite of complications arose, including the question of intergovernmental consultations before implementation and issues of national sovereignty with respect to the sites of activities pursued by NGOs. Distinct accounts of both questions emerged from the two governments.

G-MAP waited while the diplomatic communications proceeded. In the meantime, other consortia came to offer us their condolences. The word on the street was that consortia working in Brazil and other countries would have to be divided. Whereas other consortia primarily worked in one side or the other, G-MAP partners were more equally balanced across the three sides of the MAP frontier. Evidently, G-MAP would be hit the hardest. Official word eventually arrived: G-MAP would have to be divided into two parts, one managed by USAID-Brazil and the other by the USAID Regional Office for Latin America. G-MAP partners spoke of the "Solomonic division of the baby" in reference to the impending split. We wrote new funding applications, with maps of activities that ended at the frontier between Brazil and its neighbors. UF administrators again asked me, "Are you sure you want to do this?" Out of the new applications, the MABE Consortium emerged in Acre/Brazil and the M-P Consortium appeared in Madre de Dios/Peru and Pando/Bolivia.

Among other things, we made detailed, not to say tortuous, consultations about what might still be feasible by way of crossing the frontier between Brazil and its neighbors in the MAP region. Could a MABE partner maintain an activity programmed for Peru? No. The site of the activities had to be in the country in which a consortium was working. Could Brazilians participate in an M-P activity in Peru? Not a good idea, because it would be hard to justify paying for participants from outside the geographic purview of M-P. Plus, the performance reporting would be complicated. Would MABE claim the Brazilian participants, or M-P for participants in an activity in Peru?

Could a Brazilian consultant support M-P for a specific task? Yes, if it could be justified, such as in terms of the lack of someone on the M-P sides available for the task.

With the Solomonic division, the gymnastics of crossing borders thus became quite elaborate. MAPienses thus continued to think about how to cross national boundaries for collaboration in the spirit of the MAP Initiative. But the external politics stemming from concerns about consultations and sovereignty ran in the opposite direction. In the end, external politics, in this case managed via diplomacy, necessitated limitations on crossing boundaries.

In that light, it is worth noting some of the consequences for the partners. Chapter 5 outlined the collaborative structures of G-MAP, MABE, and M-P. G-MAP was designed around key MAPiense partners, such that key functions for the entire consortium were based in different organizations. The Program Manager and Financial Officer worked in Brazil; the Deputy Manager was based in Bolivia; Coordinators for specific Programs spanned all three sides of the MAP frontier. The division left MABE with the Financial Officer, which left M-P with difficulties with financial management. MABE Program Coordinators were entirely in NGOs, whereas M-P Program Coordinators were often in GAs and universities, which led to additional challenges with implementation. After a year of M-P struggling to find its feet, USAID personnel and ICAA support staff provided technical assistance. M-P eventually also hired a new Program Manager, as the original G-MAP Program Manager ended up in that role for MABE. In sum, the adjustments were neither easy nor fast, but the exigencies involved forced the partners to step up and take greater ownership of their consortia. Here it is important to make clear that the division, while traumatic, was not a locus of conflict on the part of G-MAP partners, neither among partners, nor with USAID as the funder, nor with the governments involved.

Part of the political concerns revolving around sovereignty issues stem from crossing an additional boundary, among organizational types. Vignette 6.5 provides an example of the compounded political travails when crossing multiple boundaries simultaneously. G-MAP stood for "Governance in the MAP frontier." The terms "government" and "governance" are very similar but have crucial differences with political implications. The first refers to the mandate of the state as sovereign entity over a national territory, with the legitimacy to perform various functions implied (taxation, defense, regulation of commerce, etc.). The second refers to a broader regime constituted by state as well as non-state actors, such as NGOs, to pursue shared goals for public benefit. Governance can thus be viewed as a form of spanning organizational boundaries. There is no shortage of public-private partnerships, participatory budgeting processes, and other mechanisms that could be considered examples of governance.

External political complications arise when the notion of governance, as one form of boundary crossing, becomes entangled with another form of boundary crossing, such as governance across national frontiers. Frontiers are by definition politically sensitive spaces to governments who are keen to secure their national borders against foreign incursion. As noted in Chapter 2, the history of the Amazon, including the MAP region, can be read in terms of the politics of drawing and contesting national territorial boundaries. Acre was once part of Bolivia, until Brazilian rubber barons settled sufficient numbers of rubber tappers to claim the territory. Tahuamanu province in Madre de Dios, Peru, was also a contested space with Bolivia during the rubber boom.

In that historical context, the notion of NGOs working in frontier areas on "environmental governance" became suspect. The similarity of the term "governance" with "government" left open questions as to whether NGO activities really occurred in partnerships with government or otherwise had governmental approval. Lacking governmental relationships or approval thus became a focus of concern as to what exactly NGOs were up to in frontier areas. It thus became a requirement for NGOs to be registered with government agencies in charge of international activities, activities in frontier areas, and other criteria deemed necessary for approval to work. Falling outside such institutional mechanisms was thus tantamount to informality and eventual state intervention.

Beyond the historical geopolitics of national frontiers, crossing boundaries for conservation and development in the Amazon has become sensitive due to the question of the "internationalization" of the Amazon (cf. Buarque 2002). Concerns about deforestation in the Amazon and the loss of biodiversity have led to well-known appeals for conservation measures, and at times pressure on the governments sharing the basin to act to reduce forest destruction. For their part, the governments of the countries sharing the basin have consistently reiterated that they have sovereign rights over their territories, which implies the right to promote the development of their economies for the welfare of their people. The contours of the debate over conservation versus development in the Amazon will be familiar to those cognizant of arguments concerning sustainable development in general. Particular to the Amazon is the additional discourse in Brazil about "internationalization," namely that interests in the United States wish to make the Amazon a world patrimony so that the basin would belong to all of humanity. This would provide the basis for closing the Amazon off to development, thereby conserving its forests and biodiversity.

For present purposes, I leave debates over the sources of such interests and the discourse itself to others (for reviews, see Lobo 2012; Sartre and Tavarella 2009), and focus on the ramifications for crossing boundaries for collaboration on conservation and development in the Amazon. After the end of ABCI and G-MAP, the MABE Consortium proceeded to work in Acre. In the context of internationalization, a US government entity (USAID) funding a consortium of organizations (MABE) through an agreement to a US university (UF) working with Brazilian NGOs (IPAM and SOS Amazonia) was for some a problematic arrangement. Not only were there foreign entities interested in conservation in the Amazon, but all of the organizations operating in country were non-governmental, and they were working in frontier areas. Concerns about foreign interests were thus combined with suspicions of NGOs, which were particularly problematic given the focus of MABE on a border region.

Differing noises came from different corners. The Secretary of Policy, Strategy, and International Relations of Brazil's Ministry of Defense reportedly issued a recommendation to the Ministry of Foreign Relations that all NGO activities in the Brazilian Amazon be halted (Rocha 2007). Concerns included observations that NGOs would be working in frontier areas with indigenous groups and environmentalists opposed to infrastructure projects. Nationalists in the MAP frontier spoke against ABCI as part of a broader critique of internationalization conceived as a form of neoliberal economic policy (Paula 2005). In such accounts, crossing boundaries for conservation and development was a smokescreen for subordinating local labor to global capital. The Brazilian Intelligence Agency (ABIN) sent a list of twenty-five NGOs to the Ministry of Justice for review concerning biopiracy and other criminal activity (Terra Noticias 2008). Both of the Brazilian NGOs in MABE were on the list. Both were later confirmed as "not suspect." MABE lived on, albeit in tricky circumstances.

This vignette illustrates the compounded political complexities of crossing multiple boundaries at the same time. Unease about governance prompted suspicion of NGOs; apprehension with protection of national borders fueled doubts about activities in frontier areas. Hence, NGOs working in frontier areas were cause for particular alarm. This vignette, however, also shows that even if trepidation is compounded, crossing multiple boundaries is not automatically rendered impossible.

Diplomatic relations are not always as genteel as those in the foregoing case. Vignette 6.6 reveals greater tensions among governments, and consequently greater fears among both partners and allies. The M-P Consortium worked in both Peru, where the US government had good relations, and in Bolivia, where the US government had more problematic official relations. In 2008, the government of Bolivia expelled the US Drug Enforcement Agency and the US Ambassador from the country on accusations of undermining the government. Underlying the tensions were disputes over Bolivian coca production, defended by the government of Bolivia as traditional practice among highlanders who chewed the leaves as a stimulant when working, and criticized by the US government as contributing to cocaine production for export to the United States. Deeper down were differences over economic policy, as the Morales administration came to power on a socialist-indigenist platform that flew in the face of long-standing policies friendly to foreign firms and markets. This context cast suspicion on USAID, and made it difficult for USAID to operate in Bolivia under ICAA.

As is common practice on projects reliant on external funding, especially foreign government assistance, it is important to acknowledge the funder on publications, banners at events, and other literature. Such "branding and marking" makes clear who supports collaborative projects across national boundaries. The tense diplomatic circumstances surrounding M-P made branding in Bolivia a politically tricky question. Low-key consultations by Bolivian M-P partners with allied communities, organizations, and local governments led to two conclusions: 1) there was great interest in M-P supporting municipal planning for natural resource use, but 2) there was equally great concern about publicizing that the funding for said work was coming from the US government. This left M-P between a rock and a hard place. Do the work and put consortium partners and local allies at political risk, or halt the work and risk alienating constituencies of the consortium partners.

This led to extensive discussions among M-P partners and USAID personnel. M-P proposed an exemption on branding requirements to avoid the political risks. That way, the work could be done without the political fallout. USAID personnel sagely pointed out that the work would likely call attention and potentially reveal the funder anyway. This would be worse, since the funder wouldn't be acknowledged at the outset, and catalyze something of a scandal. The Bolivian M-P partners disagreed, and positions hardened. In the end, M-P had to back off from the proposed planning activities, and apologetic explanations were offered to local constituencies.

In this vignette, broadly similar issues arose in Bolivia under M-P as in Brazil under MABE. Crossing national boundaries for governance activities led by NGOs was complicated due to the external politics at play between national governments. In both cases, the external politics had little to do with the actual content of the activities. M-P was not working in the Bolivian highlands or on issues related to coca. Nonetheless, political limitations came into play as a result of the international and inter-organizational dimensions of the boundary crossing involved. The combination of tensions between the US government and the government of Bolivia, along with proposed activities involving municipal development planning, which falls within the purview of governmental entities, proved too problematic to pursue.

External politics may pose risks to not only particular kinds of activities, but also entire collaborative ventures when crossing boundaries. Vignette 6.7 continues the story in Bolivia, four years later in 2012. Previous chapters mentioned that the Bolivian Amazon Conservation Consortium (BACC) was halted in its work just as it was ramping up. The BACC also worked with

USAID funding through the American Council on Education (ACE) via its Higher Education for Development (HED) program. In contrast to M-P, BACC focused on university capacity building through advanced training of faculty, collaborative research, and similar academic pursuits, rather than territorial development planning. The government of Bolivia approved the BACC in November of 2012, so we got to work. Initial visits by HED personnel to La Paz prompted reporting that reflected what seemed to be easing tensions between the US government and the government of Bolivia (Sol de Pando 2013).

The tensions remained, however, and after shifts within the Morales administration, on Worker's Day 2013, the government of Bolivia announced the expulsion of USAID from the country. President Morales cited conspiracies on the part of the US government and called for respect by the United States of the governments of countries in Latin America (La Razón 2013). The US government lamented the decision by the government of Bolivia, and USAID moved to close its operations. With the expulsion came a stoppage to all USAID-funded activities, including the BACC. This led to consultations as to the options, and there were none for partners in Bolivia. Then followed a flurry of budget revisions as we sorted out severance pay for Bolivian collaborators, some of whom had forgone other employment opportunities to be part of BACC, and now had no job.

As in the previous vignettes on external politics, for reasons that had little to do with the collaborative enterprise itself, crossing national boundaries proved a complicated proposition for the BACC. Governments pursue interests tied to national sovereignty, which may or may not include international cooperation. In particular, states may frown on non-governmental initiatives for international cooperation, notably if framed in terms of "governance" or if funded by another government. While it is in principle a useful policy to be proactive in engaging powerful entities to provide information as to one's motives, powerful organizations are often difficult to approach. Further, diplomatic processes among governments preclude outside overtures. Even if there are avenues, in-country partners and local allies may actively discourage overtures out of fear.

CONSEQUENCES OF INEQUALITIES AMONG COLLABORATORS ACROSS BOUNDARIES

If external politics can impose seemingly arbitrary limitations on what is possible when collaborating across boundaries, inequalities among partners can yield rather more predictable but equally intractable challenges. The opening vignettes (6.1–6.3) highlighted the role of inequalities in generating conflicts at the outset of collaborative partnerships. Here I pick up that thread by emphasizing that management of inequalities is a perennial task in collaboration, particularly when crossing boundaries. Precisely because inequalities across boundaries are often larger and more sensitive to address up front, they are often insufficiently interrogated, which can cause all manner of mischief later. Whereas acknowledging limitations, recognizing subjectivities, and managing expectations are important challenges at the outset of collaboration across boundaries, others await.

I see two reasons, beyond subjectivities and other factors already noted, that help explain why inequalities produce difficulties in collaboration, especially across boundaries, as partnerships carry on. The first is that subjectivities are colored by interests, and interests never align perfectly. Any initial understanding that partners come to as regards practices to manage inequalities will be interpreted differently based not only on unconscious habits of thought but also on their interests in the collaboration. Partners that clearly define common interests nonetheless have other interests that can come into play. In turn, those interests can play out based on the relative capacities, experience, and other sources of power possessed by those partners. While this is hardly unique to collaboration across boundaries, it is intensified by boundaries insofar as they tend to increase inequalities. Even if shared interests are the focus, private interests play a role, and they can differ significantly across boundaries. There is always a temptation to bring private interests in and argue that they be made the focus. The moment that occurs, inequality plays a large role in how the proposal is addressed. The management of inequality comes under strain, as does any hard-won accord among partners. Boundaries complicate the communication of intent in shifting the focus to emphasize private interest. Such moments often come when an initial activity has been completed (successfully or not) and the partnership is looking to define next steps. Whenever such moments occur, they constitute tests of the partners, and their inequalities and the intervening boundaries hamper their ability to negotiate.

The second reason, uncertainty and surprise, anticipates the next chapter, but deserves mention here because it can severely test the fortitude of unequal collaborators. Over time, partnerships will be tested by a parade of unexpected problems. Bureaucratic snags, personnel changes, and external political interventions, among a litany of other possibilities, will cast in harsh light the ability of partners to show grace under pressure. Inequalities across boundaries make such moments particularly educational, for better or worse, as the partner under pressure has to perform while the others in effect constitute the audience. This is when unadmitted limitations, foibles, and debilities may be put on show by the partner under duress. This is also when the other partners, who generally have to wait until the snafu is addressed, also in effect perform, revealing whether they still want to share the same stage. Inequalities across boundaries make such moments particularly hectic and difficult as partners under duress have to explain themselves and partners waiting have to decide how to respond. Large inequalities make the duress and the waiting harder to endure, and over time, nerves become frayed and patience wears thin.

Management of inequalities and the manifold moments in which they conspire with boundaries to complicate collaboration is thus a perennial challenge. I therefore offer a third suite of illustrative vignettes that highlight the complexities of inequalities in the midstream of collaboration across boundaries.

Vignette 6.8 highlights challenges stemming from attitudes of stronger partners toward weaker partners. In the context of collaborative work with partners in the Amazon, stronger partners are in my experience US-based organizations. The complications in question among US organizations and foreign partners arise from unequal institutional capacities for administering projects with external funding. The bridging personnel at the US organization doing the boundary crossing take on the dual tasks of explaining administrative protocols to the partners, and also explaining partner needs and issues to the home institution. The challenge that arises stems from the schizophrenic roles of the US organization: on the one hand, the participating institutions are partners, and bridging among collaborators is non-hierarchical; but on the other, the US institution is the lead and thus administratively has a hierarchical relationship with the partners.

The problems arise at key moments involving adjustments in budgets, personnel, activities, or other plans for the collaboration. Financial complications can stem from changes in line item costs after approval of budgets, causing a tightening of available funds for activities. Salaries, fringe benefit rates, equipment costs, plane tickets, exchange rates, and indirect rates are all subject to such adjustments. In some cases, such as changes in fringe or indirect rates, consultations with UF administrative staff led to determinations that the rates in the approved budget could be "grandfathered" and thus left unchanged. Other instances, such as plane tickets and exchange rates, could not be grandfathered, tightening budgets for partners. In these cases, consultations to reduce or otherwise adjust the UF budget to ease the pain led to varying responses, sometimes including "There's not much we can do. It's their problem." Stronger partners may thus regard problems of weaker partners as just that: their problems.

In such moments, suggestions to "take one for the team" arise, and the question is whether the large US partner or the smaller foreign partner should be the one taking the budgetary hit. Issues of justice, as well as political capital, and of course accounting and contractual protocols, all come into play. It is certainly a luxury to have other accounts on which one can draw to defray unexpected costs. This becomes more difficult when those costs are set to fall on weaker partners, since additional administrative steps are required to get them the funds from other accounts.

Here multilateral bridging, between organizations as well as between units within one's home institution, becomes priceless in managing institutional inequalities. I recall organizing the visit of the representatives of several Amazonian partner universities at UF. One part of the visit was for the representatives to meet with UF grant administrators. The UF administrators included those who had asked me whether I really wanted to pursue these complicated international projects. For their part, the UF administrators spent a chunk of the meeting showing the visitors the big numbers in UF's core budget and externally funded grant portfolio. The visitors were duly impressed. However, a by-product of the visit, unexpected for my part, was for the UF administrators to associate the names and faces of individual human beings with the universities they represented. The visitors took advantage of the opportunity to convey their interests in collaborating with UF. The UF administrators were duly impressed. Thereafter, it was not uncommon for senior administrators as well as staff to inquire about our complicated projects, especially in terms of the partner institutions, their priorities, and their representatives. That meeting unexpectedly led to a shift in perspective by UF to view the Amazonian partners as important players for their role in a significant part of the world. UF administration was happy to play a role. I have not since been asked, "Are you sure you want to do this?"

The challenges of inequalities come from both ends. Those from weaker partners include obvious issues such as limited administrative capacity, but there are also more subtle challenges. Vignette 6.9 focuses on regional universities in Latin America, or at least those which I know. Regional Amazonian universities are growing and changing rapidly, and it is exciting to watch that process. They are also political spaces, what with personnel often coming from prominent regional families, a strong identification with regional history and policy debates, and internal elections. This leads to varying degrees of internal political factionalism. Vignette 6.1 began to sketch some of the consequences in terms of internal political processes as they relate to the selection of participants in collaborative projects. Vignette 6.9 extends the example by taking up the case of another university partner with significant internal politics.

The NSF Roads project required identification of technical personnel in several disciplines to manage field data collection. In the present case, the personnel were selected by the university authority based on the strength of their vitae and their fit to the science at hand. While this process was not democratic, it did result in selection of people with solid technical skills in the areas required, something confirmed by my UF specialist colleagues. None of this, however, guaranteed that the people indicated came from the same political factions.

As it happened, they did not, and this was consequential, not only for the project but also for relations between the universities and personnel in the partner university itself. One participant in particular was a very good scientist in technical terms but also very active politically. Their performance on the project was solid in technical terms. At the same time, their political activity led to a series of confrontations with the partner university administration. I was requested to provide letters of support, putatively confirming the technical quality of the science being done, but later found out that said letters were instead being used in political confrontations. At one point there were fisticuffs over the use of funds from UF for things other than science. This included a complaint (*denuncia*) to the local courts which named UF. That necessitated some tense discussions over the project as separate from politics. It also meant that my letters of support were no longer forthcoming. Because of the internal conflicts in the partner university, personnel were increasingly selected by university authorities on the basis of not belonging to the political party repeatedly confronting the university authorities. This of course led to additional allegations, until governmental intervention gave a new executive committee the power to fire faculty for inciting protests against the university.

I recount these events not to offer a salacious tale by way of complaining about how somebody was treated or whether certain allegations were founded. Rather, I note the case of Vignette 6.9 in the context of inequalities among partners because I interpret the internal political fisticuffs in the partner university as a reflection of its history. The partner university was very young and relatively small, but growing quickly. The politics there could be interpreted in terms of a few key personalities and their politics. For present purposes, I highlight the institutional growing pains in the partner university. A young institution, with initial leadership and a newly constituted faculty body, was working to establish its identity through start-up programs and projects. This is a moment in an institution's history when structures are still being implemented and expectations delineated. It is also a moment when innovators with various goals find space to push agendas.

This stands in stark contrast with UF, a much larger, older, more established organization. Of course, if one reviews the history of UF, one can also find tales of state-level and internal politics concerning the establishment and organization of the institution. The reorganization of existing institutions of higher education in Florida, the siting of UF in Gainesville, the questions of admitting women and people of color, among others, were all historical points of contention. While the specific political issues may differ, in the broad brushstrokes, the early years of both UF and the partner institution in question begin to bear resemblances as both struggled for years to set up their programs, produce graduates, and manage growth.

Collaboration across boundaries in developing regions may thus involve organizations at very different points in their institutional histories. The boundary spanning, however, involves individuals unlikely to be aware of the parallels in their respective organizational histories. This nonetheless amounts to collaboration among unequal organizations with the consequence that the partners have rather different preoccupations. Even if this is functional in terms of complementarities, the inequalities constitute challenges, though also learning opportunities.

Whereas I highlighted the consequences of inequality among partners at the organizational level in Vignette 6.9, I want to turn to the eventualities at the individual level in Vignette 6.10. Here I pick up the thread from Vignette 6.1, on a collaborative project among universities. To refresh the memory, the partner university had engaged in a democratic process to elect the personnel to participate, though there was evidently below-the-table deal-making that influenced the outcome. The personnel elected in turn became close collaborators, though some were closer than others. Indeed, the two supervising coordinators would have very different trajectories. Whereas one was more interested in faculty group politics, and eventually left the university to become involved in regional politics, the other stayed at the university and became a point of contact for other projects. I found this to be functional since it was obvious who to contact with questions.

Subsequent projects showed this to be problematic. Having a single point of contact in a politicized organization unavoidably leads to a skewed accounting of how things work, including collaboration. When we brought in a post-doc for high-level technical support on a later project, my point of contact and the post-doc were soon in conflict. There was no further need for sleuthing, as I now had a second and very different window into the inner workings of the partner university, including the maneuvers of my point of contact. What quickly emerged was a pattern of guarding opportunities offered by UF as a collaborator, for fear they might be appropriated by others. This included selecting project participants who were not the most qualified technically, which outraged the post-doc. The issue was thrown into starker relief a few years later at the outset of a third project. A newly elected set of authorities opened another democratic process to identify key personnel on the new project. As my point of contact had cultivated ties with the side that ended up losing the election, the new authorities saw no need to favor the beneficiaries of the old guard. A question arose over whether roles on new collaborative projects should be given to people with prior experience on collaborations, or the people who received the most votes. As it happened, the election indicated that people with those two characteristics differed. The results elected new players, who seemed very good. But a fisticuff erupted, and one elected representative stepped aside and my old point of contact stepped in. Different versions of the story emerged, some rather unhappy, though all included mention that the post-election change happened because the recipient of most votes for the technical position in question was later deemed unqualified.

Vignette 6.10 thus extends Vignette 6.1 by continuing the story of how inequalities among partners may lead to political behavior in the partner with less capacity and personnel with fewer opportunities. Professionals with low pay understandably seek opportunities on collaborative projects that come with consultancies. Some individuals see collaborations in strategic terms, ostensibly for their institutions, but also for themselves, in the form of future collaborative opportunities. That of course assumes that they will get those opportunities, and not have authorities above them, or other complications like elections, get in the way. Insofar as a series of collaborations with the same individuals builds relationships and has the benefit of continuity, there is collaborative advantage. But insofar as collaboration occurs among unequal institutions such that investments by the stronger in the weaker partner go to the same few people, collaboration can become problematic as benefits are concentrated, increasing inequalities within the partner organization. What is more, jealousies and conflicts are likely to follow. Of course, the opposite problem is also possible: spreading collaborative opportunities very broadly may limit actual capacity added and create unrealistic expectations that everybody can collaborate, regardless of technical capacity, personal agenda, and so forth. When spanning boundaries, balancing these points is particularly important, but ensuring that possession of rights to collaboration is not assumed by a select few is very difficult. Explaining why somebody else needs to come in is no simple trick, even if the strategic imperatives are invoked, due to the individual interests tied to individual benefits.

A final vignette on the consequences of inequalities among partners is in order. Whereas Vignette 6.10 highlighted problematic behavior involving individuals at weaker partner organizations, Vignette 6.11 focuses on problematic behavior involving individuals at stronger partner organizations. In a word, if my collaborators are not saints, neither am I. When I visit the southwestern Amazon, many collaborators see the opportunity for lengthy discussions on complicated matters. Face-to-face talks avoid the written word when it comes to sensitive topics, and they avoid the problems of sketchy internet, as during the Amazon's rainy season. Visits obviously permit group meetings and working group sessions, when people say things publicly so everyone gets the information at the same time. This is useful for managing conflicts, and also avoids production of new misunderstandings.

Visits also constitute spaces for individualized meetings, especially right after I arrive and right before I leave. Those face-to-face meetings in turn permit requests, not to say petitions for special favors, and other forms of consultations for help with ambiguous situations. Here crossing boundaries is enormously hazardous. If UF is the lead organization and Stephen represents

UF, that makes Stephen the boss, at least in some eyes. Stephen represents a strong organization that can make things happen, including the resolution of problems on collaborative projects, even outside UF or the United States. I can no longer count on all my fingers and toes the informal entreaties brought to me during visits. I emphasize that in most cases, such requests have established channels for resolution. Equally, I admit that I know that the framing of said entreaties is colored by the strong subjectivities and interests of the petitioner. But my desire to be a good collaborator, a responsive colleague, is too great. I have therefore repeatedly assented to various petitions, putting on a brave face that I as representative of UF will make things right.

This has caused not a few problems, catalyzed a cascading administrative disaster or two, and left one or more partners with complications that they did not deserve. Rather than take the time to consult the appropriate manager or director, I took decisions or made promises on which I could not deliver. This has the effect of doing precisely the opposite of what I had intended: rather than help I have hindered, and rather than alleviate worries or conflicts I have created them or made them significantly worse. What is more, such ersatz decisions laid bare my status as the outside collaborator, and not the actual boss in the chain of command who takes care of the local logistics. The real sin in question resides in my repeat offenses: I have made such decisions even after having made the same mistake before. Whether it involves equipment purchases, hiring of personnel, or opening my mouth about decisions concerning appropriate collaborators, I have unwittingly intervened as a result of inequalities among partners and overestimated my ability to put things right. My behavior is thus the defining quintessence of stupidity.

If I may build something of a self-defense, it is worth noting that crossing boundaries makes it harder to resist informal requests since the inequalities are large, the subjectivities strong, and the difficulties of collaboration great. But overextending oneself and failing to do due diligence on behalf of one's home institution—and other collaborators—are mistakes which I have had to learn the hard way. By the words of frustrated collaborators whose more level heads and written charges I have been reminded, they'll take care of the entreaties. Respect the division of labor and the complementarities that beget collaborative advantages; don't cripple them by unilateral fiat.

INTERNAL POLITICS AMONG COLLABORATORS ACROSS BOUNDARIES

If inequalities complicate collaboration, politics among collaborators is more serious business. Collaboration across boundaries means navigation of differences for the sake of shared interests. But as was noted earlier, this does not mean that private interests evaporate. Nor does it mean that the mode of operation for one partner is compatible with that of another. Collaboration across boundaries is therefore not only a political act vis-à-vis the powers-that-be, it also may cross the interests and practices of the collaborators themselves. Consequently, the negotiations involved in collaboration across boundaries begin to resemble those among politicians with different interests. Communication skills are vital, with demands on messaging approaching great art or even divine grace in order to get things done.

My experience is that collaboration across boundaries amounts to a cascade of accumulating tensions that run risks of ruining relationships. If this sounds like hyperbole, it is worth recalling that collaboration across boundaries encompasses a large suite of tasks with significant complications that result in a panoply of problems which strategic practices can help manage, but not solve. If collaboration can be celebrated for its advantages, and collaboration across boundaries even more so due to the significant complementarities, it bears repeating that the labors involved are manifold and can be tiresome. Bureaucratic snags, unexpected problems, and difficult adjustments accumulate, and along with them, frustrations among partners. The compounding of these little difficulties with the perennial tasks and challenges of collaborating too often means there isn't time to take care of it all. As a result, the snags, problems, and adjustments build up during the course of collaboration, resulting in growing tensions, jealousies, and resentment. Eventually, somebody reaches a break point and there is an explosion, usually in the form of a complaint or accusation. Collaborators become combatants and relationships are threatened. In sum, collaboration becomes politicized.

This section therefore presents illustrative vignettes on the politics among collaborators when crossing boundaries. As in previous sections, because some unpleasant things happen and people do things they might not be proud of, I am deliberately vague about the players involved, though I self-identify when describing my own villainy. The vignettes all occurred in the midstream of projects, and usually generated or worsened tensions. Few have happy endings.

Vignette 6.12 came early in G-MAP and concerned a fight over the question of student scholarships. If many of the vignettes in this chapter have highlighted international boundaries, this case focuses on organizational boundaries. G-MAP had incurred a suite of minor altercations over various bureaucratic questions (including Vignette 6.2), when it moved to engage in final budget negotiations with USAID. The proposed G-MAP budget was rather large, so not surprisingly USAID requested a budget reduction. As negotiations proceeded, it became evident that USAID wanted to eliminate student fellowships.

This led to tense discussions among the G-MAP partners. The university partners wanted to defend the student fellowships. They cited the master logic of G-MAP as focusing on capacity building for long-term conservation in the southwestern Amazon. In that context, student fellowships were crucial as a means of freeing regional university students from working while studying. By contrast, the NGO partners cited USAID arguments that the fellowships were a rather indirect investment in conservation, especially when compared to work with rural communities and territorial management planning with governments. UF found itself caught in the middle of the conflict. As another university, UF also wanted fellowships, but as the lead organization, it had to follow USAID's mandate and make budget cuts.

Further meetings transpired among the partners. It then came out that the fellowships had been cut entirely, evidently by the NGO partners. The universities reacted negatively to this outcome and called their own meeting. Each university produced an equivalent letter to UF observing that a large portion of their budgets on the project were in student fellowships, which had been removed, and that their remaining budgets were small relative to their cost share contributions. They then demanded complete control of the capacity building program, greater respect from the other G-MAP partners, and an increased administrative overhead rate. The NGO partners responded by giving the university partners a week to tender a second and presumably more reasonable offer.

Giving the rapidly rising tensions, UF had to respond. That is, I had to respond. Taking one side over the other would detonate the consortium. I pointed out that some capacity building activities had been proposed by NGOs and GA partners, and weren't purely the purview of the universities. I noted that the overhead rate proposed by the universities was high, and in any event, raising overhead would reduce direct costs for activities. I concluded by asking for a meeting of all partners, not just NGOs or the universities, to resolve the differences.

I was traveling outside the United States and the southwestern Amazon at the time, so I missed what came next. By all reports, it was not pleasant. The universities felt jilted and the NGOs felt unfairly attacked. One of the university partners sent a letter addressed to both UF and USAID. The letter reviewed the partner university's contributions to conservation and development in the southwestern Amazon, issued the complaint that said contributions were not being recognized, lamented its budget reduction, and thereupon withdrew from G-MAP.

Then came the moment of truth for the other universities. In the event, they opted to stay on. Budget negotiations continued, and while the remaining university partners ended up with rather smaller budgets than the NGOs, they still had budgets and they were still in the consortium. The question of indirect costs was resolved, and the NGO partners also incurred reductions in their budgets. UF has since maintained positive relations with the university partner that left G-MAP, and individually we conversed about the "hot topic" of cutting the student fellowships. I fussed that the withdrawal of the university partner was a lost opportunity to collaborate and thereby learn. They were resolute in their decision, and professional on subsequent collaborations involving other activities. Key players on both sides of the G-MAP conflict remained wary of each other. One of the key arbiters in the conflict summed up the disposition of one of the NGOs by quoting Thomas Fuller: "scalded cats fear even cold water."

The organizational boundary in Vignette 6.12 was thus highly consequential, for it delineated competing proposals for handling a budget cut. The same boundary also set the field of war for the ensuing conflict, what with separate meetings, letters of recrimination, and scalded cats. Under the difficult circumstances of a significant budget cut, the organizational boundary became a fault line along which a crack appeared in the consortium. In the event, the divide could not be spanned, and while an entire bloc of partners did not march out, the consortium was nonetheless weakened by the departure of a university partner.

Also stemming from Vignette 6.2 is Vignette 6.13, concerning the political eventualities of having other partners manage funds for universities. Here again the key boundary concerned organizational types, though national boundaries also came into play. Vignette 6.2 outlined how Plan A (university foundations managing funds) and Plan B (Brazilian NGOs managing funds for a public university) and Plan C (a foreign NGO managing funds for a Brazilian public university) could not be pursued for various reasons. After G-MAP was divided into MABE and M-P, we revisited the issues under M-P. We revisited Plans A, B, and C by considering whether university foundations and Peruvian and Bolivian NGOs or GAs could manage funds for the university partners. Plan A was again shelved since the university partners either lacked foundations or they were very new and lacked sufficient experience. Plan B, Brazilian NGOs, was out for M-P due to the Solomonic division. We thus considered options for Plan C. Fortunately, there were options for both the Peruvian and Bolivian university partners, since M-P had other in-country partners on both sides.

In the event, one case worked well, whereas the other proved more complicated. The case that worked out was to have an NGO manage funds for the university. No major conflicts resulted, only the occasional bureaucratic snag, and none of those accumulated or caused tensions. The other case involved a GA managing funds for a university. Here things were more complicated, stemming largely from the fact that both were public entities with less administrative agility than an NGO. Not surprisingly, there were delays in various procedures, notably contracting personnel, clarifying requirements for invoices, and receiving payments. This led to growing tensions between the GA and university personnel.

Before long, the relationship became increasingly politicized, as the venting of frustrations morphed into allegations about unequal treatment to GA and university personnel. This crystallized during a local fair where organizations and projects had booths to present their work. As it happened, the GA had a booth and the university had a booth. The university personnel called me over to their booth. I was summarily presented a list of specific complaints against the GA. Evidently the GA personnel saw this and realized what was up, and shortly after I was called to the GA booth to receive a set of explanations. Remarkably, the explanations corresponded to most of the complaints. But strong subjectivities were at play, especially with regard to the value of the GA's administrative procedures. Visiting each booth felt like tourism to two different planets, as constructions of the same administrative procedures differed fundamentally. My resulting befuddlement prompted a visit to the consortium manager, who fortunately commanded the issues and understood the operational differences between GAs and universities. This did not magically resolve the tensions, but it did lead to meetings that mitigated the strong subjectivities at play and the resulting complaints.

Vignette 6.14 provides a case highlighting the politics of collaboration featuring boundaries as they complicate leadership and the chain of command. The M-P Consortium spanned two countries, and like other international consortia, it had a Program Manager and a Deputy Manager. The agreed-upon arrangement was thus to situate the Program Manager in one country and the Deputy in another. This ensured managerial oversight of activities in both countries, thus distributing leadership across a key boundary.

This, however, led to tensions over management in the country where the Deputy Manager was based. Whereas the roles of the Manager and Deputy are distinct, the national boundary dividing M-P fostered the tendency for the two to perform similar functions in each country. This was not acceptable to the Manager, who periodically received inquiries from USAID about activities in both countries, but had information largely limited to one. The Deputy felt the need to explain what was going on in that country directly to USAID. While it might seem obvious that the Deputy could communicate those issues to the Manager for transmission to UF and USAID, communication proved a challenge due to the subjectivities at play. Problems seen as serious issues in the context of one country were not necessarily fully understood by personnel in another. The national boundary mattered, and the challenges it posed to distributed leadership led to tensions over the question of autonomy and hierarchy in management. Those conflicts in turn required their own management for the rest of M-P.

Vignette 6.15 also provides a case of conflicts over leadership, this time featuring boundaries among organizational types. MABE avoided enduring conflicts among partners, arguably due to the lack of numerous boundaries; MABE worked in one country and involved UF and a suite of NGOs. Individuals spanned disciplinary specialties to do their work, and because they had collaborated before, there was healthy respect for interdisciplinarity.

This changed with the genesis of what I will call MABE II. A strategic priority among MAPiense partners was for UF to serve the role of lead organization for projects only on a temporary basis. Eventually, in-country partners would assume that mantle, and UF would continue in the role of a supporting partner. With the process to develop MABE II, we sought to take that step. A large Brazilian university from outside the Amazon would lead MABE II, with participation by UF and the southwestern Amazonian partners. The Brazilian university indeed had substantial experience in managing large, complicated projects with funds from various types of funders, and operated through an established foundation that was interested.

The opportunity came from USAID-Brazil, which issued a call for proposals for consortia to work on conservation in the Brazilian Amazon. Brazilian members of MABE led the organizing by inviting UF and two other new partners, including the large Brazilian university. Everyone agreed to the arrangement; the Brazilian university's foundation sent representatives for planning meetings.

Conflict arose, however, over the distribution of funds among the partners. Under MABE, UF received the agreement from USAID and administered funds via sub-agreements to the Brazilian partners. Those sub-agreements held most of the funds, since the partners led the planning and would similarly lead implementation; UF provided administrative support and participated in a technical role on specific activities. MABE, like the other boundary-crossing consortia led by UF, was thus in a material sense "a consortium for the partners." The Brazilian university partner did not have that experience and was not accustomed to managing projects in that fashion, instead preferring a more centralized budgetary model. The organizational boundary thus proved significant. There were, however, exacerbating factors, notably personal complications facing a key player at the Brazilian university. The foundation remained well disposed, but the disagreement over the organization of the budget continued.

Time was running out, forcing local partners in the southwestern Amazon to pursue alternatives. For Plan B, one Amazonian partner consulted its main office, and received replies posing additional complications. We therefore returned to UF as Plan C. This required hasty consultations with UF administration, whom I had previously informed that "this will be much simpler since UF won't be lead." I now had to beseech UF as lead again on a complicated project with a large budget, cost sharing, and large sub-agreements. This time, UF administrators didn't ask if I was sure about this, and agreed. We then had to consult with another new partner who had no prior experience in working with me or dealing with UF. The partners from MABE in effect provided the necessary references. Here the subjectivities across boundaries proved a boon and not a bane, as the local partners could explain from the point of view of their priorities and experiences how things had worked before, and—crucially—that they had worked well enough to be considered acceptable. On that basis, the new partner also agreed.

This vignette provides another case of the politics involved in arranging leadership in collaborative endeavors that cross boundaries. The disagreement between the proposed lead university and the partners over budgeting issues was a problem of crossing organizational boundaries, particularly stemming from the experience of arranging a consortium "for the partners." Here again the hierarchy of leadership and the degree of centralization of control became deciding issues. Having Plans B, C, and so on proved crucial, as has been the case in many other instances where politics stemming from crossing boundaries proved enduring.

This vignette deserves an epilogue, on two counts. First, once the personal crisis of the key player at the Brazilian university had passed, personnel from that partner later came to visit the southwestern Amazon partners. This exchange provided the opportunity for the Amazonian partners to better explain their needs and priorities as related to the allocation of the budget. At that point, it became clear to the representatives of the large Brazilian university that the large sub-agreements had a solid technical and strategic foundation, a message that calmed tensions over the budget question. The other piece of the epilogue is that in the event, MABE II was not funded. This created an unexpected positive circumstance of eased tensions along with the ongoing search for funds. Despite the problems on the first try, the door was reopened to try again.

Whereas the foregoing vignette involved a conflict that was eventually resolved, my experience is that most conflicts either persist and demand careful management, which consumes time, or end in a terminal spiral of hostilities. Vignette 6.16 therefore returns to an issue noted in Vignette 6.12, specifically a case where the politics of collaborating across boundaries led to conflicts and the end of collaborative relationships. Out of the experience recounted in Vignette 6.13, we concluded that having an NGO managing funds for a university was preferable to a GA. In a subsequent project, we got our chance to test that hypothesis for the university that had funds managed by a GA. The new consortium added an NGO with a documented track record of handling funds for large projects with governmental funding. The NGO and university both agreed to the arrangement, and got to work.

Under this arrangement, the university deemed the NGO's performance to be satisfactory. While there were minor bureaucratic snags, they did not accumulate, tensions did not rise, complaints did not ensue, and the relationship was not politicized. Management heaved a sigh of relief. However, the new relationship led to other problems which we had not anticipated. It eventually came to light that the NGO was not keeping track of documents tied to billing and expenses. While the service was faster, documentation was inadequate. As a result, it would have difficulties with a financial audit. We had thus traded one problem for another.

In-country management made several visits to the NGO to help it get the documentation ready. It became evident that the NGO was not dedicating sufficient staff time to the question, and could not produce the documents. Fortunately, management had the originals to sustain an audit, but that didn't matter: the audit would be of the NGO. We made the decision to terminate the NGO's sub-agreement. Consultations with the university partner, other consortium partners, and the UF administration ensued, resulting in a consensus about the decision. But this is not where the conflict resided.

UF issued a letter of termination of the sub-agreement. The NGO responded by asking in-country management how to resolve the mess. Management replied that the consultations had been made, the decision had been taken, and that UF's decision was final. This was the flashpoint. The NGO responded by hiring a lawyer and refusing to communicate with management. A packet of documents was sent to UF administration, which caused confusion since management was supposed to be handling the details of closeout. It became evident that the NGO wanted to manage the situation in order to delay termination. In-country management and UF sought to reassert control over the process, noting the deadlines indicated in the UF termination letter. This prompted further hostilities from the NGO, which locked personnel out of offices and seized equipment. We reasserted the deadlines and requested that the NGO reopen the offices and produce the equipment. The back and forth continued. In the event, when the deadline approached, the NGO relented from hostilities and cooperated. The accounts were balanced up and the sub-agreement terminated.

As in previous cases, boundaries made for strong subjectivities that contributed to heated conflicts. Whereas the conflicts in Vignette 6.13 stemmed from boundaries among organizational types, the conflict in Vignette 6.16 was also complicated by the national boundaries being spanned. Due to the multiple boundaries, UF and the NGO had very different understandings of what a termination letter meant, and how to respond. This made management a key player as the intermediary negotiator, working in-country but representing UF. In working with the NGO, UF spanned organizational as well as national boundaries, while management was not spanning the national boundary. This prompted the NGO to consult with management on the supposition of a shared understanding of how to escape UF's termination process. It was when management explained what the termination letter meant that the NGO realized that there were strong international subjectivities about termination. On realizing that management was indeed representing UF, the NGO responded by invoking national and organizational subjectivities in pursuing a specific course of action. The NGO felt betrayed by management and cornered by UF, and responded accordingly in its cultural and institutional context by fighting. The realization that boundaries mattered in defining how the termination would be conducted, as determined by UF, led to reification of boundaries by the NGO as it engaged in conflict to contest the terms of termination. Multiple boundaries and strong subjectivities thus complicated the situation and ruptured several relationships.

When collaborating across boundaries, the inequalities can be large and the politics complex. The foregoing vignettes offered illustrations of the problems that can arise, especially as they can hamper communication, focusing on shared goals, and so on. Taken as a whole, the vignettes show that inequalities and politics often necessitate conflict management. This occasionally permits resolution, though also in a few cases it led to meltdowns and breakups.

I conclude this chapter by offering three closing reflections on the management of conflicts arising from inequalities and politics when crossing boundaries. The first is that with regard to the consequences of inequalities, the game depends on the strategic vision that the weaker partner organizations bring to the table. Partners with clear understandings of where they are and where they want to go, and how to get there, can articulate their strategic vision, including the role of collaboration in that vision. This frames the ensuing discussion and puts stronger partners in a position to respond with ways they can provide support. This is the ideal modus operandi for any collaboration among unequals. It is particularly important when crossing one or more boundaries because of the bigger differences and stronger subjectivities at play. When the weaker partners set the tone, it valorizes their construction of the opportunities presented by the prospective collaboration. Without a clear understanding of needs and priorities, strategic goals and collaborative activities are likely to be defined by stronger partners. This may lead to conflicts with weaker partners if they resist. If they do not, goals and activities may be imposed on weaker partners, potentially resulting in conflicts later if performance delays transpire. Both paths are undesirable because they lead to conflict. None of this is to say that the outcome is determined before the outset, but it is to say that if pre-existing inequalities are not managed by inverting the roles of strong and weak partners in leading the framing of the collaborative initiative, jointly beneficial outcomes become much harder to achieve. In sum, the discussion should begin with the weaker partners articulating a strategic vision and proposing collaborative activities.

The second reflection is that the dimensions of inequalities, and the private interests that often lead to political tensions, are not always made explicit because there is often a lack of consciousness about inequalities and politics. The underlying processes involved differ somewhat. In the case of inequalities, stronger partners may not fully appreciate the challenges faced by weaker partners, and weaker partners may overestimate the capacities of stronger partners. Neither may be conscious of the implications. In the case of politics, partners are generally quite aware of their interests. Nonetheless, those interests may in turn be based on assumptions of which they are not entirely conscious, though they may entrain very deliberate practices or behaviors. In both cases, conflicts may result in a seemingly automatic fashion because of a lack of awareness, whether of the implications of inequalities or the assumptions held by one partner about the interests of another. The lack of awareness is hugely increased when crossing boundaries. Under such circumstances, the implications of inequalities and the political interests at play will usually not be sufficiently understood unless the collaborators have previous experience working together. Insofar as boundaries circumscribe cultures, it will take time to decode subjectivities and construct shared understandings. There is thus unavoidable risk in forming collaborative partnerships across boundaries, precisely because the definition of shared goals generally occurs before the implications of inequalities and private interests of the partners are fully grasped. Insofar as the shared commitment to communication and social learning is practiced, the layers of cultural assumptions will slowly be peeled back, and understanding will grow to the point where bureaucratic snags and misunderstandings will not burst into conflagrations.

A final reflection unpacks a key element in the process of deepening understanding. To the extent that inequalities are understated and interests may be hidden, it can actually be very useful for stronger partners to make themselves vulnerable in the sense of revealing more about themselves and their motives than other partners at the outset. This does several useful things at the same time. Most obviously, this exposes the proclivities, limitations, and stupidities of the stronger partner, which would otherwise be hidden. Such disclosures "humanize the machine" of the "stronger other" and thus build trust. Revealing vulnerabilities also shows that even stronger partners lack certain capacities, which may help clarify opportunities for putatively weaker partners to complement, which is tantamount to the collaborative enterprise. On a more cynical note, divulging vulnerabilities can also tempt self-interested potential partners into tipping their hand as to their motives. This can be especially useful in moments of crisis later on in collaborative endeavors. Proceeding in good faith not only constitutes ethical practice, it can also serve to uncover questionable intent. The lesson of the benefits of making oneself vulnerable for collaborative advantage is non-intuitive but nonetheless powerful. It sets a tone that collaborators are open to each other, even to the point of vulnerability, which brings with it an expectation of reciprocity. This provides a mechanism for digging down to better understand subjectivities across boundaries, including those tied to inequalities and politics. Further, it pushes all partners to stay on their toes about maintaining transparency. This strengthens the spans across boundaries.

Chapter Seven

Surprises and Uncertainty in Collaborating across Boundaries

Who learns fastest, survives. (*Quién aprende más rápido, sobrevive.*)

—Juan Fernando Reyes

The previous chapter showed that because of inequalities and politics, crossing boundaries for collaboration involves significant challenges in terms of the potential for conflicts. While some conflicts can be resolved, most require long-term management as they arise repeatedly. The upshot is that inequalities and politics underlie many of the uncertainties and surprises in collaboration (cf. Rapoport and Chammah 1965). This chapter takes up these issues by focusing on the challenges of uncertainty and surprise (e.g., Allen, et al. 2010; Bammer and Smithson 2008; Folke, et al. 2005; McDaniel and Driebe 2005), including sources beyond inequality and politics.

Because collaboration assumes predictability to permit coordinated action, uncertainty and surprise are the bane of cooperative endeavors. In this chapter I suggest that crossing boundaries multiplies the sources of uncertainty and surprise to which collaborators are exposed. Boundaries divide not only cultures but also systems of doing things, putting collaborating organizations across divides on differing clocks, such that events on one side of a boundary become surprises for collaborators on the other side. Insofar as surprises occur repeatedly and yet unpredictably, uncertainty grows. Hence, collaboration across boundaries multiplies the sources of surprises, and the accumulation of surprises constitutes increasing uncertainty.

Just as inequalities and politics can disrupt carefully designed collaborative structures, surprises, and uncertainty can necessitate unanticipated changes. In this chapter I also argue that boundaries can limit the options for responding to surprises. If a problem occurs on one side of a boundary, it can lead to vulnerabilities in collaborative structures insofar as boundaries hinder collaborators on the other side from assisting the impacted partner. Surprises and boundaries thus conspire to produce vulnerabilities in collaborative structures. This necessitates management not just for efficiency but also for resilience. This chapter therefore pays particular attention to collaborative resilience via the management of redundancies in order to avoid vulnerabilities, especially differential vulnerabilities across boundaries.

Boundaries, surprises, and uncertainty greatly tax management in collaborative endeavors. The level of uncertainty therefore influences managerial approaches to the day-to-day handling of the tasks of collaborative work. As uncertainty rises, so too does the pressure to shift from rational planning to adaptive management, scenario planning, and finally muddling through. While any collaborative endeavor ideally ought to involve rational planning, collaboration across boundaries makes this more difficult, not to say absurd.

There are numerous sources of surprise and uncertainty confronting collaborative enterprises that cross boundaries. I therefore organize this chapter around three types of sources of surprise and uncertainty. I begin by revisiting external politics. This dovetails with the previous chapter while shifting the focus to the consequences of uncertainty in collaboration. This section offers illustrative vignettes highlighting events such as elections and violence as they constitute surprises for collaborators. The second type of surprise concerns internal changes among collaborators, whether among personnel, organizations, or networks. I motivate this discussion with comments on the distinctions between management for efficiency and management for resilience. Vignettes then follow to illustrate various types of internal changes and their consequences for collaborative structures. The third type of surprise concerns extreme biophysical events as examples of rapid environmental change. This seems apropos in a book on collaboration with the substantive case of conservation and development. I focus on climate change insofar as it involves climatic variability which becomes manifest in extreme climatic events. This section provides vignettes from recent extreme climatic events in the southwestern Amazon, highlighting their ramifications for collaboration. The final section takes up the implications of surprise and uncertainty for management. Whereas the preferred modus operandi, rational planning, presumes a high level of predictability, collaboration across boundaries characteristically involves dealing with high levels of uncertainty. I therefore offer a brief review of management approaches for differing degrees of uncertainty. The section uses vignettes to illustrate the absurdity of rational planning under circumstances of high uncertainty when collaborating across boundaries. It also discusses the meaning of "deviations" from planned outcomes to reflect on alternative management approaches when crossing boundaries.

The vignettes not only provide illustrations of the complications presented by surprises and boundaries, but also the opportunities. While surprises may indeed result in catastrophes, they can also open up unexpected opportunities. Further, whereas boundaries multiply the sources of surprise, crossing of boundaries can also create new opportunities. I therefore conclude this chapter with reflections on the mixture of pains, absurdities, and joys of collaborating across boundaries due to surprises and uncertainty.

EXTERNAL POLITICS REVISITED: SURPRISES AND UNCERTAINTY

Thus far I have treated external politics as a looming threat to the prospects for collaboration across boundaries. External politics is also important due to the surprises it can bring—good and bad—for collaborators. Here I consider external politics in terms of punctuations to collaboration across boundaries, whether in terms of changes in relationships between collaborators and other groups or among partners themselves. As before, I offer vignettes to illustrate my general argument that crossing boundaries exposes collaborators to numerous sources of surprise, to make specific points as to the consequences.

Vignette 7.1 comes from an early experience when the MAP Initiative was rapidly growing and collaborative relationships were multiplying. The MAP Initiative was coming to the attention of funders and conservation organizations as a model for crossing boundaries to support participatory environmental planning. In the process, the collaborative model of the MAP Initiative came to be seen in more broadly applicable terms, whether in the MAP frontier or elsewhere, by the MAP Initiative network and various other groups.

This, however, led to problems in how the name of the MAP Initiative was used, by whom, and for what. On the one hand, the MAP Initiative was a polycentric network, and not an incorporated organization with a legal trademark. Hence, there was no legal prohibition on using the MAP Initiative name for other activities. On the other hand, the MAP Initiative comprised a grassroots movement built by local organizations, and grown by expanding participation by civil society, which lent legitimacy to the activities and proposals generated. Hence, the terms "MAP" or "MAP Initiative" carried legitimacy when invoked to refer to the grassroots movement itself.

In early 2005, MAP Initiative coordinators learned of a proposed consortium of large international NGOs that invoked the MAP name in relation to conservation and sustainable development in the southwestern Amazon. By itself, this was not cause for concern, though use of the MAP name called attention. MAP coordinators began conferring as to who in the MAP Initiative was involved. They discovered that nobody in the MAP Initiative was involved in the "MAP" consortium. This was a surprise.

MAP Initiative coordinators received a copy of the proposal. The project contemplated the support of conservation initiatives in the southwestern Amazon, particularly in Bolivia. This goal the MAP Initiative coordinators wholeheartedly endorsed, along with the significant investment represented by the seven-figure project budget. That said, the organization of the project, the focus of the project, and the degree of local participation all raised flags. First, whereas the MAP Initiative worked in an inclusive, participatory mode to produce proposals for action that were legitimate before local organizations and civil society, the "MAP" proposal came from large international NGOs. This raised concerns that the "MAP" project would be run in a top-down, hierarchical manner, in contrast to the horizontal networking that characterized the MAP Initiative. Second, whereas the MAP Initiative highlighted four co-equal priorities—environmental conservation, economic development, social equity, and public policies—the "MAP" proposal focused only on the first. This was cause for worry that the multifaceted priorities of the MAP Initiative would be reduced to biological conservation, potentially alienating other constituencies in the MAP Initiative. And third, whereas the MAP Initiative was a grassroots movement involving various local organizations and communities, the "MAP" proposal listed no such entities as participating members.

MAP Initiative coordinators, particularly those in Pando/Bolivia, therefore held a series of meetings, which led to the writing of *La Carta* ("The Letter"), a seventeen-page indictment of the "MAP" proposal. *La Carta* came with a table of contents to orient readers, as it was organized into multiple sections recounting the manifold problems associated with the "MAP" proposal. The sins were many, and they reflected a lack of attention to the exigencies of crossing boundaries for collaboration. Problems included a lack of communication with local organizations, differences in goals, inequalities in control and budgets, and lack of participation by local organizations. *La Carta* reviewed the relations of the participating international NGOs with local organizations up to that point. While it noted crucial support to MAP Initiative forums, it concluded that the record was mixed. *La Carta* was signed by numerous MAP Initiative coordinators and participants, translated into multiple languages, and sent to the government funder of the "MAP" proposal.

These events transpired shortly after publication of Chapin's (2004) critique of international environmental NGOs and their mixed record of working with indigenous and other local peoples. Chapin's critique surprised some conservationists, and led to spirited responses by international NGOs and other stakeholders (WorldWatch 2005). In that context, *La Carta* may or may not have come as a surprise to the organizations involved. In any event,

La Carta prompted the government funder to require changes in the "MAP" proposal. Project implementation went forward with an Oversight Committee representing coordinators active in the MAP Initiative in Pando/Bolivia. After the first few months of the "MAP" project, the MAP Initiative members of the Oversight Committee concluded that the issues raised had not been dealt with, and resigned. This was not a decision taken lightly, given the considerable funding behind the "MAP" project for activities in Pando. Additional meetings among MAP Initiative coordinators ensued, and another, shorter letter went to the governmental funder. The letter concluded that the "MAP" project and the MAP Initiative, whatever resemblances there might be in name usage, were following distinct paths.

The case of *La Carta* illustrates the difficulties posed by surprises when crossing boundaries. The network of the large NGOs was not at the time highly articulated with the MAP Initiative's network. This made proposals for projects surprising to the MAP Initiative network. In turn, *La Carta* posed significant challenges for the "MAP" consortium, and likely came as a surprise, as well as for the governmental funder. The subsequent twists and turns in the "MAP" consortium's situation from proposal to implementation were significant, and reflected the boundaries at play, both between countries and networks with different operational cultures. Because the boundaries could not be effectively spanned in light of the surprises, articulation with the MAP Initiative proved unworkable.

Surprises and uncertainty due to politics can also involve violence. Vignette 7.2 offers one such case, stemming from the *autonomista* (autonomy) conflict in Bolivia from roughly 2006 to 2008 (cf. Barndt 2012). Despite having a predominantly indigenous population, Bolivia's first indigenous president, Evo Morales, was only recently elected, in 2005. His political party, MAS (Movement Toward Socialism) pursued a series of policies promoting social equality. This was popular primarily in the highlands, and was increasingly resisted by political opposition in the Amazonian lowlands, including in Pando. The tensions grew over time and led to political crisis in September 2008. *Prefectos* (governors) of the *Media Luna* ("Half-Moon") departments openly declared their autonomy and supported street demonstrations and acts of civil disobedience such as roadblocks. The protests led to police crackdowns on opposition supporters as well as attacks by armed Media Luna militias on MAS supporters.

These events transpired during scheduled fieldwork for the NSF Roads project. We had planned to visit twelve rural communities along primary roads in Pando for interviews about rural livelihoods. On September 11, sixteen MAS supporters were found dead in the forest in Pando (Garcia 2008). The government of Bolivia declared a state of emergency in Pando. In the days following, more bodies were discovered, for a total of thirty, and the count of other people still missing rose to one hundred (*El Clarín* 2008). This came to be called the "Massacre of Porvenir." Meanwhile, one of the participating UF graduate students was still in the field. After an anxious twenty-four hours of e-mails and phone calls, we received confirmation that she was unharmed. However, we also learned that the community where she was working had families with MAS supporters who left to protest, and never returned. For obvious reasons, we cancelled planned visits to the remaining communities on the interview schedule.

News also came out that the *autonomista* governor of Pando, Leopoldo Fernandez, had ordered the massacre (Garcia 2008). The governor responded that the MAS protesters were armed. The government of Bolivia sent in national troops to secure the airport in Pando's capital, Cobija. The governor ordered armed militias to keep national troops out of Pando. The conflict became ever more complicated. The government of Bolivia expelled the US Ambassador on accusations of supporting the *autonomistas*, and the US government put Bolivia on its "black list" of drug-exporting countries. The Union of South American Nations (UNASUR) called a summit that issued a declaration supporting the government of Bolivia over the *autonomistas*. Legal proceedings were duly instituted against the governor, who was apprehended on September 16 (*Página 12* 2008).

The events of September 2008 are important for many reasons, most of all the loss of life and the damage done to society and polity. For present purposes, I underscore three other points. The first is that political dynamics in one country can greatly affect collaboration with partners from other countries. Political conflict in Bolivia between the MAS government and the *autonomistas* came to be determinant of what collaborators from the United States and Bolivia could accomplish together. Sampling of communities was unavoidably curtailed. Collaborating in Bolivia in 2008 thus incremented sources of uncertainty. The second point is that violence manifests extreme uncertainty that stops any collaborative activity. In the heated rush of hostilities, not only is there hardly time to communicate, everything beyond safety becomes irrelevant. The best-laid plans have little value without contingencies for evacuation of personnel. The third reflection is that the events of September 2008 colored political perceptions thereafter concerning collaboration across boundaries in Bolivia. Governor Fernandez had been a supporter of the MAP Initiative. After his apprehension, President Morales appointed a new governor. Since then, as noted in vignettes in the previous chapter, collaboration with Bolivian partners has become more complicated, as well as relations between the MAP Initiative and governments in Pando. Hence, one difficult circumstance driven by external politics can beget others; surprises can cascade, perpetuating uncertainty and undermining collaboration.

All that said, there are very different consequences that may arise due to political violence. Rather than a cascade of hostilities, violence can also beget unlooked-for pleasant surprises, which bring their own challenges. Vignette 7.3 provides a useful illustration. Also in 2008, there was political violence in Madre de Dios/Peru, next door to Pando/Bolivia. In March 2008, the Garcia administration promulgated a series of decree laws to facilitate resource exploitation in the Peruvian Amazon by domestic as well as foreign companies. The decrees came as part of the negotiations for a Free Trade Agreement with the United States in order to harmonize regulations among trading partners.

Regional society in Madre de Dios interpreted the decree laws as attacks on property rights, cultural traditions, and traditional resource use, and dubbed the decrees the *Ley de la Selva* ("Law of the Jungle"). The Alliance of Federations of Madre de Dios, representing ten regional federations including indigenous peoples, small farmers, forest extractivists, miners, and mothers issued their own decree announcing a seventy-two-hour general strike from July 7 to 9. Estimates varied but a common figure was that roughly 5000 people took to the streets of the capital of Madre de Dios, Puerto Maldonado, in peaceful protest. A key demand from the Alliance was to the regional government of Madre de Dios (GOREMAD), calling on the *Presidente Regional* (Governor) to oppose the decree laws promulgated by the government of Peru.

It came out that the governor had instead fled to Brazil to avoid having to respond. The protest turned angry, and GOREMAD's main compound was set afire. Firefighters were unable to control the fire due to water scarcity. The fire spread out of control and GOREMAD buildings, office equipment, and vehicles were lost. News reports indicated twenty-one police injured and over two hundred protesters jailed (*El Comércio* 2008a).

The day after, the finger-pointing began in earnest. President Garcia accused the protesters of setting the fires; protesters responded by blaming goons planted in the crowd by the government of Peru. More national police and troops were sent in, and roughly fifty leaders of federations were apprehended (*Servindi* 2008). By July 11, roughly thirty "vandals" were in police custody, accused of setting fire to GOREMAD (*El Comércio* 2008b). At the same time, meetings between Alliance leaders and Peru's Minister of Environment were under way to discuss possible compromises over the decree laws and the need to protect natural resources and local livelihoods. By September, the talks had collapsed due to disagreements over the roles of governmental entities and civil society in implementing the decrees.

For all the surprises in the turns of events noted thus far, there was another in store, and it was eminently relevant to collaboration across boundaries. At the time, the M-P Consortium was implementing the conservation and development model discussed in Chapter 2, which combines scientific research with stakeholder participation for regional environmental planning. GOREMAD had limited relations with M-P, even though M-P included PEMD, a GA. After the burning of the GOREMAD compound, GOREMAD GA Directors realized that M-P partners had geodata and other information relevant to regional environmental planning. In particular, GOREMAD had been working on the second phase of the ecological-economic zoning plan for Madre de Dios. M-P suddenly became an important partner to support GOREMAD in compiling databases to minimize the time lost for developing the zoning plan. Thus, because of political violence with a catastrophic outcome for GOREMAD facilities, the relationship of a regional government was strengthened with a consortium with a predominance of universities and NGOs.

Like the previous vignette, this case shows how crossing national boundaries adds sources of surprise due to political conflicts. However, this vignette also reveals that surprises, even disastrous ones, can in turn beget other surprises, including unexpected opportunities. While it would obviously be preferable to have avoided the burning down of GOREMAD, external politics are generally well beyond the control of collaborators crossing boundaries. Consequently, we took the opportunity, completely unlooked-for, to engage GOREMAD GAs in common purpose. In developing regions with political contention, surprise, and uncertainty are rife, but that is not a reason to withdraw, especially from trusted partners; just the opposite. The bottom line is to be ready for an opportunity at all times.

When crossing boundaries, political surprises can come from either side of a divide, but affect all collaborators. External political surprises are in my experience usually unpleasant, whether for proposed collaborations (Vignette 7.1) or established collaborations (Vignettes 7.2–7.3). There is considerable uncertainty as to the outcome of external political surprises, as they may prevent collaborators from advancing in their work (Vignette 7.2), or they may lead to unexpected opportunities (Vignette 7.3).

INTERNAL CHANGES IN COLLABORATIVE STRUCTURES

The vignettes thus far amount to tales of largely unpleasant surprises driven by external politics. More common if more mundane are surprises and uncertainties caused by internal changes in the personnel and organizations among partners. Chapter 5 spent time on the issue of designing collaborative structures to ensure the performance of key functions; Chapter 6 observed that in spite of the best of intentions in collaborative design, there can nonetheless be conflicts. Here I raise additional concerns stemming from surprises due to changes in collaborative structures themselves, whether due to turnover in personnel, modifications in organizations, or realignments of networks. Such changes often come as surprises to partners across boundaries, and thus constitute uncertainties. Indeed, changes in collaborative structures can induce significant complications for partners, even those that a collaborative structure was specifically designed to avoid.

Changes in personnel and institutions pose particular risks to collaborative structures designed only with efficiency in mind. The design of collaborative structures must account not only for efficiency in the performance of functions, but also for redundancy in personnel with regard to functions, as a means of avoiding vulnerability. Especially for individuals and organizations with capacities that are difficult to replace, there is a need to incorporate redundancy in anticipation of change. Uncertainty thus makes managing for resilience via redundancy a priority.

Managing for resilience is especially important in the context of collaboration across boundaries. This is because administrative systems on different sides of such boundaries have distinct rhythms and moments for possible change points. For example, calendars vary among organizational types: universities have academic years, governments have fiscal years, and so on. Further, academic years differ among universities in different countries. The consequence is that partners on one side of a divide may experiences changes in a partner on the other side as surprises if they follow different calendars.

At the same time, boundaries make management of surprises and uncertainty due to partner changes more difficult. Whereas the design of collaborative structures across boundaries often calls for personnel performing a function on both sides of a boundary, changes may reduce the availability of personnel on one side but undermine the viability of the corresponding function for both. Simple changes such as the loss of staff responsible for invoicing may lead to delays for payments if the partner does not receive invoices. The presence of a boundary complicates communication and thus coordination of handling of invoices in the meantime, unless collaborators specifically dedicated to bridging are there to backstop. On that note, the loss of personnel involved in bridging undermines bridge strength, with the result that bridges across boundaries may be undermined or even fail, with potentially large consequences for the entire collaborative structure.

Here I offer something of a typology of types of internal changes in collaborative structures. I begin with micro-level changes involving individuals, before moving up to larger-scale changes in organizations and networks. Along the way, I offer "Mini-Vignettes" to illustrate particular types of changes. The larger point of the Mini-Vignettes is that there are numerous possible internal personnel changes, and all may come as surprises.

Changes at the individual level are manifold. Individuals may change positions within an organization: students become faculty, and people get promotions. Such changes are institutionalized and thus programmed and predictable, so they may not individually constitute surprises. Uncertainties stem from the fact that such changes may be frequent, which does pose challenges insofar as promotions impose greater demands on an individual by the organization. This raises questions about time for one's home organization as opposed to a collaborative network, which puts into play their contributions to the network.

More consequential at the individual level is when individuals change organizations, especially types of organizations. My experience is that people change organizations quite frequently. Students graduate and go work for NGOs or GAs; university faculty may leave for an NGO or GA; NGO and GA staff may similarly move to another organizational type. While such changes may come as surprises and create uncertainties, this is usually a good thing for collaborative enterprises seeking to cross boundaries. A consortium of universities seeking stronger outreach is likely to be served if a faculty or student member relocates to an NGO or GA. An individual from one organizational type relocating to another is itself the act of building a bridge that spans a boundary. Whereas collaboration across boundaries poses challenges, having individuals who have worked on "both sides" of a divide is extremely valuable for navigating those challenges.

Here "Mini-Vignette" 7.4 is illustrative. This concerns the case of an individual from the unhappy experience in G-MAP of the fight over the student fellowships, noted in Vignettes 6.2 and 6.12. The individual, then in an NGO, became very wary of universities. She later entered a PhD program run jointly with one of the universities involved in the G-MAP fellowships fight. This provided the opportunity to "see how the other side works" across the organizational divide. Her wariness of the uncertainties of working with universities was mitigated by the importance she attached to gaining an advanced degree. Another example is Mini-Vignette 7.5, the case of two colleagues who switched places between their roles directing a local NGO and holding a key technical management position in a GA. The switch was made possible by mutual knowledge and trust. In turn, the switch served to strengthen understanding between the GA and NGO, since both organizations had personnel who had worked in both organizational types.

Having discussed changes in individuals, I now consider changes in organizations. One key institutional change at the organizational level concerns elections, which affect universities and GAs. Elections often change the politics of the relevant administration, which is a potentially large source of uncertainty. New governmental administrations require something of a reset with collaborative networks in which they are situated, as via proactive efforts to introduce the network's members, communicate goals, and so on. The same is true in universities, as new authorities seek to set their own tone in relations with faculty and students.

Mini-Vignette 7.6 offers a case of high uncertainty stemming from frequent changes in university leadership. A partner university with significant political infighting exhibited rapid turnover in leadership: eight *rectores* (university presidents) in nine years. Each new president that came in was subjected to certain political litmus tests by a vocal minority group. When the new president failed to capitulate and instead asserted their authority, the vocal group brought allegations and engaged in various protest actions including strikes. The result was turnover. Eventually the government intervened, sending an executive committee. The same pattern ensued. Finally an executive committee arrived with the authority to fire anybody who questioned its decisions. Finally, calm ensued. But during the multi-year interim, the lack of a stable institutional leadership made collaboration difficult. Even establishing cooperative agreements was hard since signed agreements required institutional approval that entrained processes over periods which spanned multiple administrations. The lack of institutional memory due to frequent leadership changes made sustaining collaborative activities arduous due to the periodic need to re-explain everything. This is especially complex coming in from "outside" across a national boundary, due to the additional challenges of communication.

Elections and other leadership turnover may also result in substantial personnel turnover, another source of uncertainty. In many parts of the world, governmental elections sometimes also lead to some degree of staff turnover. We have thus adopted a "non-aligned policy" with respect to elected representatives, and primarily seek to work with their technical staff, who in many cases exhibit partial turnover in cases of leadership change. Relations with authorities of organizations with elections can nonetheless be very delicate to the extent that elections are tied to patronage, and insofar as boundaries become politicized in elections. Patronage may extend only to a selected set of leadership positions, in which case appointments at the management level will not affect technical staff. Not all incoming authorities have that vision.

That comment motivates Mini-Vignette 7.7. A collaborating partner coming out of an election received new leadership which declared a thorough turnover in its staff. This not only involved changes in administrators and staff paid out of the organization's core budget. The leadership also called for replacement of personnel paid out of an externally funded agreement, which was not allowed under its sub-agreement. This circumstance created large uncertainties for collaboration. Not only could participating personnel not continue, greatly undermining ongoing technical work, but the funder raised serious questions about the partner, which affected the entire consortium. These unfortunate changes in the partner led to termination of its sub-agreement, which launched a cascade of very messy events which I will recount in the concluding chapter.

Crucial in this mini-vignette is to recognize that the uncertainties were magnified due to the lack of understanding by the new leadership of the boundaries being crossed as reflected in the rules in the agreement. On the one hand are the national boundaries and resulting complications of differing administrative systems; in addition are the organizational boundaries between organizational types, since universities as well as GAs are involved in the regulations behind the agreements. In seeking wholesale staff changes, including the personnel funded by the agreement, the leadership of the partner failed to recognize the national and organizational boundaries as they manifest in differences in operational standards. This constituted a surprise to management of the consortium, and required extended communications over the question of personnel changes. At the same time, the partner's lack of understanding of the importance of boundaries made those communications very difficult.

Another form of changes in organizations concerns modifications to organizational structures. Universities create and close programs, NGOs dependent on projects have activities that come and go, and GAs incur policy changes that may greatly alter the capacity of specific technical units. Many such structural changes can be anticipated, but many cannot, especially when partners are subject to external evaluations. Funders evaluating partners occasionally require modifications in partner structures. When collaborating across boundaries, structural changes required of one partner create uncertainties for others when bridges are affected.

Mini-Vignette 7.8 concerns one such case. Public universities in Latin America are coming under increasing scrutiny by governmental overseers, and Amazonian universities face particularly rigorous challenges since expectations are often defined in comparison to older, larger national universities or those in developed countries. Partner universities in the MAP frontier thus face uncertainties around times of program evaluation. Such evaluations can come with significant consequences, up to and including closure of programs with performance deemed below par by evaluators. In the case of one university partner, this transpired, and it threw into doubt the viability of continued collaboration with colleagues in the threatened program. Those colleagues then sought a reevaluation, and when that was turned down, they moved to close their program in order to open a new one. The first part involved considerable uncertainty as to the outcome, and the second required considerable time and effort by the collaborators. These events in turn undermined collaborator participation in joint activities. This mini-vignette provides another example of how crossing boundaries exposes one partner to the exigencies and uncertainties facing another. It also adds the observation that the ability of one partner to assist another in distress is hampered by the national boundary separating them.

Beyond changes in organizational personnel and structures, organizations themselves may come and go. Organizational courtships are periods of uncertainty as potential partners learn about each other. Of course, such courtship periods are time-limited and may or may not yield all of the relevant information necessary to make fully informed decisions. Boundaries of course impede access to information among partners, as well as comprehension of information provided. This implies that partners often consummate collaborative relationships, as by signing contracts, by assuming risks to the sustainability of the relationship. As a result, surprises can arise, including those which lead to changes in partners.

Mini-Vignette 7.9 offers such a tale, and expands the story from Vignette 6.16, which involved hostilities over termination of an agreement with an NGO partner. As it happens, that organization was just one in a parade. The consortium in question had originally incorporated another organization during the design phase of the project. During preparation of agreements, certain facts about that organization's past management of money came to light that made signing a bad proposition. That discovery prompted a hasty search for another partner to fill the void. As noted in Vignette 6.16, we went with an organization well known to a key player in the consortium. Management later discovered problems in documentation that led to termination of the second new partner. That led to negotiations with a third new partner. That partner proved to have insufficient capacity for the ensuing administrative demands, which led to the search for a fourth new partner.

At each step, major administrative complications arose for the other partner organizations in the consortium. Each new partner had slightly different protocols for handling specific procedures for hiring, processing invoices, and making payments. What is more, the partner most affected was a university, thus of a different organizational type. The many changes and resulting adjustments constituted unpleasant surprises. As each new partner issued its procedures and requirements, they were increasingly questioned by the university partner. The bureaucratic hassles were compounded by the repeated negotiations, which constituted an unwelcome distraction. Rapid courtships led to unhappy surprise discoveries that necessitated further changes, leaving established partners increasingly ill-disposed to each new partner, regardless of its merits on paper. This vignette illustrates how crossing organizational boundaries increases uncertainties, as in knowing how to appraise the capacities of a potential new partner. It also shows that surprises caused by one type of organization are more difficult to manage if there are negative consequences for organizations of another type. Building trust in new partners gets harder with each change in partner.

Beyond frustration with new partners, the frequent changes in the consortium lineup constituted a significant source of uncertainty in the entire collaborative structure. On that note, I shift focus from uncertainties in organizations to those at the level of collaborative structures. One key type of change at the level of collaborative structures involves shifts in network goals. This may result from alterations in the lineup of partners. Shifts in focus can also transpire as a consequence of modifications to collaborative relationships in the structure. More radical are changes in entire collaborative structures.

A final Mini-Vignette, 7.10, is therefore in order. Two previous Vignettes, 6.7 and 7.2, told portions of the story concerning the tensions between the US government and the government of Bolivia, and the consequences for various collaborative activities in Pando. One result was the stoppage of the BACC in 2013. The stoppage meant that funding for collaboration with all Bolivian partners would be limited to phone calls and e-mail, rather than joint activities on the ground. This did not mean that all ties with our Bolivian collaborators were severed, but rather that the hunt for new resources got under way. Then came the surprise: USAID and HED came back with a proposal to form a new consortium, between UF and partners outside of Bolivia in the Andean Amazon. This led to consultations with partners in Brazil, Colombia, Ecuador, and Peru. The new partners indicated interest in a new suite of activities, and the funders approved. Rather than work with one group of organizations in one country, we worked with another group in several other countries. The Andes Amazon Rim Consortium (AARC) was born. The focus of collaborative activities among universities suddenly spanned several national borders. While some of the partners were known entities, many others were not. The introductions started anew, along with the other trappings of collaborating across boundaries.

This final mini-vignette illustrates that wholesale replacement of entire consortia can occur, and involves big surprises. The surprises constituted extraordinary circumstances involving direct action by multiple governments. All-important in these consortium-level changes were national boundaries, which defined the geographic options for delineating the new collaborative structure. Also significant in this vignette is the positive eventuality of forming a new consortium following the imposed termination of another. Here previous experience in spanning national boundaries proved crucial. While crossing a specific national boundary exposed one consortium to the risk of termination, it also permitted the pleasant surprise of getting the opportunity to form a new consortium by crossing several others.

In sum, internal changes to collaborative structures often come as surprises when crossing boundaries (Mini-Vignettes 7.6–7.10). Such changes can undermine necessary functions that support collaboration, as they are frequently difficult to manage and they require dedication of time and energy to replace personnel or partner organizations. Changes reconfigure relationships, and when those relationships span boundaries, bridges must often be reinforced, lest they collapse (Mini-Vignette 7.9). Of course, internal changes can also build or reinforce bridges across boundaries (Mini-Vignettes 7.4–7.5).

The foregoing mini-vignettes return us to the questions of management of collaborative structures for resilience. Surprises and uncertainty due to internal changes can only be managed for resilience up to a point. The mini-vignettes indicated cases where redundancy was viable, as in instances of personnel changes (Mini-Vignette 7.5). There were, however, cases where redundancy was more difficult, such as among organizations (Mini-Vignette 7.9) and entire consortia (Mini-Vignette 7.10).

That said, any discussion of managing for resilience that focuses purely on structural considerations runs the risk of missing other aspects of resilience, such as adaptation. Here crossing boundaries can be turned from a collaborative stumbling block into an asset. Individuals who change organizations among organizational types in effect cross boundaries and build adaptive capacity in collaborative networks (Mini-Vignette 7.4). Individuals thus learn to perform bridging functions. Similarly, organizations inhabited by individual boundary crossers can become bridging organizations (Mini-Vignette 7.5). Hence, a means of managing collaboration for resilience is training in bridging. In my experience, this largely occurs on the job rather than as a distinct capacity-building activity.

EXTREME BIOPHYSICAL EVENTS

Thus far the discussion of surprise and uncertainty has focused on human decisions and institutional arrangements. It is, however, apropos in a book about collaboration for conservation and development to recognize that rapid environmental change due to biophysical processes can also provide surprises.

Extreme biophysical events readily cross boundaries based on human institutions. The most obvious instance concerns biophysical events that cross national borders. Examples from the southwestern Amazon include the drought that spans national borders or the flood that begins upriver in one country and then moves downstream to another. Also important is the demand placed by extreme biophysical events on various types of organizations. While GAs have an obvious role in public safety, there is recognition of the need for inclusive models of coordinated disaster response (Cutter, et al. 2008; Gaillard and Mercer 2012). Examples include community-based (Allen 2006) and decentralized disaster response (Hultman and Bozmoski 2006; Gopalakrishnan and Okada 2007). Finally, the complex causation behind biophysical events has stimulated interdisciplinary rethinking about disaster risk reduction. Of particular importance have been integrative frameworks highlighting vulnerability and resilience (Cutter, et al. 2008; Fatti and Patel 2013; Hultman and Bozmoski 2006). Surprises and uncertainty due to unprecedented biophysical events are thus catalysts for collaboration across boundaries.

At the same time, extreme biophysical events pose challenges to extant collaborative structures. An obvious reason is that disasters may affect the partners and their infrastructure, hampering their ability to get work done.

Even more important is the circumstance where a disaster prompts organizations and communities to approach a partner of a collaborative network for assistance. The request may be urgent, and the need significant. Alternatively, the request may be strategic, focusing on preparing for the next disaster, specifically to avoid the negative outcomes of the last one. In both cases, there are significant challenges for collaborators, which come with the opening of a major opportunity to have an impact.

As noted in Chapter 2, the southwestern Amazon has recently incurred an unprecedented series of droughts and floods. While weather records for the basin show earlier such events, their frequency and severity have increased. There is growing attention to "precipitation anomalies" in the Amazon (Marengo, et al. 2008). Whereas the Amazon historically had well-defined seasons marked by changes in rainfall, the cessation and onset of rains has become less predictable (Davidson, et al. 2012). Whereas the end of the dry season was the time to burn recently cut vegetation, in years of drought, setting fires is a dangerous proposition. Further, whereas rains during the wet season can be heavy, there have been several sustained precipitation events that have caused record flooding. I therefore offer two vignettes, one on a drought and the other on a flood, and consider their implications for collaboration across boundaries.

Vignette 7.11 concerns the drought of 2005. This drought had its epicenter in the southwestern Amazon (Brown, et al. 2011; Marengo, et al. 2008; Lewis, et al. 2011). It is significant for several reasons: it led to a series of unexpected problems, it prompted rapid response that required extensive collaboration across multiple types of boundaries, and it stimulated ongoing collaboration in preparation to mitigate future disasters.

The 2005 drought became evident due to the unusual length of the dry season that year, and manifest in reduced forest greenness in satellite images and water availability in soils (Brown, et al. 2011; Marengo, et al. 2008). When farmers set fire to cut vegetation, it burned hotter and faster than usual, and spread out of control. Fires proceeded to damage houses and other installations on farms, causing an estimated \$100 million in damage (Brown, et al. 2011). Drought conditions also permitted the entry of fires into forests, burning an estimated 400,000 ha of mature forest in the MAP frontier. If one assumes that the value of losses is roughly equal to fines applicable for burning forests in Acre (approximately \$500/ha), then the forest fires in Acre in 2005 led to more than \$200 million in impoverishment beyond property damage. Beyond the countryside, the fires became a serious health problem for the urban population due to the smoke. Particulate matter concentrations reached ten times the level considered the maximum allowable by the WHO. The result was numerous visits to hospitals and then political pressure to do something.

The drought and fires thus constituted a disaster and prompted a rapid organizational response (Brown, et al. 2011). In Acre/Brazil, the state government and municipal civil defense forces set up a "situation room." Data collection on fire risk and actual fires was implemented to support coordination of civil defense teams as they targeted specific locations to fight fires. This involved collaboration between GAs from the national to the local levels in Brazil and work with university students on fire monitoring and local peoples on firefighting in their communities.

Disaster risk reduction (DRR) has traditionally been understood as having four phases, namely preparedness, response, mitigation, and recovery (Kapucu 2008; UN/ISDR 2005). The 2005 drought and fires revealed limitations with respect to all four. This helped define the strategy for improving future DRR. The experience of 2005 prompted creation of the mini-MAP in Civil Defense across the MAP frontier, along with other collaborative initiatives to build capacity for DRR. Here the collaborative network across the MAP frontier proved invaluable for proposing and disseminating ideas to improve various phases of DRR. Under the M-P Consortium, Bolivian partners invested in environmental monitoring capacity and produced weather bulletins for distribution to GAs and communities. Meanwhile, Peruvian partners collaborated with communities on disaster preparation by creating community fire brigades, setting up radio communication networks as an early warning system, and conducting disaster response simulations. Similarly, under the MABE Consortium, Brazilian partners continued to work with various GAs to sustain DRR, including environmental monitoring.

The classic cycle of attention following a disaster leads to relaxation of preparedness if another extreme event does not soon follow. However, the southwestern Amazon had no such opportunity, for six months after the 2005 drought, the region incurred an historic flood in early 2006. Additional floods followed in 2009 and 2010. In 2010, the southwestern Amazon was again in the epicenter of a major drought. Only months later in early 2011, the region experienced another flood, followed by floods in 2012, 2014, and 2015. While the occurrence of extreme climatic events is no longer surprising, each event is distinct and still poses challenges to DRR.

Vignette 7.12, the 2012 flood, provides a case in point. Each of the floods between 2005 and 2011 caused economic losses on the order of $10 million in Acre's capital city, Rio Branco. The 2012 flood, however, was significantly worse; it caused roughly $100 million in damage in Rio Branco (Xavier 2012). Depending on location, the Acre River crested at or above the highest levels recorded previously. Smaller towns on the Acre River were largely inundated; thousands of people had to flee their homes. In Rio Branco alone, 57,000 were affected (*A Gazeta* 2012). The state government declared the flood the worst disaster in the state's history, up to that point. (The 2015 flood turned out to be worse.)

Around the time of the 2012 flood, additional initiatives emerged to mitigate the effects of extreme climatic events. Mini-MAP Civil Defense held meetings on a tri-national early warning system. Collaboration with ham radio operators advanced, as cell phone service often went out during heavy precipitation events. River gauges were installed in remote locations upriver from key towns in order to provide data on water levels. Under the MDD Consortium, GOREMAD was to coordinate a collaborative process for a regional adaptation plan for climate change.

These vignettes show that extreme climatic events constitute surprises that have major effects on collaboration across boundaries. For one thing, such events impel boundary crossing and formation of new collaborative ties, as via Mini-MAP Civil Defense. At the same time, extreme events drive repeated shifts in collaborative activities under DRR, as worse disasters occur and new debilities become evident. Consortia collaborating across borders, which originally formed due to concerns about the impacts of the Inter-Oceanic Highway, increasingly focused on the impacts of extreme climatic events.

In turn, extreme climatic events constitute shocks to collaborative enterprises themselves. Extreme climatic events constitute a severe test of adaptive capacity, including the plasticity of collaborative structures and their ability to respond effectively via new activities. The challenge of responding to surprise biophysical events is to understand that DRR must include preparation for the unimaginable: the disaster bigger than all previous disasters.

THE BEST OF INTENTIONS: PLANNING UNDER UNCERTAINTY

Surprise and uncertainty place significant demands on management in collaborative enterprises. This has led to a variety of management approaches, differentiated on the basis of the level of uncertainty. Here I briefly consider four: rational planning, adaptive management, scenario planning, and muddling through. Each is considered appropriate under a distinct level of uncertainty, and makes different assumptions about decision making.

Rational planning is the type case in management where information is abundant, certainty is relatively high, decisions can be programmed, and outcomes are predictable. These conditions are greatly facilitative of collaborative decision making. Planning can occur on relatively long timescales. Consequently, planning can be detailed as to overall strategy, the actions to be taken, the processes that will unfold, and the outcomes. Indeed, the rationality of planning under conditions of high certainty is reflected in the fact that outcomes should be not only predictable but measurable in quantitative terms with a high level of accuracy.

Adaptive management (Holling 1978; Lee 1993) was developed for contexts where certainty is lower but information is still available. The type cases of adaptive management usually highlight the importance of collaboration among diverse stakeholders to jointly make decisions. A big reason for this is that the information necessary for those decisions is not available from any single source. Starting with exchanges of available information and continuing with joint fact finding, stakeholders make programmed decisions for management over a specified period. They then observe the results to see if the anticipated outcomes in fact occur. Due to uncertainty, there is an expectation that adjustments will be necessary; hence, the designation of "adaptive" management.

Under conditions of greater uncertainty, scenario planning may be necessary (Shoemaker 1991; Peterson, et al. 2003). Scenario planning responds to significant uncertainty by bringing together stakeholders to design several probable scenarios of future change. Each scenario explicitly states assumptions about key drivers of change, outlines in detail key change processes that then ensue, and envisions particular outcomes, usually in qualitative terms. Planning then attends to each scenario and highlights actions that can be taken to mitigate the negative eventualities. Scenario planning thus shifts from programmed to non-programmed decisions insofar as decisions are programmed for scenarios, without knowing which scenario is most likely to unfold.

Muddling through (Lindblom 1959; Wollenberg, et al. 2007) is appropriate under conditions of very high uncertainty. When change is sufficiently rapid, predictability becomes so low that any form of planning, even with scenarios, becomes unwieldy since the futures that seem probable change quickly. The fundamental assumption of the rational instrumentality of programmed decisions—that decisions made now will yield predictable outcomes in the future—is thus violated. Under such conditions, muddling through instead demands a focus on goals, and exploitation of means when they become available. Muddling through thus seeks to take advantage of the very unpredictability that characterizes conditions of high uncertainty: what may seem impossible now may suddenly become viable tomorrow. The key is to be ready with a suite of possible actions, in the hope that one will be workable, if only temporarily.

The importance of boundaries for surprise and uncertainty bears implications for approaches to management by collaborators. When spanning divides, there are likely to be more disruptions to the working routine, whether due to external politics, internal changes, or the occasional disaster. Under conditions of high uncertainty, plans to do X and thereby obtain Y products for an outcome by Z date look rather dubious. Despite the best of intentions, rational planning becomes more difficult and other management approaches become more important. The higher the level of uncertainty due to boundaries and their challenges, the stronger the pressure to shift to adaptive management, scenario planning, or even muddling through. If collaborators are crossing multiple boundaries, and if collaboration is focused on a region with a high intrinsic level of uncertainty, muddling through becomes harder to avoid.

Higher levels of uncertainty become manifest in different ways. Surprises from different sources arise of their own logics on independent calendars. Consequently, surprises may come regularly as would fall a steady rain, or they might instead occur at the same time in an unholy conjunction. The diversity of surprises colors perceptions about the ambient level of uncertainty. Big surprises that confound fundamental assumptions call particular attention (cf. Kagan 2002). Clustered surprises likely have a similar effect.

In practice, the adoption of the appropriate management approach generally transpires in response to surprises as they occur. The ideal is to maintain the commitment to rational planning, since it is a powerful model for management. Large or numerous surprises thus enforce departures from rational planning. Shifting to another approach is, however, a difficult decision. It is often less clear how radical the shift in management should be. Especially if surprises come in clusters, the level of uncertainty is difficult to assess on anything other than the very short term. At the same time, shifts in management models also take time.

The upshot of these considerations is that high levels of uncertainty are difficult to manage. On the one hand, all four management models share a commitment to rationality in decision making. An implication of this commitment is that their application under a level of uncertainty contrasting with their designed intent is irrational. Better put, management for one level of uncertainty becomes inefficient if uncertainty is higher or lower. On the other hand, the experience of implementing a management protocol under conditions of uncertainty that contrast with the design intent takes on the feeling of the absurd. Crossing boundaries in rapidly changing developing regions by pursuing rational planning is absurd because it is unworkable.

On that note, two illustrative vignettes are in order. Both come with apologies to funders. Vignette 7.13 highlights the issue of work plans and technical performance reports. As noted earlier, there has been something of a revolution in the philosophy and methodology for evaluating project performance via the application of assessment tools. These tools are important and useful, and funders are right to require them for project implementation.

Nonetheless, there are complications when collaborating across boundaries or in other situations where uncertainty is high. Work plans are detailed documents that spell out what will be done, by whom, with whom, in a specific location, over a particular period of time, with an explicit amount of financial resources, in order to generate some quantity of a product, and, per a theory of change, yield an outcome of value. Technical performance reports derive directly from work plans by reporting in the same terms what in fact happened. In sum, work plans and technical performance reports are tools of rational planning. These tools therefore come with the caveats stemming from assumptions about uncertainty noted above for rational planning. They can be seen as members of a broader family of tools such as results frameworks, on which work plans are based, along with their financial cousins, namely project budgets, anticipated accruals, and financial reports. In all cases, the assumptions are the same: uncertainty is low, work plans can be detailed, implementation will follow as expected, and performance reports will permit evaluation in light of plans.

The occurrence of all manner of surprises, such as those recounted in the foregoing sections, makes the implementation of work plans an honorable goal as well as an irrational, absurd exercise. Crossing boundaries means the cascade of surprises comes faster, and decision processes other than that presumed by the tools thus become necessary. Under such conditions, the focus is on avoiding absurd situations by identifying a decision process appropriate for the level of uncertainty. This is hard when the surprises are big and frequent.

Rational planners could rightly respond that results frameworks which underlie work plans do seek to account for changes in circumstances due to surprises. The focus in such responses, however, tends to reside with external surprises and not internal surprises. The problem is that when crossing boundaries, both types multiply.

The upshot is that managing high levels of uncertainty in collaboration across boundaries becomes an adventure in the absurd. This is due to the tendency to persist with rational planning tools or shifting to adaptive management by requesting changes in the details of plans. That becomes unworkable when the pace of new surprises is faster than the pace of adjustments in plans. There thus comes a choice: manage aggressively or sit and wait. In the first, collaborators advise funders of what is going on and then take decisions to respond, often acting on modifications in plans (regardless of whether there is time for approval) in order to keep up with events. In the second, per the tradition of muddling through, collaborators stick to their plan but wait for an opening to pursue established goals.

There is thus an awkward dance among management models under conditions of high uncertainty. Surprises occur in semi-random fashion, with the consequence that there may be few surprises for a long time and then several all at once. Variation in the rate of surprise does not necessarily mean that one merely shifts from rational planning in tranquil spells to muddling through for, say, a week, and then shift back. While different management protocols are designed for differing levels of uncertainty, rational planning, adaptive management, and scenario planning all feature fairly long timescales. Hence, shifting back and forth per se is not very workable, especially when working across boundaries, since different partners may be facing contrasting challenges at a given point in time. Rather, the recommendation is to manage uncertainty via a combination of planning and decision methods from the outset. Rational planning may be fine as the point of departure, but there should also be contingency planning for adaptation when surprises occur, ideally along the lines defined by a scenario planning process as to what the probable changes might look like. In other words, if one lacks scenarios from the outset, it may be too late if surprises occur to shift in project midstream. What is more, scenario planning at the outset needs to include possible scenarios with changes larger than those seen before. Examples from earlier in this chapter, such as external politics and extreme climatic events, offer cautionary tales. In sum, when crossing boundaries, making provision for all four management protocols from the outset is wise. Rather than being prepared to shift from one to another, the expectation should be to use one to get to another. In moments of rapid change, muddling through should be deployed to return to rational planning; if that is not possible, then partners can try adaptation, possibly by revisiting available scenarios.

These reflections prompt the question of the ramifications for the bottom line: the comparison of expected versus actual performance. That question motivates Vignette 7.14. Under the low level of uncertainty assumed by rational planning, expected and actual performance should correspond. "Deviations" from planned targets are thus flagged and funders require explanations.

Here another form of absurdity appears, as conditions of high uncertainty imply that deviations are more likely to be large. Surprises that contribute to high uncertainty may result in reductions or increments in performance. External political interventions and biophysical disasters constitute obvious explanations for work delays or stoppages, and thus shortfalls in performance. Other surprises, however, can yield increments in performance, such as greater than expected societal demand for a training activity or participation in programs investing in new technologies. The tables in Chapter 2 show significant disparities between performance targets and achievements.

Deviations from planned targets raise important questions, especially when technical performance is related to budgets, as in work plans. A performance target of X products for Y dollars that actually results in X+N products for Y dollars can thus raise questions in the minds of accountants. To the extent that the explanations are simple, such as "the families in the community invited their friends in the neighboring community," there are likely no worries. But when crossing boundaries, communication challenges and other issues become more complex, and so too can the explanations.

The intervention of a complex surprise, possibly both positive and negative in their ramifications, complicates explanation. The fires of 2005 and the flood of 2006 made DRR a priority that was not in the original work plan for

G-MAP. Societal demand nonetheless rose for a suite of such activities on DRR, and adaptive management duly got under way, under MABE and M-P. Disciplinary, organizational, and national boundaries were crossed in the process. Subsequent floods in turn revealed oversights and limitations in initial activities, prompting adjustments and new initiatives. Surprise and uncertainty can thus impede performance or catalyze a chimerical proliferation of activities.

What is more, crossing boundaries may prove effective and catalyze growth in demand. Positive experiences with communities participating in territorial planning prompted neighboring communities to approach the M-P Consortium for similar activities. Similarly, extreme climatic events led to a proliferation of communities and local governments seeking participation in DRR activities. In both cases, the outcome was the same: performance rose in quantitative terms beyond the original target.

In sum, surprises and uncertainty can prompt not only qualitative changes in plans and activities, but also a decline or a rise in the values of performance metrics. Working across boundaries may indeed expose partners to risks from surprises that cause impediments. Alternatively, crossing boundaries may make certain activities possible for which there is latent demand which then rises rapidly beyond planned targets. While deviations from planned targets might be a preoccupation to some, under conditions of high uncertainty, deviations themselves should hardly be a surprise. In particular, incremented performance deserves positive attention rather than the same treatment as inadequate performance. What is more, whether deviations result in decrements or increments in performance, they should also be viewed as opportunities for learning. The key lesson is that surprises demand learning and adaptation in order to maximize downstream performance and impact.

Crossing boundaries has a multifaceted relationship with surprise and uncertainty. I first argued that crossing boundaries multiplies the sources of surprise, and thereby results in a higher level of uncertainty for collaborators. Various vignettes offered diverse examples, ranging from external politics (including violence) to internal changes (even entire consortia) to biophysical events (like fires and floods). Surprises may occur in one discipline, organizational type, or country but not another. The consequence is that surprises may cause delays or disasters (in unpleasant cases) or bring new opportunities and challenges (in pleasant cases). In any case, surprises require adjustments which affect all partners across boundaries and thereby complicate collaboration.

I also suggested that spanning divides makes management of surprises and uncertainty more difficult due to the complications boundaries pose for collaboration. Trying to explain a crisis within a discipline, organization, or country is less complicated than attempting the same across a boundary. Reminders about some previous crisis in a partner that now has to wait can be instructive; snags can happen anywhere. The Amazon has floods, and Florida gets hurricanes. Both cut the lights. Similarly, the use of irony can also be useful. Chapter 4 suggested that NGOs might be less stable than GAs since the former often live from one project to the next. Moments of contentious elections thus provide an opportunity to observe with irony that GAs also face instability.

Surprises and uncertainty need not be catastrophes. Surprises, even disasters, can also lead to unexpected opportunities. In such moments, spanning boundaries can be particularly advantageous precisely because partners are likely to have complementary capacities that offer a wider response latitude. This is not to say that the numerous opportunities across boundaries are any less difficult to manage.

Finally, I argued that management of high levels of uncertainty that tend to characterize collaboration across boundaries requires multiple management approaches. Rational planning may become irrational under circumstances of rapid change and appear absurd to collaborators. Come ready to implement the alternatives, even if it means following the mercurial genius of muddling through. It may be painful to watch destruction unfold, but catastrophes may present opportunities. Like relationships in general, and partnerships across boundaries in particular, surprises and opportunities alike come of their own accord.

Chapter Eight

Crossing Boundaries and the Self

> To be honest is the weakness that makes you strong. (*Ser honesto es la debilidad que te hace fuerte.*)
>
> —Graffiti on the Inter-Oceanic Highway, Poetry Action, Puerto Maldonado, Madre de Dios, Peru

Previous chapters considered various challenges in collaborating across boundaries. The design of collaborative structures, inequalities and politics, and surprises and uncertainties all pose complications to partners when working across one or more divides. This chapter, however, goes beyond issues of collaborative structures, organizational protocols, and cultural expectations to focus on the additional challenges of individual proclivities. Just as boundaries constitute differences among partners in different disciplines, organizational types, and countries, individuals vary considerably, across as well as within boundaries.

This adds complexity to collaboration across boundaries. For one thing, the selection of individual collaborators can greatly affect the prospects for the collaborative relationship. Depending on an individual's personality traits, collaboration can be substantially complicated or greatly facilitated. Some individuals are characteristically disposed to be effective collaborators: they are gregarious, sympathetic, and good listeners and communicators. Further, individuals also face a specific institutional and cultural context, which they may or may not be effective at navigating, depending on their personality traits. Hence, the personality traits of individuals matter, not only with respect to their collaborating partners, but also to their home context.

Any consideration of individual characteristics requires reflexivity with regard to one's own personality. Collaboration requires accountability all around, and personality traits that leave one indisposed to give as well as receive are cause for concern. In particular, crossing boundaries tests the emotional preparedness of all collaborators to move out of their comfort zone. Everybody has one, and all comfort zones have their limits. The first question then is whether one is sufficiently self-aware before crossing a boundary to meet the "other." My experience is that most are not; I wasn't. The second question is whether crossing a boundary carries the collaborator beyond their comfort zone. The implication is that crossing boundaries becomes a potentially severe test of the self, due to the typically steep learning curve. This takes strength and commitment, not just disposition and talent, though they certainly help.

The remainder of this chapter considers these two perspectives—personality traits and the reflexive self—on the importance of individual characteristics as they affect collaboration across boundaries. One way to distill the contrast in the two perspectives is to characterize the focus on personality traits as an "exterior view" of individuals whereas the focus on the self is a more "interior view" (cf. Wilber 2001). The first part of this chapter therefore considers models of personality traits, drawing from literature in psychology and management. I also briefly consider the concepts of emotional intelligence and cultural cognition. The second section then draws out the implications of personality traits for collaboration across boundaries. I provide vignettes from collaborative projects to illustrate the importance of personalities when crossing boundaries. The third section of the chapter then takes up issues of collaboration and the self, offering a more reflexive view of the experience of collaboration and the implications for the individual. There I draw on insights from ethnography, where the self becomes the instrument of observation, and thus requires reflexivity to evaluate the validity of the information acquired and interpretations derived therefrom. In light of these considerations, the chapter discusses two strategic practices which are especially relevant to collaboration and the self: servant leadership and intentional vulnerability. I conclude with some final reflections on subjectivities and ethics in collaboration across boundaries.

PERSONALITY TRAITS AND COLLABORATION

Collaboration in general, and collaboration across boundaries in particular, leads to critical moments for which one cannot specifically prepare. In such moments, personality traits, here understood as "behavioral dispositions," tend to come to the fore in determining how one chooses to respond. These critical moments constitute proverbial forks in the road where one is in effect tested—sometimes quite intentionally—by partners. Knowing oneself and one's collaborators certainly makes responding less difficult. When crossing boundaries, such knowledge is generally more limited, and the exigencies are more emotionally taxing. Collaboration thus becomes an effective if occasionally unpleasant teaching tool as collaborators learn more about each other, for better or worse. But even if a decision and its immediate fallout are unhappy, in the bigger picture, learning is an improvement, for collaborators gain understanding of each other's cognitive styles and modes of expression.

There are sundry ways to differentiate personalities. One way is to view personalities of collaborators in structural terms. This permits a focus on how well individual dispositions function in particular collaborative structures. In turn, various categories of behavioral dispositions can be distinguished. Whereas some individuals may be natural networkers who tend to see collaborative structures in terms of teams managed with a horizontal organizational style that highlights collegiality, others may be more disposed to see collaborative structures in hierarchical terms where authority is clearly differentiated and relationships are highly vertical. Plunking someone with a horizontal management style in a highly vertical collaborative structure may lead to tensions and confusion.

Another perspective on personalities, practiced more than admitted, is to engage in stereotypes based on various social and organizational designations. All stereotypes are dubious at best (racism, sexism, and nationalism all spring to mind), and serve to hamper mutual understanding. For present purposes, I focus on stereotypes about personalities stemming from boundaries. Crossing disciplinary boundaries requires navigating past stereotypes tied to the supposition that different disciplines attract particular personalities. Here I confess my struggles to overcome conceits that social scientists (such as myself) are neurotic, whereas field biologists are humorless, and engineers are dictatorial. Crossing organizational boundaries is hardly any less fraught with stereotypes about personalities. University faculty (such as myself) with their job security have been characterized as academic prima donnas, whereas NGO personnel who live from project to project resemble sharks when they smell the lifeblood of donor money, and GA staff subject to the vagaries of political winds are scrupulously noncommittal and apolitical. Perhaps the most colorful stereotypes, however, are tied to nationalities. Using the countries represented in the projects in which I have participated in the southwestern Amazon, the Americans (such as myself) are of course the gringos, which roughly translates as capitalist pigs. National stereotypes like others are selective in the traits they acknowledge, and are often simplistic, negative, and inaccurate.

This is not, however, to suggest that personality traits are not relevant or important to consider in the context of collaboration. Numerous insights on personality traits are available from the behavioral sciences, particularly work in psychology. This research has found manifold applications in business administration via management practices. There are various models of individual personalities, stemming from Jung, Allport, and many others. Personalities are generally viewed as groups of traits defined as more or less stable behavioral dispositions observable in individuals. One approach to personalities is the Myers-Briggs Type Indicator (MBTI). Derived originally from Jung, the MBTI posits a set of traits defined by polar opposites, such as introversion and extroversion. This permits measurement of personalities based on where an individual falls along a continuum between the two extremes. More recently, the "five-factor" model of personality came into vogue (Digman 1990; Wiggins and Pincus 1992). On that account, the "big five" dimensions of personality were extroversion, agreeableness, conscientiousness, emotional stability, and openness. This drove a large research literature, including comparisons of the MBTI to the "big five" (Furnham 1996; McCrae and Costa 1989) and proposals for a six-dimensional model that added humility (Ashton and Lee 2007). Research also inquired into the proposition that personality traits underlie personality "styles," as in cognitive styles, problem-solving styles, and so on. Multidimensional models therefore found applications in management for hiring decisions. The ideal for many positions is to hire candidates who score highly on all of the five dimensions.

Lists of personality traits led to revisionist efforts especially relevant for collaboration, such as the concept of "emotional intelligence" (Goleman 1995; Salovey and Mayer 1989). Also called "social and emotional intelligence," these terms refer to a set of four alternative dimensions of personality traits: self-awareness, self-management, social awareness, and relationship awareness. These traits focus on the ability of individuals to manage themselves with respect to others via social relationships. In particular, there is emphasis on self-mastery in moments of ambiguity, crisis, and other forms of external threat. Insofar as there is variation in self-awareness, scores on emotional intelligence will differ, with implications for behavioral dispositions to work with others.

Also relevant to a discussion of personality traits in the context of crossing boundaries is the concept of "cultural cognition" (Tomasello, et al. 2005). Cultural cognition explicitly relates individual characteristics to their social context, which by definition varies across boundaries. Research on cultural cognition comes out of studies on symbolic cognition as related to social interaction. Symbolic cognition is defined as situated in a culture, which in turn informs every nuance of interaction, whether via words, gestures, or exchanges. Crossing boundaries thus requires learning the symbolic code and interactional rules of another culture. This occurs by reading and understanding the intentions behind symbolic cues being offered by partners, in order to develop a "shared intentionality." The challenge stems from the fact that while interpretive frameworks arise from cultural contexts, they are nonetheless interpreted by individuals. Because individuals bring their own cognitive frameworks, the interpretation of cultural symbols varies individually. The travails of cultural cognition across boundaries are evident in research literature, such as on communication of science to lay audiences (Kahan, et al. 2011), assessment of risk (Kahan 2012), and water management planning (Overdevest and Christiansen 2013).

The foregoing discussion of personality traits offers a suite of insights for collaboration across boundaries. First, analyses of personality traits are applicable to identification of individuals disposed to be effective collaborators. The management literature points to the use of the "big five" personality dimensions as helpful indicators for use in hiring and placement decisions, in part out of recognition of the importance of forming teams of compatible coworkers. Similarly, prospective partners might benefit from knowing about personality profiles of the individuals at hand. High scores on the five dimensions would denote behavioral dispositions favorable to establishing and maintaining collaborative relationships. Second, work on emotional intelligence highlights some key individual characteristics relevant to collaboration. The emphasis on traits like self-management and relationship awareness is eminently germane to effective communication, conflict management, and the other issues, tasks, and challenges of collaboration outlined in Chapter 3. Individuals with low scores on emotional intelligence will likely have more problems should they embark on collaborative enterprises. And third, cultural cognition skills are eminently relevant to crossing boundaries for collaboration. The ability to grasp cultural symbols and interactional rules in other disciplines, organizational types, and countries is obviously advantageous when spanning divides.

All that said, the focus on personality traits carries a tendency to essentialize behavioral dispositions. This implies an understatement of the potential for individual learning and growth to influence behavioral dispositions, as emphasized in humanistic psychology (e.g., DeCarvalho 1990; Maslow 1970). Given the importance of learning in collaboration, I therefore suggest "derived" personality dispositions as being especially relevant to collaboration. I used the term "derived" to denote personality characteristics that stem from the types of traits in the major frameworks. As derived dispositions, they must be cultivated, and can be learned during the process of collaboration. As dispositions that are learned, the derived characteristics in turn lend themselves to application via strategic practices. Derived personality dispositions can thus facilitate collaborative advantage. I offer a short list of derived behavioral dispositions.

The first derived disposition concerns the orientation of an individual to collective goals. A strong collective orientation derives from personality traits including extroversion and relationship awareness; all constitute various forms of an outward orientation by an individual. But while a collective orientation may in some sense derive from personality traits, it constitutes a more specific behavioral disposition. A collective orientation is not merely an outward orientation, but a disposition to seek out and identify shared goals, and to form commitments in order to achieve those goals. A collective orientation underlies the identification of points of consensus, the brokering of relationships in networks, and the building of bridges to sustain relationships. By definition, a collective orientation amounts to a disposition to subordinate one's self-interest for the sake of larger goals. This is not to say that individuals with a collective orientation have no self-interest, nor to suggest that they are not proactive, competitive, or non-conformist. It is, however, to assert that they will work to identify shared goals and then advocate for those goals. A collective orientation thus also encompasses a strong element of empathy, which permits understanding of another's priorities, as well as dedication to resolving conflicting goals. This in turn leads directly into various strategic practices including communication about goals, venting of grievances, and identification of jointly held objectives. Individuals who lack this derived disposition will struggle in collaboration. Individuals who learn from those frustrations by developing a collective orientation can become more effective collaborators.

A second derived disposition is the tolerance of ambiguity. This disposition derives from personality traits like openness and self-mastery, and is related to cultural cognition. As with collective orientation, tolerance of ambiguity is more specific than the personality traits that make it possible, and it too requires development via learning. In my experience, collaborating across boundaries has led to circumstances where things I thought were polar opposites occur together. Examples include the experience of learning and bafflement, advances with complications, and even conflict with love. Such "joint dualities" constitute ambiguities that are likely to stem from the particularities of the partners involved, along with the cultures of the boundaries being crossed and the inequalities at play. During collaboration, specific ambiguities arise that demand careful attention in interpretation and decision making. The lack of tolerance of ambiguities is likely to result in hasty decisions that may prove regrettable. Tolerance of ambiguity is particularly vital when crossing boundaries, for until foreign cultural codes are understood, conclusions about symbols and rules are likely to be naïve and mistaken. Indeed, a lack of tolerance of ambiguity is likely to be problematic at best and toxic at worst for sustaining collaboration across boundaries. Because it reflects understanding of cultural symbols and interactional rules, tolerance of ambiguities affords pursuit of various strategic practices that involve recognition and management of relationships, as in calling out assumptions and biases and in making explicit how benefits of collaboration will be distributed.

A third and final derived disposition concerns equanimity. Equanimity comes from personality traits like emotional stability and self-awareness. Collaboration across boundaries requires significant commitment, and can run risks of becoming all-consuming of the time, effort, and focus by participating individuals. While I do not question the importance of commitment among collaborators, at the same time, there remains a need to maintain a level of dispassion about the collaborative endeavor. Full-on commitments become unitary foci, and unitary foci present great risk of blocking out information that permits more considered analysis. When conflicts arise, individuals overly focused on a technical goal may be blind to issues of collaboration and the need to manage them. Surprises pop up, and individuals who cannot step back to evaluate may not respond effectively. Panicking never helps. Except in the most obviously dire of circumstances dangerous to life and limb, dropping everything and adopting a new focus is also likely to do more harm than good. Equanimity, the ability to step back from a complicated situation, permits cool, calm evaluation of the situation from various angles. This becomes especially important when crossing boundaries due to the added complexity, which may necessitate taking multiple proverbial steps back to appraise. Dispassion about conflicts, surprises, and other crises is crucial to the sustainability of collaborative relationships. Partners need to know they can rely on each other to not act rashly when difficulties arise. Equally, partners need to know that decisions taken reflect consideration and care, especially when said decisions are unpopular. Like the other derived dispositions, building equanimity takes time and experience. Supportive personality traits certainly help, but the specifics of particular situations that arise are also crucial. In turn, equanimity permits more careful implementation of strategic practices, as in the adoption of non-aligned policies among conflicting parties and the disciplined enforcement of deadlines.

Derived dispositions thus provide a framework for thinking about individual personalities in a way that is eminently germane to dynamics of collaborative relationships and the self. Because they are derived characteristics, these three dispositions stem to a greater or lesser degree from personality traits in established frameworks. In turn, as dispositions that require cultivation, they also necessitate learning from experience. Finally, insofar as they are consciously cultivated, they contribute to the intentional implementation of strategic practices to manage collaboration. In these respects, derived dispositions do not replace but instead complement models of personality, particularly by moving beyond essentialist traits and making the link to strategic practices.

I selected these particular dispositions—a collective orientation, tolerance of ambiguity, and equanimity—for additional reasons. For one thing, they permit management of strong subjectivities likely to be at play in collaboration across boundaries. These derived dispositions serve in complementary ways to support collaborative relationships among individuals, particularly in moments of conflict and crisis. A collective orientation maintains a focus on what is shared rather than what divides, which provides a basis for questioning the assumptions behind conflictive subjectivities. Tolerance of ambiguity is instrumental to acknowledging differences of opinions and selections of facts while avoiding the temptation to jump to hasty conclusions. Equanimity is crucial to avoid taking sides, or at least not taking sides without very clear reasons.

A second reason for this particular combination of derived dispositions is that each recognizes in different ways the distinctions between personal dispositions and shared priorities. This is especially relevant to collective orientation, which is not the same as following the crowd or lacking agency. Individuals with a collective orientation have self-interests, but recognize the greater value of shared interests, and acknowledge that the two are not always aligned. This is vital to avoid imposing personal interests on a group, while at the same time also requiring recognition by the group of one's viewpoint. In the context of collaboration, tolerance of ambiguity often refers to ambiguities stemming from differences of opinion between oneself and one's partners. Such tolerance is crucial in joint problem solving and management of potential conflicts. Equanimity presumes that one has preferences but is prepared to set them aside. Awareness of the distinction between self-interest and shared priorities are central issues in collaboration, for they underlie the challenges outlined in the foregoing chapters.

Collective orientation, tolerance of ambiguity, and equanimity are, however, found in varying degrees among collaborators. I therefore offer a suite of vignettes related to projects in the southwestern Amazon as illustrations of the importance of these three dispositions. Vignette 8.1 concerns the formation of collaborative ties at UF which eventually helped birth the NSF Roads project. The "Roadies" group emerged organically as a working group of faculty and graduate students from diverse disciplines with shared interests. We anticipated the subjectivities stemming from disciplinary differences in intellectual commitments. The group devoted substantial time to discussion of the assumptions underlying various theoretical frameworks as a means of learning about disciplinary cultures. However, there emerged the additional challenge of navigating the individual subjectivities as to the purpose of the group. Some participants wanted free-wheeling intellectual discussions which they took as the end purpose; others saw the discussions as the means to formulating proposals for projects.

Subjectivities concerning the purpose of the group became the larger issue. Debates increasingly revolved around what to do with the ideas at hand, and the differences became less oriented to disciplinary backgrounds than subjectivities about goals. Eventually some participants broke off from the main group and composed a theoretical framework that was published (Cumming, et al. 2005). This was possible because the subgroup retained a collective sense of shared purpose that they wanted to see through to a final product. In turn, the subgroup began to produce proposals for funding. Here again something of a division appeared as to the specific applications of the theoretical framework. Another instance of ambiguity occurred as the participants decided what to do. A sub-subgroup eventually coalesced around the proposal that resulted in the NSF Roads project.

In both instances, a collective orientation, tolerance for ambiguity, and equanimity were required for individuals to address the intellectual differences at play in a constructive manner. In both moments of group fluidity, individuals varied in terms of their willingness to pursue goals even if only shared by others (collective orientation), engage opposing arguments (tolerance of ambiguity), and step back to reconsider their positions (equanimity). Individuals who refused to consider or failed to recognize the path favored by an emerging group tended to fall outside the subsequent collaboration. By contrast, individuals who grasped the goals of the group, even if distinct from their particular interests, often came along because they recognized the distinction and set their individual interests aside out of commitment to the shared interests of the group.

To provide a strong contrast, I offer Vignette 8.2, of another group of faculty at UF who also sought to cross disciplinary boundaries to produce a proposal for funding, with very different results. While participating in the Roadies group, I was invited by another group working on a different topic. Rather than engage in regular meetings to understand each other, I was told what my role on the proposal would be. When I asked who else was participating, I was given names and departments but very little else. When I inquired about the purpose of the proposal, I received a similarly terse reply. I expressed concerns about these issues in order to raise doubts that I still wanted to participate. I was told that my role was defined: the team needed a social scientist, and I was thereafter cajoled to participate so they could check a box, and promised that my time commitment would not be large.

The proposal went out. I confess to ardently wishing that it be rejected, out of concerns about the way it was developed. In the event, the project was not funded. I was informed that this was due to a lack of integration of the project team. My abiding memory is that the PI called to give me the news, and asked how long it takes to constitute a true interdisciplinary team. I replied that it takes at least a solid year of regular meetings. The silence at the other end of the line was deafening. I never heard back from the group in question.

In this vignette, the work of crossing a boundary was underestimated, and key individual dispositions, notably of the PI, played a central role. While at first glance one might suppose there was a collective orientation, it soon became evident that the input of at least some of the co-PIs was not of shared interest. The proposal goals were set, which foreclosed on a collective process of proposal development. There was precious little tolerance of ambiguity. With the proposal goals defined, there was no further discussion, regardless of the issues that might arise. The process was presumed to be logical and linear, rather than open and creative, and thus potentially ambiguous. Finally, there were strong intellectual commitments to the project but there was a lack of equanimity on the part of key players, including myself. Equanimity would have permitted dispassionate reflection. By itself, this was not enough to improve the teaming behind the proposal; but without equanimity, there was no impetus to stimulate such reflection.

Vignette 8.3 provides a third illustration. Here I call attention to subjectivities, across boundaries as well as among individuals within boundaries, during the process of defining objectives for a collaborative project. I draw on the case of the HED project when it was beginning in Bolivia. That project focused on investments by a consortium of universities and NGOs in a focal partner, the Amazonian University of Pando (UAP), to build its research and advanced training capacity. The question then focused on the specific content of the training objectives. Here different personality dispositions emerged among the Bolivians, in two key moments.

The first concerned the moment when personnel in the key department at UAP needed to come together to make joint decisions about what exactly they wanted. At that moment, it became evident that some personnel were not going to participate. This stemmed from past conflicts, but in particular, how behavioral dispositions influenced how those conflicts had unfolded, how they were interpreted, and how they influenced subsequent departmental functions. Salient in the most recent such conflict was a pronounced difficulty on the part of the two sides, namely the chair and most everyone else, to find common ground. One rendition was that the chair wanted a more thorough review of faculty in order to renew their contracts, and the faculty rebelled; the other rendition was that the chair was making demands that would be impossible to meet, which gave the faculty the options of failing to get their contracts renewed or rebelling. By both accounts, rebellion ensued. By both accounts, neither side could find middle ground. In other words, neither exhibited a collective orientation, tolerance for ambiguity, or equanimity.

The outcome of the conflict had ramifications for the second key moment, when the department faculty who chose to participate on the project had to define the objectives of the training program. Here again a split emerged, along a different fault line than in the previous conflict. But the same behavioral dispositions emerged, and attitudes began to harden. However, the discussion was moderated by visiting consultants who demonstrated all three of the key dispositions: they pointed out that the faculty would have to jointly define the objective (collective orientation), they observed that there were advantages and disadvantages, and thus risks and uncertainties, in both options put forward (tolerance of ambiguity), and they made it clear that they were not getting involved in the debate over the merits of the two proposals (equanimity). Interestingly, both sides increasingly adopted those dispositions as they negotiated. The faculty sought a consensus on the understanding that everyone would have to contribute for the program to be workable (collective orientation), they all recognized that both proposals had risks and unknowns (tolerance of ambiguity), and at least some put themselves in the position of those arguing the other side (equanimity). It became evident that the faculty recognized that they could not advance via the style of debate that had transpired in the previous conflict.

To be sure, the behavioral dispositions were facilitative though not entirely determinant. After all, the former chair who had refused to negotiate in the first conflict did not participate. Further, the decision eventually taken left open some important questions about the details of implementation. But the shift focused the entire consortium on a shared goal, as the partners were prepared to support UAP in their decision. What is more, the demonstration of behavioral dispositions important to reach a consensus by the UAP faculty left a decidedly positive impression on the partners, which was vital for moving forward.

Collective orientation, tolerance for ambiguity, and equanimity are not by themselves sufficient to guarantee effective collaboration. Nothing is. They have, however, proven to all be necessary, particularly when crossing boundaries. Fundamentally, they constitute behavioral dispositions that permit management of self and others in the context of collaboration, which involves delicate relationships rife with subjectivities and other complications.

CROSSING BOUNDARIES AND THE SELF

The discussion thus far has focused on an "exterior" view of personalities, understood in terms of observable behavioral dispositions. This approach, however, raises additional issues about an individual's personal experience of the rigors of collaboration. This is the focus of the second half of this chapter, where I highlight an "interior" view of the self in the context of collaboration across boundaries.

The experience of collaboration across boundaries is likely to be confusing, frustrating, and exhausting. I have lost no end of sleep due to waking up in the middle of the night worrying about the challenges of collaboration, especially across boundaries. The subconscious operations of my brain seem to sound alarms over ongoing ambiguities and unresolved problems. Evidently, there is not sufficient time in the day to think through and put out all the fires when collaborating across boundaries.

Bridging divides and working jointly are likely to force into the open one's preferences and blind spots. One's proclivities will eventually go on display before collaborators, consciously or unconsciously. Those behavioral dispositions therefore require self-reflection. This is not for the sake of learning how to hide one's limitations. On the contrary, reflection permits personal growth in order to better meet the challenges at hand, as well as those as yet unrecognized by oneself. In sum, collaboration across boundaries is transformative of the self, and thus challenges the self to change. The key is to go prepared to reflect and enact said change.

Crossing boundaries for collaboration is not unlike ethnography or studying abroad. For all three, it is wise to embark on self-reflection as a part of preparation (e.g., Agar 1996; Jackson 2006). Auto-evaluation involves reflection, not just on one's professional preparation and the substantive reasons for crossing boundaries, but also on one's emotional maturity, cultural awareness, and self-knowledge. I suggest that self-reflection on one's profile with respect to the three behavioral dispositions underscored earlier—collective orientation, tolerance of ambiguity, and equanimity—can be especially useful.

Initial self-evaluations are typically optimistic and generous, and should be treated as provisional hypotheses. Frustrations and conflicts arise and constitute tests, not only in the harsh light of experience in front of one's partners, but also of one's personal hypotheses. Initial self-evaluations thus provide a baseline that permits further self-reflection later on. Re-evaluations then have to confront one's initially rosy self-perceptions with the potentially ugly facts of a collaborative situation and its travails. The rite of passage is to admit and accept one's proclivities and limitations in order to manage them and by extension one's collaborative practice. The healthy result is some modicum of attitude adjustment and humility. On that note, a few of my own personal failings serve as examples: poor language skills ("I thought I understood what they said . . . " and "That's not what I meant . . . "), naïve idealism ("Everyone will act in the name of the greater good, right?"), impatience ("Let's just vote on this . . . "), an idiotic savior complex ("I can fix this problem, even though I have no idea how; I'm sure I won't make it worse . . . "), and the kicker, a hegemonic white male problem ("I'll just explain it to them and surely they will all agree . . . "). These failings are based on a mix of self-reflection and discrete comments by sincere but blunt collaborators.

Any instance of an unproductive meeting, an activity that doesn't come out as expected, a crisis handled poorly that worsens, and conflicts among individuals, are moments for self-reflection and thus re-evaluation. Before pointing fingers at others, I look in the interior mirror and ask myself what I might have done differently. Such moments are also auspicious for consultations with partners. Subjectivities are likely to be strong, but an external appraisal of one's behavior can usefully complement a self-evaluation by identifying potential blind spots.

Consultations with partners are also useful for calling out behavioral disorders ("neuroses"). Such disorders are behavioral dispositions that reveal limitations in one's ability to adapt to a social environment, especially a changing one, and thus develop a more actualized personality. Obsessive-compulsive disorders, fears and anxieties, and similar disorders often head the list, as defined in recent Diagnostic and Statistical Manuals of Mental Disorders (DSMs). Disorders constitute blind spots that limit or distort what individuals perceive in others and how they interpret social situations. Not surprisingly, disorders present stumbling blocks in collaboration, particularly across boundaries, especially in moments of crisis. Self-oriented behavior in moments of conflict may indicate compulsions to withdraw from social situations with the hope of ensuring control. Fear can lead to aggression, stimulating conflict. Both cases undermine collaboration.

This discussion prompts Vignette 8.4. I do not consider myself a highly gregarious person. As a sociologist, I like to watch more than participate in social situations. I collaborate out of a recognition of the need to support partners to achieve shared goals. This carries several implications involving my limitations which pose hazards to collaborative activities. I tend to see collaboration as a professional activity, despite all of the personal travails and shared experiences. Regardless of the similarities between preparing to cross boundaries and do ethnography, I resist full cultural immersion; the "post-dinner-chat-over-a-beer" is my preference. Interestingly, this is not the preference of most of my collaborators, who in many cases see the stakes of our joint goals as being too great to stop working. For me, after a beer, it's time to retire. After a long day of meetings, little problems that lead to side meetings, unexpected snags that lead to side-side meetings, various naïve questions about how something gets done around here, and an accumulation of unresolved issues, I crave personal space. When a serious crisis arises that requires a solution that night, I become crabby and impatient and various compulsive behaviors begin to emerge. Rather than use the restroom next to the conference hall, I will race back to my hotel room, even if only to wash my hands, in order to breathe for just a minute. Then I wash those hands again. After the resolution, even if it is late, I need a silly movie, or some such cognitively light fare. Whether due to attention deficit disorder, a need for private space, or some more disturbing complex, I compulsively collaborate by day, and equally obsessively feel the need to recoil at night. Perhaps this explains the sleep interrupted by my burdened subconscious, still working on the day's unresolved problems, which frequently awakens me around 2 AM.

With that general vignette, I offer two more, each with more specific instances of experiences involving personality clashes that motivated self-reflection. Vignette 8.5 involves the near detonation of a project. We had one project and then got another, with some of the same partner organizations. This led to discussions of how the two projects might coordinate. I called a meeting of the program managers to discuss the question. The two managers differed markedly in various respects: their ages, national origins, and professional backgrounds. Consequently, their cognitive styles, preferences about project management, and proclivities regarding appropriate processes also differed. This led to strong subjectivities in how the managers saw the question of coordination, and how to manage processes to arrange specific instances of coordinated activities. As a result, it was clear to each person how to proceed, but it was less clear that the other understood what exactly was meant. Further discussion ensued, on a more operational level. The subjectivities became more evident in the form of frustration as one project manager felt he had to repeat what he had already said. I found myself moderating the conversation by gingerly trying to ask questions in order to clarify, feeling like I was attempting to defuse a bomb. Finally the questions were answered and the conversation seemed to be drawing to a conclusion. The frustrated manager slyly smiled and thanked me for the meeting. The other manager, who had been asking for the clarifications, still seemed tense. The first manager departed for another meeting. The second manager then asked me what to do. We reached a tentative consensus to make a priori consultations before any coordinated activity. Despite the preoccupations with a potential bomb detonating, no collaborative explosions occurred. Evidently we got the first manager's real message—show deferential respect—and played the game correctly.

With that happy ending, a final vignette is in order. Vignette 8.6 recounts a collaborative detonation and the personal fallout that ensued. I was contacted by e-mail by collaborators at a partner organization with an urgent request for a virtual meeting. Wanting to be a good collaborator, I agreed, and the day and time were set. The collaborators thereupon provided a list of concerns about an unfolding drama over the case of one of their staff whose contract the project management had decided not to renew. This was explained as an example of a larger pattern of dictatorial management, something that left them unhappy. Consultations with the management revealed significant limitations on the part of the staff person in question, who was having others do their work for them. Finding no concrete reason to intervene, I said as much to the partners with the concerns.

That decision not only failed to put out the fire, it led to another e-mail requesting a meeting to discuss the underlying issue of dictatorial management. Given the first case, I became doubtful that such a meeting was necessary. As it happened, I was set to visit the partners seeking the meeting on my way to visit other partners elsewhere. When word of that got out, the e-mailing partners insisted on a meeting. I advised the manager of the meeting in question and went to see what the issue was. The partners were visibly nervous and began to recount not one but a series of various allegations about the management. I expected a one-hour meeting; I called a halt after two and a half. Rather than list the allegations, it became evident that there were communication problems between partners and management, and management was making demands of the partners with which they were unhappy. They were also unanimous in their conclusion: fire the manager.

I followed up with the manager, whereupon I admitted the allegations had left me with some doubts. The manager became very offended and demanded a meeting with the partners to respond to the allegations. Discussion of a couple of allegations made it evident to me that the allegations themselves were not that serious; there were procedures to put out these fires. What is more, the ensuing discussion led to one of the most dizzying reversals in the tone of the conversation I have ever experienced. The manager was overtaxed because the person indicated as deputy manager was not doing their job. This was especially problematic as the project was under growing pressure to spend money and deliver results. Hence, it was not surprising that the manager was being demanding. Everyone was under stress, and the partners had personalized the demands. The solution became obvious: get a new deputy manager to better support the project in handling the pressure. This was not only acceptable to the manager, it would come as a godsend.

After the visit, I followed up with a long e-mail. The situation was very complicated due to the external pressures, the internal structural problem, and the hostilities among the players involved. The subjectivities were exceptionally strong, and what is more they occurred along a boundary where I'd not expected such difficulties: between local partners and management with international experience. I proposed the hiring of a new deputy manager and invited comments on the proposal. Virtually everyone repeated their grievances and almost no one commented on the proposal. This was bad as the repetition indicated that not only did those with grievances evidently feel that they had not been heard, they still felt there was no way out except for me to favor one side or the other in the dispute. I concluded that removing the manager, as well as not hiring a real deputy manager, would be disastrous as it would ensure that the problems would continue and the project would not advance. The external pressure and the internal limitations left no choice but the unpopular one: increase management capacity.

The new deputy manager embarked on a suite of actions to clarify the chain of command, the accounts of the partners, and various procedures from hiring to firing. The ensuing flurry of accusatory e-mails, from all around, was nothing less than frightening. I was sure I'd destroyed the project, and a panoply of collaborative relationships in the process. I kept telling myself that all of this stress was to get the project back on track. What is more, the point was to help ensure the strategic viability of the partners to secure future funding to continue their work.

Then some crucial decision points approached. I held my breath. Clarify the accounts, or no more money. The accounts were suddenly forthcoming. Respond to requests from the funder, or face removal from the project. Requests were largely met, and in those cases where they were not, personnel changes indeed transpired. It was strong medicine, and some collaborative relations were ravaged. Some players left the consortium. But in the process, the operational capacity of the group rose.

The whole episode constituted a trial by fire for a suite of individuals, myself included, which revealed various behavioral dispositions, personal limitations, and behavioral disorders. Most everyone involved lapsed into self-preservation, and/or sought black/white clarity, and/or passionately defended values they deemed inviolable. Most everyone involved pursued allegations against others while erecting defenses against anticipated attack. In sum, there was rather little self-reflection going on. That said, I began pointing out problematic behavioral tendencies of others, as others did of me, notably the idiotic savior complex. At the same time, most everyone involved came to recognize that there was still a collective project to advance. While it became evident to others that I was not going to reverse my decision, the lamentations petered out and people got back to work. My next visit was rather less intense than the one with the detonation, but I went equipped with the preparation of past experience—the knowledge of how others behaved and how I responded under significant personal stress, along with self-reflection on it all.

Collaboration across boundaries thus tests the participating individuals, not only in terms of their understanding of others, but also of themselves. The first part of this chapter noted how individuals may respond differently to the challenges of collaboration, even individuals within a given side of a divide. The second part highlighted the personal travails that make up the experience of collaboration.

I conclude this chapter with two strategic practices for managing personalities, including one's own, in the context of collaboration across boundaries. The first practice is servant leadership (cf. Greenleaf 2002; Sendjaya and Sarros 2002). As noted in Chapter 3, servant leadership highlights investments by an individual in order for others to succeed. For present purposes, servant leadership can be seen as an ethic that encompasses the behavioral dispositions highlighted in the first half of this chapter. Servant leadership extends beyond merely one's collaborative relationships with partners to co-workers in one's home institution, as well as friends and family. This carries an important implication. As a highly generalized ethic, servant leadership imposes limitations on the self, including limitations on the acceptance of responsibilities to prospective collaborators. Effective servant leadership thus entrains reflection on the behavioral dispositions of others as well as one's own behavioral limitations. Participation in collaborative enterprises must be highly selective to be practiced effectively. The compulsion to collaborate requires moderation even as one seeks to respond to the call of new opportunities.

The other strategic practice is intentional vulnerability, discussed earlier in Chapter 6. Here I invoke the concept in light of the personal limitations of others and oneself. Such limitations mean that personal vulnerability is endemic to collaboration, since one invests trust in collaborators and vice versa. Of course, the investment can be risky, as when a partner engages in unethical behavior, such as misappropriation of funds. However, such behavior reveals who is trustworthy. Further, unethical behavior can reveal what people will not openly admit about their circumstances, notably in cases of highly unequal partners spanning divides. This leads to better understanding of others as well as oneself. As with servant leadership, intentional vulnerability has definite limitations, since some risks have more dire consequences than others. The fundamental challenge remains: for collaborators to better understand their partners, and themselves.

Chapter Nine

Conclusion: The Seasons of Collaboration

Sometimes collaboration requires a strong stomach and gallows humor.

—Foster Brown

Previous chapters in this book took up the issues, tasks, challenges, and complications of collaborating across boundaries. Much of the discussion thus far focused on initiating and then working to sustain collaborative efforts when spanning multiple types of divides. Such a focus leaves open the question of how collaborative relationships end. This calls attention to the issue of the "life cycle" of collaborative relationships, from birth to death.

I therefore begin this chapter with a discussion of the life cycles of collaboration. I suggest that collaborative relationships have stages or "seasons" like other social relationships. Collaborative relationships begin with 1) a courtship or initiation, then 2) they evolve via maintenance and investments in sustainability, and along the way, 3) they experience crises that lead to adjustments or transformations, and, in many instances, 4) they end, whether via mutually agreed parting of ways or unilateral termination.

Of particular interest in the context of collaboration across boundaries is the issue of crises which can lead to breakups. As previous chapters have shown, relationships spanning boundaries incur various strains and repeated shocks that can lead to crisis. Even if partners have a smooth courtship and make significant investments in maintenance, internal dynamics and external shocks pose considerable threats to sustainability. The key point is that repeated crises put relationships at risk of slipping into the final stage. Collaborative relationships across boundaries are highly dynamic and their trajectories are hard to predict.

Because of those characteristics, this chapter offers a discussion of bifurcation theory, which stems from thought on system complexity and chaotic dynamics. Bifurcation theory might seem an abstruse point of departure for thinking about the end of collaborative relationships. I suggest that bifurcation theory is germane here for at least two reasons. The first is that bifurcation helps account for non-linear changes in relationships among components of complex systems, which is useful when watching for signs that collaborative relationships in crisis may experience ruptures and end badly. The second reason is that bifurcation highlights how systems at critical points may go in one of two radically different directions. In my experience, collaborative relationships similarly tend to end in one of two very different fashions. Following one path, there is a mutually agreed conclusion as shared goals are met and partners merrily go their separate ways. Along the other, there are hostile confrontations involving most anything ranging from expletives to legal action. The second case is my primary concern. It is important to identify signs that crises may result in hostilities and end relationships irreparably. This can pose strategic problems for estranged partners to respond jointly to future needs and opportunities.

The remainder of this chapter proceeds as follows. I first outline basic tenets of bifurcation theory. The chapter then applies insights about bifurcations to the case of collaborative relationships. I offer a selection of vignettes on unpleasant breakups between partners collaborating across boundaries. Despite the downbeat tone, the intent is not to end this book with sad stories of breakups and woe. Rather, I provide concluding reflections on the transformative power of collaboration across boundaries, highlighting the strategic implications of this experience. Partners need each other; they change each other. Collaboration across boundaries can bring unlooked-for positive changes in partners, even after unhappy breakups. This is not only made possible through collaboration itself, but by the uplifting character of the experience of crossing boundaries in its own right.

COMPLEX SYSTEMS AND BIFURCATION THEORY

Bifurcations constitute one key concept among others in the broader confines of complex systems theory (Iooss and Joseph 1997; Sun and Luo 2006). This body of thought highlights non-linear dynamics in systems, whether in cases of stability, oscillation, or chaotic behavior. There has been particular attention devoted to moments where heretofore stable or predictable systems suddenly exhibit unprecedented and rapid change (Gunderson and Holling 2002; Scheffer 2009). Popularly referred to as “tipping points,” such moments are also called “critical transitions,” and the need to understand the conditions of such transitions has driven substantial research. This is because bifurcations at critical transitions can lead a system into an undesirable state, from which it may be difficult to escape. Predicting whether a system will stay in its current state or shift to another state has proven difficult, though some indicators have been identified (Scheffer, et al. 2009, 2012).

Work on topics like complexity, chaos, and bifurcations emerged from physics and mathematics in the 1960s, but has since been applied in various other fields. Where change is non-linear, there is the potential to bring complex systems thought to bear via a focus on chaotic system behavior, including bifurcations. This is true in the social sciences, as when post-modern theory became popular by the 1990s (Young 1991). Setting the post-modern trappings aside, ideas about chaos and the problems of bifurcations found applications to topics relevant to understanding collaborative relationships. One example is the application of bifurcation theory to industrial psychology and the management of relationships and productivity in work environments (Guastello 1995). Another concerns the problems presented by crises in the context of public relations (Murphy 1996), where crisis management must account for low predictability in how publics interpret information (Seeger 2002). A third application concerns bifurcations in family relations, especially in light of shifting inequalities of power, whether due to changing gender roles or cognitive development in children (Ward 1995). In all of these applications, bifurcations that lead to new system states are generally considered system breakdowns in social relations, as in workplaces, communities, or families.

In such work, there is a shared focus on several key issues which I highlight in this chapter. The first is the understanding of a set of social relationships as a system with internally generated non-linear dynamics. Individuals or organizations relating to one another may collaborate harmoniously or contentiously; in either case, what one does affects how the others respond. In complexity thought, internal dynamics in relationships can yield a range of system behavior, from predictable oscillations among system states to unpredictable changes among states. Collaborators may understand each other well and know what to expect, resulting in predictable interactions and outcomes, or one may surprise the other, leading to responses which are harder to anticipate.

The second issue concerns the importance of initial conditions as determinants of outcomes. Complexity theorists often focus on "path dependency," where initial conditions or early changes delineate a predictable trajectory thereafter. Initial impressions or early crises may color enduring perceptions among collaborators. Negative responses to initial crises may leave other collaborators already planning their exits from the collaborative enterprise.

The third is the low predictability of outcomes when a system is punctuated by an external shock. Perturbations to established systems with known relationships may result in unexpected behavior, including bifurcations.

Changing circumstances can render heretofore stable relationships much more difficult to sustain if the stresses on one or all collaborators rise. In particular, bifurcation theory calls attention to moments where a partner reaches its limit, whether emotionally or in terms of organizational capacity, and greatly changes its behavior. The resulting emotional outburst or unexpected refusal to comply with a bureaucratic requirement or spend money on something comes as a big surprise to partners.

Fourth is the concern with critical thresholds, where bifurcations occur and a system may shift into an undesirable state. Even incremental changes in a key parameter may push the system over a critical threshold. How partners respond in moments of crisis is obviously crucial to the outcome. If a partner declares it is unable to continue with things as they are, somebody else may have to step forward. If such declarations are, however, seen as illegitimate, conflict and breakdown are much more likely. Perceptions matter greatly in crisis moments, and differing perceptions, as across boundaries, can lead to very divergent responses and thus contention and rupture.

CRISES, BREAKUPS, AND BIFURCATIONS IN COLLABORATION

Bifurcation theory provides two useful insights when applied to the question of how collaboration across boundaries may end. Here I focus on how crossing boundaries makes crises more likely, and consider how crises in collaborative relationships may result in unpleasant breakups. With regard to the first insight, collaborative structures that are larger and more complicated exhibit more complex internal dynamics. Further, collaborative structures that cross more boundaries are subject to more sources of external shocks. Given the additional exigencies of crossing boundaries—greater structural complexity, greater inequalities, more sources of surprise, and stronger individual subjectivities—collaborative relationships are frequently pushed from maintenance into crisis. Collaborations across boundaries thus have numerous critical moments where the players have to make difficult decisions, frequently over the distribution of benefits, often with the proposition that somebody is going to lose something.

Concerning the second insight, on how relationships can lurch from crisis to breakups, three particular points become salient. First, in crisis moments, which can be understood as critical thresholds, partner behavior can change dramatically, in ways for which other partners may be unprepared. Such situations may reflect external shocks that cause panicked responses by partners. Unanticipated behavior can also ensue when tensions among partners have been incrementally growing over time, and finally reach a critical threshold. Second, per the concept of bifurcation, partner behavior in critical moments can lead to contrasting outcomes. A collaborative structure may be preserved, or it may be altered, possibly radically. Of particular concern is the departure of one or more partners. And third, what follows either outcome is often stabilization in a new state. The question then is whether the state reached is desirable in light of partnership goals.

The behavior of partners during crises thus becomes a central question. Recent work on system behavior leading to bifurcations has suggested some predictors, including a slowdown in responsiveness and increased variability in responses (Scheffer, et al. 2009). While direct translation to collaboration is hazardous, there are some noteworthy parallels. A reduction in responsiveness applied to collaborative structures might be evident in limited response capacity to a shared problem. An external shock may simply be too big to manage, such as a major flood or direct governmental intervention. Alternatively, the explanation may reside with internal dynamics in a collaborative structure, as when partners become mutually antagonistic and refuse to cooperate with each other. Similarly, increased variability as a precursor to bifurcation exhibits a frightening parallel to the case of collaboration. Whether due to an external shock or slowly rising internal tensions, increased variability could become evident in the form of very different partner responses to a shared problem. Alternatively, rather than varying responses among partners, partners change their responses in rapid fashion, whether in panic or in angry response to decisions by other partners. Both cases are extremely difficult to manage.

In turn, partner behavior during crises bears important implications for the outcomes of bifurcations. Whereas one bifurcation point would be whether an unhappy partner departs or not, that is often followed by additional bifurcations with regard to the implications for the remaining partners. Breakups in collaborative structures might be good or bad, depending on the devils in the details. Partner departure may lead to additional bifurcations with outcomes defined by three vital factors that affect the viability of collaboration. The first concerns the retention of shared goals. Partner departure may occur precisely because somebody does not in fact share goals with the others in the collaboration; in this case, the departure may be a good thing. By contrast, if the departed was instead crucial to keeping the other partners together and leaves out of frustration with the lack of cohesion, the rest of the collaborative structure is likely to disintegrate. A second key factor is the performance of key functions supporting the collaboration. Partner exodus may in turn catalyze additional crises for the remaining partners in the collaborative structure if the departed partner was uniquely capable of performing a key function. Of course, if there is redundancy and the departed partner only caused problems, the change might be welcome. The third vital factor is the capacity to continue with activities in order to achieve objectives. As with key functions, the departure of a partner uniquely qualified to lead an activity of central importance for achieving objectives is little short of disastrous. By contrast, breaking up with a partner in a supporting role may be less consequential, though it may nonetheless reduce the overall capacity of the collaborative structure.

The contribution of bifurcation theory and the great worry in managing collaborative relationships in moments of crisis concerns the low predictability of partner behavior at critical thresholds. Whether they arise due to internal dynamics or external shocks, crises constitute a form of critical threshold for the viability of collaborative enterprises. During crises, for various reasons, partners may change their behavior, whether due to the stresses involved or due to conflicts. As a result of crises, if a critical threshold is crossed, a bifurcation may occur, resulting in the exit of one or more partners and possibly the dissolution of the collaborative structure.

Due to the complications of boundaries noted in previous chapters, managing crisis moments at bifurcation points is more difficult and more consequential when spanning divides. Boundaries multiply the sources of external shocks and thus the potential number of crises. Boundaries strengthen subjectivities and thereby hamper effective communication. Boundaries increase complementarities and consequently the pain if somebody leaves. When crossing boundaries, managing hostilities in crisis moments becomes nerve-wracking because of the dynamics and consequences of bifurcations.

"BREAKING UP IS HARD TO DO" . . . BUT CAN EASILY HAPPEN

With that theoretical context, I turn to a suite of vignettes involving crises and breakups, drawing on direct experiences in collaborative projects that crossed boundaries in the southwestern Amazon. These vignettes serve as illustrations of what can happen, and provide the basis for a discussion of collaborative crisis management. While the details of a given case may be context-specific in some respects, the larger issues at play, concerning external shocks and internal conflicts, as well as the struggles of partners to deal with critical thresholds and breakups, are of general interest.

For Vignette 9.1, I return to Vignettes 6.2 and 6.12. In Vignette 6.12, the G-MAP consortium experienced internal conflict over the question of student fellowships. The division between the NGO and university partners helps to explain the tensions, as the NGOs were willing to go along with the funder's request to cut the fellowships, while the universities were not. Fundamental to the unfolding dynamic was the political question of who was in charge of the fellowships. For the NGOs, this was a responsibility to be shared across the consortium. For the universities, the fellowships were theirs to manage as they saw fit. The back and forth between the two sides increased the tensions as each side hardened its position. Hence, there was both increasing divergence between the two sides as well as decreasing responsiveness, both of which are indicators of an oncoming bifurcation.

With the adversarial positions thus adopted, the universities organized their universities-only meeting to issue their demands. When they were not met, the critical threshold was reached. In the event, one of the universities departed; this represents one bifurcation point. Several processes then followed in rapid succession. One concerned what the other two universities would do. In this case, there was not a second bifurcation, as they softened their positions and chose to stay on, thereby accepting the cuts in the fellowships. One bifurcation involving a departing partner thus did not catalyze another, but that did not necessarily mean that other consequences would not follow. Yet another process was for the remaining partners of the consortium to evaluate the implications. The consortium therefore analyzed the technical and financial responsibilities in light of the reduced overall capacity of the consortium. One of the NGOs took on a training program, and another NGO incremented its cost share obligation. While these were significant adjustments, they did not catalyze further crises.

Vignette 9.1 illustrates a case of an external shock (a request for a budget cut) which catalyzed an internal dynamic (tensions between partners). Partners exhibited several behavioral shifts, first toward hostility, which constituted the crisis and led to the critical threshold (whether to accept the cuts in fellowships). A bifurcation then resulted (the departure of a university partner), along with additional behavioral changes (the other universities backed off from their demands and opted to stay on). After the bifurcation, the consortium re-stabilized by making adjustments in the allocation of technical and financial responsibilities. Nonetheless, as a result of the bifurcation, the consortium and the departed university went their separate ways.

Vignette 9.2 provides another case of a breakup in the form of the exit of a partner, under rather different circumstances. This case stems from a portion of the story recounted in Vignette 7.9, which involved a series of changes in partners. This involved a project that included some new partners. One was an NGO that was to manage funds for a university partner. When it came time to prepare for an external financial audit, management sat down with the NGO to review the accounts. The documentation was not in order, so it could not be reviewed. Management gave the NGO a time limit to get the documents in order. Over several months, it gradually came out that the NGO was not paying attention to the documents, and would not be ready in time for the audit. As the lead organization with the contractual obligations, UF would have to make a determination as to the next steps. Management recommended termination of the partner due to the inattention to the accounts. UF concurred and initiated a termination protocol with a letter to the NGO outlining the procedures and deadlines.

This catalyzed a crisis, as the NGO changed its behavior from inattention to hostility. It responded by hiring a lawyer and making accusations of management. It locked consortium personnel out of its office. It issued requests for considerable additional funding, beyond what was needed for closeout, according to management and UF. The larger question that arose concerned who was in charge of the process. As far as UF was concerned, that was UF; but as far as the NGO was concerned, it was the NGO. The NGO then sought to circumvent management by contacting UF administration directly. I was then told by UF administration to tell the NGO to communicate with UF via management. Hence, a collaborative relationship became a contentious relationship, with a pronounced switch in behavior by the NGO.

As stated deadlines drew near, another behavioral shift occurred. The accusations ceased, offices were unlocked, financial reports arrived, and the final closeout was completed. The NGO left the consortium, and as in the previous vignette, other partners were left to manage the technical and financial responsibilities still in play.

In Vignette 9.2, an internal dynamic (inattention to financial documents) led to tensions (over preparation for an audit) and a critical threshold at a bifurcation point (the decision to terminate). The crisis in this case was prolonged by the change in behavior of the partner (hiring lawyers, locking offices), presumably hoping to avoid the termination. Given the basis for the termination, avoidance was not contractually viable, and the bifurcation occurred under more or less managed circumstances as circumscribed by the termination letter. As in the previous vignette, the exit of a partner reduced the overall capacity of the consortium. As noted in Vignette 7.9, that departure prompted an ongoing search for a replacement partner due to additional travails in securing sufficient management capacity. In that sense, the bifurcation led to other problems.

Vignette 9.3 is an especially painful bifurcation because it involves the removal of two long-time collaborators. This contrasts with the two previous cases, where departures either occurred at the outset of a project (Vignette 9.1) or involved a new partner (Vignette 9.2). While the collaborators were of long standing, they also presented known limitations. External circumstances delayed their removal due to institutional instabilities, which hindered discussions to secure the support of institutional representatives. Meanwhile, a series of mistakes and missed opportunities accumulated over time. There was a lack of leadership in terms of strategic vision. This led to a tendency to manage activities in an ad hoc rather than programmatic fashion. The upshot was a series of failures to respond to opportunities, including those presented by crises, or to use installed resources more effectively, such as new labs for research programs. Lessons were not being learned. At the same time, a series of problems due to inattention, such as the lack of insurance on equipment that was then stolen, caused growing logistical headaches. The collaborators had a strong orientation to their institution rather than the consortium or the region. Despite the limitations that resulted for the consortium, the collaborators felt their membership was inviolable.

Tensions between management and the collaborators in question gradually rose. These tensions became entangled with the fight reported in Vignette 8.6, as the collaborators in question were the same individuals who contacted me with the intent to change management. Not long after that episode, the decision was finally made, with my participation, to remove the collaborators in question. Management in country called a meeting to inform the collaborators of their removal, with the approval of representatives of their home institution. Subjectivities were exceptionally strong, as the collaborators interpreted the timing of the move as revenge for their having had the temerity to question management. It became clear in angrily-worded e-mails that I had disappointed them for not taking their side to continue our collaboration. Perhaps the reasons they were given in the meeting with in-country management were not clearly explained; they were definitely not accepted.

In this case, a long-running internal dynamic (tensions with two collaborators) and complicated institutional circumstances (instability) delayed a decision to remove collaborators, which led to disagreements as to the adequacy of their performance (the collaborators felt it was satisfactory, management did not). When the critical moment finally came (relative institutional stability) and brought the critical threshold (the decision to remove), the bifurcation proved especially unpleasant. Here two rapid changes in behavior transpired, and interpretations revealed very strong subjectivities. First, management finally took the decision to remove the collaborators. There had been noises about this before, but no action had followed, so the collaborators presumed this time would be no different. To their surprise, it was. Second, the collaborators responded angrily, and the bridge burning began in earnest. Vignette 9.3 is a cautionary tale of the importance of communicating critiques of partner performance, lest it be presumed adequate when it is not.

One last story. Vignette 9.4 picks up where Vignette 7.7 left off. A GA partner coming out of regional elections incurred widespread changes in leadership and staff. We had collaborated with the GA for many years, and worked through numerous previous changes in leadership and staff. Each time, management met with the new leadership and explained the contractual requirements of the funder and UF. Those requirements had always been respected, on the understanding that the relationship with UF was strategically important, and having the financial resources added to GA capacity for activities in support of local communities. Consultations with the new government suggested that this time would be no different. We were proven wrong.

The new government began issuing public statements criticizing international collaboration, and the GA leadership followed suit. This came shortly after passing a new funding increment to the GA. The GA responded by making declarations that it could spend the money as it saw fit, and replace project staff with personnel of its choice. As both would constitute breaches of contract, management requested meetings and we wrote e-mails and letters trying to explain the contractual requirements to the new government and GA leadership. They reiterated their declarations and moved to remove project personnel, who were contractually responsible for the project funds and performance of activities. Another round of requests for meetings, e-mails, and letters ensued. Still no movement. These were telltale signs of a critical threshold; a bifurcation point loomed close.

The removal of project staff constituted breach of contract, and UF took the decision to terminate. I was in the southwestern Amazon at the time, and so requested a meeting with GA leadership to personally deliver the termination letter. Everyone in the room was visibly angry, and the subjectivities were extraordinarily strong in the recriminations that followed. I began by reviewing what makes a good collaborator, in terms of communication and the rest, before recounting the GA's problematic declarations, lack of responsiveness, and decisions that constituted breach of contract, whereupon I presented the termination letter, including translation. The GA leadership thereupon reasserted its inalienable right to make decisions about its funds and personnel. They then declared that the previous management, which included members of our project team, had violated the GA's operating protocols. Papers with questionable cell phone recharges were brought out to justify removal of the project team. Project management and I retorted by declaring that such misdeeds paled when compared to the actions in breach of contract. The GA leadership spoke dismissively of the contract, noting that they hadn't actually reviewed it yet. They reasserted their superior management skills and made further allegations. I tried switching gears, arguing that unless the GA recognized the contract and followed the terms of termination, there could be no way to collaborate in the future, and thus no further resources. Whatever the reasoning, the GA leadership toed the new government's party line and said that would be just fine. Vignette 9.4 illustrates an externally induced shock (an election) that led to a critical threshold (a complete change in behavior of a partner) made more difficult to manage by the presence of multiple boundaries (organizational types and countries), which resulted in hostilities and a bifurcation (the GA's exit).

We left the meeting stunned, but already working out the many implications. The GA's exit catalyzed other crises. Because the GA complemented other partners as a distinct organizational type, its departure in many ways crippled the entire consortium, and required a complete overhaul of the collaborative structure in order to maintain collaborators expelled from the GA in order to rescue key functions and sustain activities. The overhaul led to more critical thresholds as we made decisions to shift functions among partners and bring in new partners. The funder wanted more information, UF administrators asked hard questions, and surviving partners became nervous. Word came down that the funder might just shut down the project. It felt like keeping a patient on life support, not knowing if they would recover. This vignette shows how rupture in one collaborative relationship can affect others when crossing multiple boundaries.

THE STRATEGIC IMPORTANCE OF COLLABORATING ACROSS BOUNDARIES

Whether one refers to the termination of collaborative relationships using the language of bifurcations, it is important to recognize that separations constitute an important eventuality. Collaborative relationships have a life cycle, with positive energy at the outset in light of what might be achieved, followed by the onset of exigencies of maintaining the relationships. It is vital to acknowledge that in any collaborative enterprise, but especially when crossing boundaries, complications will arise, and they bring the possibility of breakdowns and thus breakups. For all the effort required, collaborative relationships may be long-lived or short term. What is more, in both cases, they can end suddenly, unexpectedly, and disagreeably.

After the exigencies catalogued in the foregoing chapters, the new batch of warnings just recounted might serve as the final straw to drive prospective collaborators back to the comfort and protection of the familiar. Here it is worth briefly revisiting the arguments for collaboration with which this book opened. Chapter 1 posed the "unmet need" argument. Acting in a collaborative mode on conservation and development challenges is badly needed and must be scaled up. The costs of acting in terms of the drudgery of collaborating are far outweighed by the more frightening prospect of the ecological and economic costs of failing to act. We should choose our nightmares carefully. Chapter 1 also presented the "constant improvement" argument, noting that a variety of models for conservation and development have been implemented and found wanting. Unitary models that center on one type of social actor, whether the state or communities or others, have noteworthy strengths but come with limitations. Those limitations have increasingly motivated approaches that are collaborative, particularly across various types of boundaries.

There is one other argument I now wish to add, in order to further undermine the temptation to avoid collaboration. In light of the previous chapters, and especially the agonies of breakups discussed in this chapter, I want to add the "strategic practices" argument. Collaboration across boundaries is itself a strategic practice with long-term benefits. While the daily slog of collaboration across boundaries might obscure such benefits, it is crucial that they be appreciated. Collaborative relationships generally form on the basis of tactical discussions focusing on shared objectives. Participants in discussions during collaborative courtships can rarely foresee the larger significance of the relationship being contemplated. Partly this is due to the complexities of relationships across boundaries, which serve to obscure strategic vision, and partly this stems from the intervening politics, surprises, and other distractions that can change the game along the way. But beyond the strategic importance of a specific collaborative relationship is the strategic importance of collaborating across boundaries by any partner in the first place.

This further unpacks the promise of collaborative advantage. In addition to the notion that complementarities in capacities and experiences can permit greater achievement together than apart, there is the process of social learning. Collaborative relationships have a larger strategic importance for collaborators because partners stand to learn enormously from each other. For that reason alone, they need each other. Collaborative relationships, especially across boundaries, impose a rigorous learning process that forces all partners to "break the routine" and rethink their assumptions about their goals, capacities, and potential contributions. There is no choice: regardless of the shared interests, collaboration, particularly across boundaries, imposes a rigorous schedule for learning. Collaboration across boundaries accelerates learning, which is vital in contexts where change is rapid. Networks are fine for casually passing around information as via an e-mail discussion list, but the flow runs faster if you are all in the same boat as in a project with deadlines.

Of course, arguing that partners need each other and should therefore collaborate is not enough. Some collaborators fail to see the strategic importance of partnering, or may not share the same long-term goals. This is a recipe for conflicts and bifurcations, and begs questions about how to avoid breakups. One obvious answer is to just say no, at least at a given juncture. Not all partners offer complementarities, and not all have the same depth of strategic vision, particularly if they face high levels of uncertainty. That, however, is not to say that the courtship ends. On the contrary, the courtship should continue. New crises and thus opportunities are likely to arise, which may offer occasion to revisit the question of collaboration as it squares with evolving interests and capacities of the courting parties. Saying no forever is a mistake; but so is saying yes when the puzzle pieces do not fit together. Another mistake is to fail to maintain ties with a possible partner in a fluid context.

Further, partnerships that end in breakups without achieving shared goals also present opportunities for learning. There is of course the obvious set of lessons stemming from the basis for disagreements, the tactical exchanges that ensue, and the manner of the breakup itself. All of this is cause for reflection on what one might have done differently. The tactical considerations of a specific situation should not, however, be confused with the larger strategy of pursuing collaboration. The failure of one relationship need not poison prospects to form others. When crossing boundaries, the additional vagaries become sources of learning and open avenues for improvements in practice in the future. The ultimate strategic practice in collaborating across boundaries is to learn from mistakes. By extension, one stands to learn most from unhappy breakups.

A key implication here is to learn how to get along with ex-partners. Saying no to prospective partners for now provides the basis to keep doors open for future collaborations. Similarly, breakups stemming from a particular situation can serve as a springboard for reflection, learning, and change among ex-partners. It is unwise to completely burn bridges. All of the cases of conflicts and even the breakups reported here do not necessarily prohibit future collaboration, provided there is an acknowledgment of the disagreements, mistakes in decision making, and bad behavior. For the right reasons, bridges can be repaired. I hold open the possibility of future relations with once-upon-a-time partners where we failed to get a project or where we got the project but then had it stopped for external reasons. I also keep doors open with partners that left collaborative projects due to conflicts. In the last case, the key question is whether they have reflected and learned.

These remarks might smack loudly of naïve optimism about the power of collaboration. Here it is worth remembering that even in the context of competitive enterprise, collaboration is highly valued as a means of impelling innovation (Khanna, et al. 1998; Teece 1992). Rather than idealism, the philosophy underlying collaboration, including collaboration across boundaries, is pragmatism in the search for the means to achieve goals greater than one can attain on one's own. Despite the toil and trouble, we pursue collaboration across boundaries because the need is there, the alternatives have limitations, and the experience teaches and strengthens partners. We pursue collaboration across boundaries because we recognize we are ignorant of vital aspects of shared problems, and acknowledging our ignorance opens us to learn from partners. We pursue collaboration across boundaries out of the admission that we must step forward in hope, and not back in fear, if shared problems are ever to be resolved.

Bibliography

Adger, W. Neil. 2000. "Social and Ecological Resilience: Are They Related?" *Progress in Human Geography* 24: 347–364.

Agar, Michael H. 1996. *The Professional Stranger*. San Diego: Academic Press.

Agarwal, Bina. 2001. "Participatory Exclusions, Community Forestry, and Gender: An Analysis for South Asia and a Conceptual Framework." *World Development* 29(10): 1623–1648.

Agrawal, Arun and Clark Gibson. 1999. "Enchantment and Disenchantment: The Role of Community in Natural Resource Conservation." *World Development* 27(4): 629–649.

Ainlay, Stephen, Gaylene Becker, and Lerita Coleman, eds. 1986. *The Dilemma of Difference: A Multidisciplinary View of Stigma*. New York: Plenum.

Albert, Mathias, David Jacobson, and Yosef Lapid, eds. 2001. *Identities, Borders, Orders: Rethinking International Relations Theory*. Minneapolis: University of Minnesota Press.

Aldaba, Fernando. 2002. "Philippine NGOs and Multistakeholder Partnerships: Three Case Studies." *Voluntas* 13: 179–192.

Allen, Katrina. 2006. "Community-based Disaster Preparedness and Climatic Adaptation: Local Capacity Building in the Philippines." *Disasters* 30(1): 81–101.

Allen, Peter M., Liz Varga, and Mark Strathern. 2010. "The Evolutionary Complexity of Social and Economic Systems: The Inevitability of Uncertainty and Surprise." *Risk Management* 12: 9–30.

Aragão, Luis E.O.C. 2012. "The Rainforest's Water Pump." *Nature* 489: 217–218.

Armitage, Derek, Ryan Plummer, Fikret Berkes, Robert Arthur, Anthony Charles, Iain Davidson-Hunt, Alan Diduck, Nancy Doubleday, Derek Johnson, Melissa Marschke, Patrick McConney, Evelyn Pinkerton, and Eva Wollenberg. 2009. "Adaptive Co-management for Social-Ecological Complexity." *Frontiers in Ecology and Environment* 7(2): 95–102.

Arraut, Josefina Moraes, Carlos Nobre, Henrique Barbosa, Guillermo Obregon, and Jose Marengo. 2012. "Aerial Rivers and Lakes: Looking at Large-Scale Moisture Transport and Its Relation to Amazonia and to Subtropical Rainfall in South America." *Journal of Climate* 25: 543–556.

Ashton, Michael C. and Kibeom Lee. 2007. "Empirical, Theoretical and Practical Advantages of the HEXACO Model of Personality Structure." *Personality and Social Psychology Review* 11(2): 150–166.

Assies, William. 2002. "From Rubber Estate to Simple Commodity Production: Agrarian Struggles in the Northern Bolivian Amazon." *Journal of Peasant Studies* 32(3–4): 83–130.

Austin, James E. 2000. *The Collaboration Challenge: How Nonprofits and Businesses Succeed through Strategic Alliances*. San Francisco: Jossey-Bass.

Bakx, Keith. 1988. "From Proletarian to Peasant: Rural Transformation in the State of Acre, 1870–1986." *The Journal of Development Studies* 24(2): 141–160.

Baldersheim, H. and L.E. Rose, eds. 2010. *Territorial Choice: The Politics of Boundaries and Borders*. Basingstoke: Palgrave Macmillan.

Balint, Peter, Ronald Stewart, Anand Desai, and Lawrence Walters. 2011. *Wicked Environmental Problems: Managing Uncertainty and Conflict*. Washington, DC: Island Press.

Bammer, Gabriele and Michael Smithson. 2008. *Uncertainty and Risk: Multidisciplinary Perspectives*. London: Earthscan.

Baraloto, Christopher, Paula Alverga, Grenville Barnes, Sufer Baez Quispe, Grenville Barnes, Nino Bejar Chura, Isaias Brasil da Silva, Wendeson Castro, Iracema de Souza Moll, Jim del Alcazar Chilo, Hugo Dueñas Linares, Jorge Garate Quispe, Dean Kenji, Herison Medeiros, Skya Murphy, C. Rockwell, A. Shenkin, Alexander Shenkin, Marcos Silveira, Jane Southworth, Guido Vasquez, and Stephen Perz. 2014. "Tradeoffs among Forest Value Components in Community Forests of Southwestern Amazonia." *Ecology & Society* 19(4): 56–66.

Barkley, Elizabeth F., K. Patricia Cross, and Claire Howell Major. 2005. *Collaborative Learning Techniques: A Handbook for College Faculty*. New York: Jossey-Bass.

Barndt, William T. 2012. "Destroying the Opposition's Livelihood: Pathways to Violence in Bolivia since 2000." *Journal of Politics in Latin America* 4(3): 3–37.

Barth, Frederik, ed. 1998. *Ethnic Groups and Boundaries: The Social Organization of Cultural Difference*. Long Grove, IL: Waveland.

Barzelay, Michael with Babak J. Armajani. 1992. *Breaking through Bureaucracy: A New Vision for Managing in Government*. Berkeley: University of California Press.

Batterbury, Simon P.J. and Jude L. Fernando. 2006. "Rescaling Governance and the Impacts of Political and Environmental Decentralization: An Introduction." *World Development* 34(11): 1851–1863.

Bebbington, Anthony and Graeme Thiele, eds. 1993 *Non-governmental Organizations and the State in Latin America*. New York: Routledge.

Becher, Tony and Paul R. Trowler. 1989. *Academic Tribes and Territories: Intellectual Enquiry and the Cultures of Disciplines*. Bristol, PA: Society for Research into Higher Education / Open University Press.

Beekman, Christopher S. and William W. Baden. 2005. *Nonlinear Models for Archaeology and Anthropology*. Aldershot, UK: Ashgate.

Belmont Forum. 2011. "The Belmont Challenge: A Global Environmental Research Mission for Sustainability." Retrieved January 1, 2012 (www.igfagcr.org/sites/default/files/documents/belmont-challenge-white-paper.pdf).

Bercovitch, Jacob and Jeffrey Rubin, eds. 1992. *Mediation in International Relations: Multiple Approaches to Conflict Management*. New York: St. Martin's Press.

Berger, Peter and Thomas Luckmann. 1966. *The Social Construction of Reality: A Treatise on the Sociology of Knowledge*. Garden City, NY: Anchor Books.

Berkes, Fikret, ed. (1989) *Common Property Resources: Ecology and Community-based Sustainable Development*. London: Belhaven Press.

Biermann, Frank and Philipp Pattberg. 2008. "Global Environmental Governance: Taking Stock, Moving Forward." *Annual Review of Environment and Resources* 33: 277–294.

Birkinshaw, Julian M. and Allen J. Morrison. 1995. "Configurations of Strategy and Structure in Subsidiaries of Multinational Corporations." *Journal of International Business Studies* 26(4): 729–753.

Blumer, Herbert. 1969. *Symbolic Interactionism: Perspective and Method*. Berkeley: University of California Press.

Bonnell, Victoria E. 1980. "The Uses of Theory, Concepts and Comparison in Historical Sociology." *Comparative Studies in Society and History* 22: 156–173.

Borgatti, Stephen P., and Martin G. Everett. 1992. "Notions of Position in Social Network Analysis." *Sociological Methodology* 22: 1–35.

Brando, Paulo, Daniel Nepstad, Eric Davidson, Susan Trumbore, David Ray, and Plínio Camargo. 2008. "Drought Effects on Litterfall, Wood Production, and Belowground Carbon Cycling in an Amazon Forest: Results of a Throughfall Reduction Experiment." *Philosophical Transactions of the Royal Society* B 363: 1839–1848.

Bretell, Caroline B., ed. 2007. *Constructing Borders, Crossing Boundaries: Race, Ethnicity and Immigration*. Lanham, MD: Lexington Books.

Brick, Philip, Donald Snow, and Sarah Van de Wetering, eds. 2001. *Across the Great Divide: Explorations in Collaborative Conservation and the American West.* Washington, DC: Island Press.

Brown, I. Foster, Silvia Brilhante, Elsa Mendoza, and I. Ribeiro de Oliveira. 2002. "Estrada de Rio Branco, Acre, Brasil aos Portos do Pacífico: Como Maximizar os Benefícios e Minimizar os Prejuízos para o Desenvolvimento Sustentavel da Amazônia Sul-ocidental." Pp. 281–296 in *La Integración Regional Entre Bolivia, Brasil y Peru,* edited by A. Wagner Tizón and R.S. Gadea Duarte. Lima, Peru: CEPEI.

Brown, I. Foster, George Pereira Santos, F. Ferreira Pires, and C. Batista da Costa. 2011. "Brazil: Drought and Fire Response in the Amazon." World Resources Report, Washington, DC. Retrieved January 1, 2012 (www.worldresourcesreport.org).

Brown, Valerie A., John A. Harris, and Jacqueline Y. Russell, eds. 2008. *Tackling Wicked Problems through the Transdisciplinary Imagination.* London: Earthscan.

Brunet-Jailly, Emmanuel. 2005. "Theorizing Borders: An Interdisciplinary Perspective." *Geopolitics* 10(4): 633–649.

———. 2011. "Special Section: Borders, Borderlands, and Theory: An Introduction." *Geopolitics* 16(1): 1–6.

Buarque, Christovam. 2002. "A Internacionalização do Mundo." *Revista Diálogo Educacional* 3(5): 1–2.

Buck, Louise E., Charles C. Geisler, John Schelhas, and Eva Wollenberg, eds. 2001. *Biological Diversity: Balancing Interests through Adaptive Collaborative Management.* Boca Raton, FL: CRC Press.

Buckles, Daniel, ed. 1999. *Cultivating Peace: Conflict and Collaboration in Natural Resource Management.* Ottawa: International Development Research Centre.

Burr, Vivien. 2003. *Social Constructionism.* Psychology Press.

Burton, John Wear and Frank Dukes. 1990. *Conflict: Practices in Management, Settlement and Resolution.* New York: St. Martin's Press.

Burton, Richard M., Børge Obel, and Gerardine DeSanctis. 2011. *Organizational Design: A Step-by-Step Approach, 2nd ed.* Cambridge: Cambridge University Press.

Calderón, César, and Luis Servén. 2004. "The Effects of Infrastructure Development on Growth and Income Distribution." Central Bank of Chile, Working Paper No. 270.

Carlsson, Lars and Fikret Berkes. 2005. "Co-management: Concepts and Methodological Implications." *Journal of Environmental Management* 75: 65–76.

Carpenter, S.L. and W.J.D. Kennedy. 1988. *Managing Public Disputes: A Practical Guide to Handling Conflict and Reaching Agreements.* San Francisco: Jossey-Bass.

CEPEI (Centro Peruano de Estudios Internacionales). 2002. *La Integración Regional entre Bolivia, Brasil y Peru.* Lima, Peru: CEPEI.

Cernea, Michael, ed. 1991. *Putting People First: Sociological Variables in Rural Development, 2nd ed.* New York: Oxford University Press and World Bank.

Chambers, Robert. 1994. "The Origins and Practice of Participatory Rural Appraisal." *World Development* 22(7): 953–969.

———. 1997. *Whose Reality Counts? Putting the First Last.* London: Intermediate Technology.

Chapin, Mac. 2004. "A Challenge to Conservationists." *World Watch* 17: 17–31.

Chasek, Pamela S., ed. 2000. *The Global Environment in the Twenty-first Century: Prospects for International Cooperation.* New York: CAB Direct.

Chavez, A. 2009. *Public Policy and Spatial Variation in Land Use and Cover in the Southeastern Peruvian Amazon.* PhD Dissertation, University of Florida.

CHS Alternativo. 2012. *La Trata de Personas en la Región Madre de Dios.* Lima: CHS Alternativo.

El Clarín. 2008. "Dudas y Rumores en Torno a la Masacre de Pando." September 15. Retrieved January 1, 2015 (edant.clarin.com/diario/2008/09/15/elmundo/i-01760625.htm).

Cleaver, Frances. 1999. "Paradoxes of Participation: Questioning Participatory Approaches to Development." *Journal of International Development* 11: 597–612.

Coffin, Alisa W. 2007. "From Roadkill to Road Ecology: A Review of the Ecological Effects of Roads." *Journal of Transport Geography* 15: 396–406.

Coleman, Gilroy. 1987. "Logical Framework Approach to the Monitoring and Evaluation of Agricultural and Rural Development Projects." *Project Appraisal* 2(4): 251–259.

Colfer, Carol, ed. 2010. *The Complex Forest: Communities, Uncertainty, and Adaptive Collaborative Management.* Washington, DC: Resources for the Future.

El Comércio. 2008a. "Queman la Sede del Gobierno Regional de Madre de Dios." July 9. Retrieved July 9 (www.elcomercio.com.pe/ediciononline/html/2008-07-09/queman-sede-gobierno-regional-madre-dios.html).

———. 2008b. "El Gobierno Regional de Madre de Dios Denunciara Penalmente a Vandalos." Retrieved July 11 (www.elcomercio.com.pe/ediciononline/html/2008-07-10/el-gobierno-regional-madre-dios-denunciara-penalmente-vandalos.html).

Constantino, C.A. and C.S. Merchant. 1995. *Designing Conflict Management Systems: A Guide to Creating Productive and Healthy Organizations.* San Francisco: Jossey-Bass.

Cooke, B. and U. Kothari, eds. 2001. *Participation: The New Tyranny?* London: Zed Books.

Cossio Solano, Rosa. 2009. *Capacity for Timber Management among Private Small-Medium Forest Enterprises, Madre de Dios, Peru.* PhD Dissertation, University of Florida, Gainesville.

Courpasson, David, Damon Golsorkhi, and Jeffrey J. Sallaz, eds. 2012. *Rethinking Power in Organizations, Institutions and Markets.* Bingley, UK: Emerald.

Cropper, Steve, Mark Ebers, Chris Huxham, and Peter Smith Ring, eds. 2008. *The Oxford Handbook of Inter-Organizational Relations.* Oxford: Oxford University Press.

Cumming Graeme, Grenville Barnes, Stephen Perz, Marianne Schmink, Katie Sieving, Jane Southworth, Michael Binford, Robert D. Holt, Claudia Stickler, and Tracy Van Holt. 2005. "An Exploratory Framework for the Empirical Measurement of Resilience." *Ecosystems* 8: 975–987.

Cumming, Graeme, David Cumming, and Charles Redman. 2006. "Scale Mismatches in Social-ecological Systems: Causes, Consequences, and Solutions." *Ecology and Society* 11(1): article 14.

Cutter, Susan, Lindsey Barnes, Melissa Berry, Christopher Burton, Elijah Evans, Eric Tate, and Jennifer Webb. 2008. "A Place-based Model for Understanding Community Resilience to Natural Disasters." *Global Environmental Change* 18: 598–606.

Daft, Richard L. 2005. *Management, 7th ed.* New York: Thomson Southwestern.

Dale, Reidar. 2003. "The Logical Framework: An Easy Escape, a Straightjacket, or a Useful Planning Tool?" *Development in Practice* 13(1): 57–70.

Davidson, Eric, Alessandro C. de Araújo, Paulo Artaxo, Jennifer Balch, I. Foster Brown, Mercedes Bustamante, Michael Coe, Ruth DeFries, Michael Keller, Marcos Longo, J. William Munger, Wilfrid Schroeder, Britaldo Soares-Filho, Carlos Souza, and Steven Wofsy. 2012. "The Amazon Basin in Transition." *Nature* 481: 321–238.

DeCarvalho, Roy J. 1990. "A History of the 'Third Force' in Psychology." *Humanistic Psychology* 30(4): 22–44.

Dickinson, Robert E., ed. 1987. *The Geophysiology of Amazonia: Vegetation and Climate Interactions.* New York: Wiley.

Digman, John M. 1990. "Personality Structure: Emergence of the Five-factor Model." *Annual Review of Psychology* 41: 417–440.

Dourojeanni, Marc. 2005. "Futuro de las Relaciones entre las Organizaciones No Gubernamentales Ambientales Transnacionales y Nacionales." Paper presented at the seminar, "Estratégias para o Século XXI para Reduzir a Pobreza e Conservar a Natureza na América Sul-Occidental: Rumo a Modelos de Gestão Participativa," Cobija, Bolívia, September 30–October 2.

Dourojeanni, Marc, Alberto Barandiaran, Diego Dourojeanni. 2010. *Amazonia Peruana en 2021.* Lima, Peru: SPDA.

Easterby-Smith, Mark, Mary Crossan, and Davide Nicolini. 2000. "Organizational Learning: Debates Past, Present, and Future." *Journal of Management Studies* 37: 783–796.

Edmunds, David and Eva Wollenberg. 2002. "Disadvantaged Groups in Multistakeholder Negotiations." CIFOR Program Report. Retrieved June 30, 2007 (www.cifor.org/publications/pdf_files/Strategic_Negotiation_report.pdf).

Ehrenreich, R.M., C.L. Crumley, and J. E. Levy, eds. 1995. *Heterarchy and the Analysis of Complex Societies*. Arlington: American Anthropological Association.

Ehringhaus, Christiane. 2005. *Post-Victory Dilemmas: Land Use, Development, and Social Movements in an Amazonian Extractive Reserve*. PhD Dissertation, Yale University, New Haven.

Elden, Stuart. 2005. "Missing the Point: Globalization, Deterritorialization and the Space of the World." *Transactions of the Institute of British Geographers* 30: 8–19.

Escobar, Arturo. 1995. *Encountering Development: The Making and Unmaking of the Third World*. Princeton: Princeton University Press.

Ewing, Katherine Pratt. 1998. "Crossing Borders and Transgressing Boundaries: Metaphors for Negotiating Multiple Identities." *Ethos* 26(2): 262–267.

Faminow, Merle. 1998. *Cattle, Deforestation and Development in the Amazon: An Economic, Agronomic and Environmental Perspective*. New York: Cab International.

Fatti, Christina Elizabeth and Zarina Patel. 2013. "Perceptions and Responses to Urban Flood Risk: Implications for Climate Governance in the South." *Applied Geography* 36: 13–22.

Fearnside, Phillip M. 2012. "Brazil's Amazon Forest in Mitigating Global Warming: Unresolved Controversies." *Climate Policy* 12: 70–81.

Ferguson, Kathy E. 1993. *The Man Question: Visions of Subjectivity in Feminist Theory*. Berkeley: University of California Press.

Folke, Carl, Thomas Hahn, Per Olsson, and Jon Norberg. 2005. "Adaptive Governance of Social-ecological Systems." *Annual Review of Environment and Resources* 30: 441–473.

Forman, Richard T.T., Daniel Sperling, John Bisonette, Anthony Clevenger, Carol Cutshall, Virginia Dale, Lenore Fahrig, Robert France, Charles Goldman, Kevin Heanue, Julia Jones, Frederick Swanson, Thomas Turrentine, and Thomas Winter. 2003. *Road Ecology: Science and Solutions*. Washington, DC: Island Press.

Freeman, Linton C. 2004. *The Development of Social Network Analysis: A Study in the Sociology of Science*. Vancouver: Empirical Press.

Frodeman, Robert, Julie Thompson Klein, and Carl Mitcham, eds. 2010. *The Oxford Handbook of Interdisciplinarity*. New York: Oxford University Press.

Furnham, Adrian. 1996. "The Big Five versus the Big Four: The Relationship between the Myers-Briggs Type Indicator (MBTI) and the NEO-PI Five Factor Model of Personality." *Personality and Individual Differences* 21(2): 303–307.

Gaillard, J.C. and Jessica Mercer. 2012. "From Knowledge to Action: Bridging Gaps in Disaster Risk Reduction." *Progress in Human Geography* 37(1): 93–114.

Gaillard, Jacques F. 1994. "North-South Research Partnership: Is Collaboration Possible between Unequal Partners?" *Knowledge and Policy* 7: 31–63.

Galat, David L. and Jim Berkly. 2014. "Introduction to Exploring Opportunities for Advancing Collaborative Adaptive Management: Integrating Experience and Practice." *Ecology and Society* 19(2): article 40.

Garcia, E. 2008. "Bolivia Split in Two by Protests; Deaths Mourned." *Reuters UK Edition*. Retrieved September 12 (uk.reuters.com/article/2008/09/12/bolivia-idUKN124237492008 0912).

Garfinkel, Harold. 1967. *Studies in Ethnomethodology*. Englewood Cliffs: Prentice-Hall.

Gatti, L.V., M. Gloor, J.B. Miller, C.E. Doughty, Y. Malhi, L.G. Domingues, L.S. Basso, A. Martinewski, C.S.C. Correia, V.F. Borges, S. Freitas, R. Braz, L.O. Anderson, H. Rocha, J. Graced, O.L. Phillips, and J. Lloyd. 2014. "Drought Sensitivity of Amazonian Carbon Balance Revealed by Atmospheric Measurements." *Nature* 506: 76–80.

A Gazeta. 2012. "Cheia do Rio Acre já Atinge 57 Mil Pessoas em Rio Branco." Retrieved February 23 (agazetadoacre.com/noticias/cheia-do-rio-acre-ja-atinge-57-mil-pessoas-em-rio-branco-2/).

Giddens, Anthony. 1984. *The Constitution of Society: An Outline of the Theory of Structuration*. Cambridge: Polity Press.

Gittell, Jody Hoffer, and Leigh Weiss. 2004. "Coordination Networks within and across Organizations: A Multi-level Framework." *Journal of Management Studies* 41: 127–153.

Gittell, Jody Hoffer, Rob Seider, and Julian Wimbush. 2010. "A Relational Model of How Higher-Performance Work Systems Work." *Organization Science* 21: 490–506.

Globo. 2015. "Maior cheia da história do Rio Acre desabriga milhares; saiba como ajudar." Retrieved March 15 (g1.globo.com/fantastico/noticia/2015/03/maior-cheia-da-historia-do-rio-acre-desabriga-milhares-saiba-como-ajudar.html).

Goffman, Erving. 1959. *The Presentation of Self in Everyday Life.* New York: Anchor Books.

Goleman, Daniel. 1995. *Emotional Intelligence.* New York: Bantam Dell.

Gomes, Carlos Valerio Aguiar. 2009. *Twenty Years after Chico Mendes: Extractive Reserves' Expansion, Cattle Adoption and Evolving Self-Definition among Rubber Tappers in the Brazilian Amazon.* PhD Dissertation, University of Florida, Gainesville.

Gomez-Mejia, Luis, David Balkin, and Robert Candy. 2008. *Management: People, Performance, Change.* Boston: McGraw-Hill.

Goodland, R.J.A., and H.S. Irwin. 1975. *Amazon Jungle: Green Hell to Red Desert?* New York: Elsevier.

Gopalakrishnan, Chennat, and Norio Okada. 2007. "Designing New Institutions for Implementing Integrated Disaster Risk Management: Key Elements and Future Directions." *Disasters* 31(4): 353–372.

Governo do Acre. 2012. *Plan Estadual de Recursos Hídricos.* Rio Branco: Governo do Acre.

Grann, David. 2005. *The Lost City of Z: A Tale of Deadly Obsession in the Amazon.* New York: Vintage Books.

Gray, Barbara. 1989. *Collaborating: Finding Common Ground for Multi-Party Problems.* San Francisco: Jossey-Bass.

Greenleaf, Robert K. 2002. *Servant Leadership: A Journey into the Nature of Legitimate Power and Greatness.* Mahwah, NJ: Paulist Press.

Groves, Leslie and Rachel Hinton, eds. 2004. *Inclusive Aid: Changing Power and Relationships in International Development.* New York: Earthscan.

Guastello, Stephen J. 1995. *Chaos, Catastrophe and Human Affairs: Applications of Nonlinear Dynamics to Work, Organizations and Social Evolution.* New York: Taylor & Francis.

Gulati, Ranjay, and Martin Gargiulo. 1999. "Where Do Inter-Organizational Networks Come From?" *American Journal of Sociology* 104: 1439–1493.

Gunderson, Lance. 2000. "Ecological Resilience—In Theory and Application." *Annual Review of Ecology and Systematics* 31: 425–439.

Gunderson, Lance and C.S. Holling, eds. 2002. *Panarchy: Understanding Transformations in Human and Natural Systems.* Washington: Island Press.

Haggett, P., A.D. Cliff, and A. Frey. 1977. *Locational Analysis in Human Geography.* New York: John Wiley.

Handel, Michael J., ed. 2003. *The Sociology of Organizations: Classic, Contemporary and Critical Readings.* Thousand Oaks, CA: Sage Publications.

Hardy, Cynthia and Nelson Phillips. 1998. "Strategies of Engagement: Lessons from the Critical Examination of Collaboration and Conflict in an Interorganizational Domain." *Organization Science* 9(2): 217–230.

Hall, Anthony. 2008. "Better REDD than Dead." *Philosophical Transactions of the Royal Society B* 363: 1925–1932.

Handelsman, Jo, Sarah Miller, and Christine Pfund. 2007. *Scientific Teaching.* New York: W.H. Freeman.

Hansen, Morton T. 2009. *Collaboration: How Leaders Avoid the Traps, Create Unity, and Reap Big Results.* Boston: Harvard University Press.

Harary, Frank. 1969. *Graph Theory.* Reading: Addison-Wesley.

Heath, Renee Guarriello. 2007. "Rethinking Community Collaboration through a Dialogic Lens." *Management Communication Quarterly* 21: 145–171.

Heckscher, Charles, and Paul S. Adler. 2006. *The Firm as a Collaborative Community: Reconstructing Trust in the Knowledge Economy.* Oxford: Oxford University Press.

Heemskerk, M., K. Wilson, and M. Pavao-Zuckerman. 2003. "Conceptual Models as Tools for Communication across Disciplines." *Conservation Ecology* 7: article 8.

Hemmati, Minu, with Felix Dodds, Jasmin Enayati, and Jan McHarry. 2002. *Multi-stakeholder Processes for Governance and Sustainability: Beyond Deadlock and Conflict.* New York: Earthscan.

Herod, Andrew. 2011. *Scale.* New York: Routledge.

Hickey, Samuel and Giles Mohan, eds. 2004. *Participation: From Tyranny to Transformation? Exploring New Approaches to Participation in Development.* London: Zed Books.

Hindess, Barry. 1986. "Actors and Social Relations." In M.I. Wardell and S.P. Turner, eds., *Sociological Theory in Transition.* Boston: Allen and Unwin.

Hirsch Hadorn, Gertrude, Holher Hoffmann-Riem, Susette Biber-Klemm, Walter Grossenbacher-Mansuy, Domonique Joye, Christian Pohl, Urs Wiesmann, and Elisabeth Zemp, eds. 2008. *Handbook of Transdisciplinary Research.* Bern: Springer.

Hockings, Marc, with Sue Stolton and Nigel Dudley. 2005. *Evaluating Effectiveness: A Framework for Assessing the Management of Protected Areas.* Cardiff: IUCN / Cardiff University.

Holling, C.S. 1973. "Resilience and Stability in Ecological Systems." *Annual Review of Ecology and Systematics* 4: 1–23.

———, ed. 1978. *Adaptive Environmental Assessment and Management.* New York: Wiley.

———. 1996. "Engineering Resilience versus Ecological Resilience." Pp. 31–44 in P.C. Schulze, editor, *Engineering within Ecological Constraints.* Washington, DC: National Academy Press.

Holstein, James A., and Jaber F. Gubrium, eds. 2008. *Handbook of Constructionist Research.* New York: Guilford.

Horton, John. 1966. "Order and Conflict Theories of Social Problems as Competing Ideologies." *American Journal of Sociology* 71(6): 701–713.

Hultman, Nathan and Alexander Bozmoski. 2006. "The Changing Face of Normal Disaster: Risk, Resilience and Natural Security in a Changing Climate." *Journal of International Affairs* 59(2): 25–41.

Huxham, Chris and Siv Vangen. 2005. *Managing to Collaborate: The Theory and Practice of Collaborative Advantage.* New York: Routledge.

IIRSA (Iniciativa para la Integración de la Infraestructura Regional de America del Sur). 2008. Retrieved January 20, 2008 (www.iirsa.org).

Iooss, Gérard and Daniel D. Joseph. 1997. *Elementary Stability and Bifurcation Theory.* Berlin: Springer Verlag.

Jackson, Jane. 2006. "Ethnographic Preparation for Short-Term Study and Residence in the Target Culture." *International Journal of Intercultural Relations* 30: 77–98.

Jasanoff, Sheila. 1997. "NGOs and the Environment: From Knowledge to Action." *Third World Quarterly* 18: 579–594.

Kagan, Jerome. 2002. *Surprise, Uncertainty and Mental Structures.* Cambridge: Harvard University Press.

———. 2009. *The Three Cultures: Natural Sciences, Social Sciences, and the Humanities in the 21st Century.* Cambridge: Cambridge University Press.

Kahan, Dan M., Hank Jenkins-Smith, and Donald Braman. 2011. "Cultural Cognition of Scientific Consensus." *Journal of Risk Research* 14(2): 147–174.

Kahan, Dan M. 2012. "Cultural Cognition as a Conception of the Cultural Theory of Risk." Pp. 725–759 in *Handbook of Risk Theory,* edited by S. Roeser, R. Hillerbrand, P. Sandin, and M. Peterson. New York: Springer.

Kainer, Karen, Marianne Schmink, Arthur Cezar Pinheiro Leite, and Mário Jorge da Silva Fadell. 2003. "Experiments in Forest-Based Development in Western Amazonia." *Society and Natural Resources* 16(10): 869–886.

Kainer, Karen, Marianne Schmink, Hannah Covert, John Richard Stepp, Emilio Bruna, Jonathan Dain, Santiago Espinosa, and Shoana Humphries. 2006. "A Graduate Education Framework for Tropical Conservation and Development." *Conservation Biology* 20: 3–13.

Kapucu, Naim. 2008. "Collaborative Emergency Management: Better Community Organising, Better Public Preparedness and Response." *Disasters* 32: 239–262.

Kearney, M. 1998. "The Local and the Global: The Anthropology of Globalization and Transnationalism." *Annual Review of Anthropology* 24: 547–565.

Keck, Margaret E., and Kathryn Sikkink. 1998. *Activists beyond Borders: Advocacy Networks in International Politics.* Ithaca: Cornell University Press.

Keim, Gerald D. 1978. "Corporate Social Responsibility: An Assessment of the Enlightened Self-interest Model." *Academy of Management Review* 3(1): 32–39.

Keller, Michael, Mercedes Bustamante, John Gash, and P. Silva Dias, eds. 2009. *Amazonia and Global Change.* Washington, DC: American Geophysical Union.

Kennedy, Paul. 2007. "Global Transformations but Local, 'Bubble' Lives: Taking a Reality Check on Some Globalization Concepts." *Globalizations* 4(2): 267–282.

Khanna, Tarun, Ranjay Gulati, and Nitin Nohira. 1998. "The Dynamics of Learning Alliances: Competition, Cooperation, and Relative Scope." *Strategic Management Journal* 19: 193–210.

Killeen, Timothy J. 2007. *A Perfect Storm in the Amazon Wilderness: Development and Conservation in the Context of the Initiative for the Integration of Regional Infrastructure of South America (IIRSA).* Arlington: Conservation International.

Kim, Yeonjoo, Ryan Knox, Marcos Longo, David Medvigy, Lucy Huytra, Elizabeth Pyle, Steven Wofsy, Rafael Bras, and Paul Moorcroft. 2012. "Seasonal Carbon Dynamics and Water Fluxes in an Amazon Rainforest." *Global Change Biology* 18(4): 1322–1334.

Klein, Julie Thompson. 1990. *Interdisciplinarity: History, Theory, and Practice.* Detroit: Wayne State University Press.

———. 1996. *Crossing Boundaries: Knowledge, Disciplinarities, and Interdisciplinarities.* Charlottesville: University of Virginia.

———. 2010. *Creating Interdisciplinary Campus Cultures: A Model for Strength and Sustainability.* San Francisco: John Wiley and Sons.

Knight, Andrew, Richard Cowling, and Bruce Campbell. 2006. "An Operational Model for Implementing Conservation Action." *Conservation Biology* 20(2): 408–419.

Knoben, J. and L.A.G. Oerlemans. 2006. "Proximity and Inter-organizational Collaboration: A Literature Review." *International Journal of Management Reviews* 8(2): 71–89.

Kraus, W.A. 1980. *Collaboration in Organizations: Alternatives to Hierarchy.* New York: Human Sciences Press.

Lank, Elizabeth. 2006. *Collaborative Advantage: How Organizations Win by Working Together.* New York: Palgrave MacMillan.

Lauber, T. Bruce, Richard Stedman, Daniel Decker, and Barbara Knuth. 2011. "Linking Knowledge to Action in Collaborative Conservation." *Conservation Biology* 25(6): 1186–1194.

Laurance, Susan G.W., Ana Andrade, and William F. Laurance. 2010. "Unanticipated Effects of Stand Dynamism on Amazonian Tree Diversity." *Biotropica* 42(4): 429–434.

Laurance, William F., Thomas Lovejoy, Heraldo Vasconcelos, Emilio Bruna, Raphael Didham, Philip Stouffer, Claudio Gascon, Richard Bierregaard, Susan Laurance, and Erica Sampaio. 2002. "Ecosystem Decay of Amazonian Forest Fragments: A 22-Year Investigation." *Conservation Biology* 16: 605–618.

Laurance, William F., Harko Koster, Monique Grooten, Anthony B. Anderson, Pieter A. Zuidema, Steve Zwick, Roderick Zagt, Antony Lyman, Matthew Linkie, and Niels Anten. 2012. "Making Conservation Research More Relevant for Conservation Practitioners." *Biological Conservation* 153: 164–168.

Lee, Kai N. 1993. *Compass and Gyroscope: Integrating Science and Politics for the Environment.* Washington, DC: Island Press.

Lélé, Sharachchandra, and Richard B. Norgaard. 2005. "Practicing Interdisciplinarity." *BioScience* 55: 967–975.

Lemos, Maria Carmen and Arun Agrawal. 2006. "Environmental Governance." *Annual Review of Environment and Resources* 31: 297–325.

Leonel, Mauro. 1992. *Roads, Indians and the Environment in the Amazon: From Central Brazil to the Pacific Ocean.* Copenhagen: International Working Group for Indigenous Affairs.

Levin, Simon. 1999. *Fragile Dominion: Complexity and the Commons.* Reading, MA: Perseus.

Lewis, Simon L., Paulo Brando, Oliver Phillips, Geertje van der Heiden, and Daniel Nepstad. 2011. "The 2010 Amazon Drought." *Science* 331: 554.

Lima, Alexandre. 2013. "Após um Ano da tragédia, Brasiléia Ainda Tenta Se Recuperar da Enchente." *O Alto Acre.* Retrieved February 17, 2013 (www.oaltoacre.com/index.php/apos-um-ano-da-tragedia-brasileia-ainda-tenta-se-recuperar-da-enchente/).

Lindblom, Charles E. 1959. "The Science of 'Muddling Through.'" *Public Administration Review* 19(2): 79–88.

Lloyd, Chrisopher D. 2014. *Exploring Spatial Scale in Geography*. Hoboken: Wiley.

Lobo, Rafaella. 2012. *A Amazônia e as Relações Internacionais: O Mito da Internacionalização e os Desafios da Conservação*. MA Thesis, Catholic University of Goiás, Brazil.

Locke, Edwin A., ed. 2000. *The Blackwell Handbook of Principles of Organizational Behavior*. Oxford: Blackwell.

Long, Norman. 2001. *Development Sociology: Actor Perspectives*. New York: Routledge.

Ludwig, Donald, Marc Mangel, and Brent Haddad. 2001. "Ecology, Conservation, and Public Policy." *Annual Review of Ecology and Systematics* 32: 481–517.

M-P (Madre de Dios-Pando) Consortium. 2011. Fiscal Year 2011 and Life-of-Project Report. Submitted to the US Agency for International Development by the University of Florida, with Herencia, Proyecto Especial Madre de Dios, Universidad Amazónica de Pando, Universidad Nacional Amazónica de Madre de Dios, and Woods Hole Research Center.

MABE (Manejo Ambiental de Bacias e Estradas) Consortium. 2011. Final Performance Report. Cooperative Agreement 512–A-00–08–00003–00. Submitted to the US Agency for International Development by the University of Florida, with the Instituto de Pesquisas Ambientais da Amazônia, SOS Amazônia, and Woods Hole Research Center.

MacMynowski, Dena P. 2007. "Pausing at the Brink of Interdisciplinarity: Power and Knowledge at the Meeting of Social and Biophysical Science." *Ecology and Society* 12(1): article 20.

Mahapa, Sabina M., and Mac Mashiri. 2001. "Social Exclusion and Rural Transport: Gender Aspects of a Road Improvement Project in Tshitwe, Northern Province." *Development Southern Africa* 18: 365–376.

Mahar, Dennis J. 1989. *Government Policies and Deforestation in Brazil's Amazon Region*. Washington, DC: World Bank.

Malhi, Yadvinder, J. Timmons Roberts, and Richard Betts. 2008. *Climate Change and the Fate of the Amazon*. Special Issue, *Philosophical Transactions of the Royal Society B* 363(1498).

Manring, Susan. 2007. "Creating and Managing Inter-Organizational Learning Networks to Achieve Sustainable Ecosystem Management." *Organization and Environment* 20: 325–346.

Mansbridge, Jane J., ed. 1990. *Beyond Self-interest*. Chicago: University of Chicago Press.

Marengo, Jose, Carlos Nobre, Javier Tomasella, Marcos Oyama, Gilvan Sampaio de Oliveira, Rafael de Oliveira, Helio Camargo, Lincoln Alves, and I. Foster Brown. 2008. "The Drought of Amazonia in 2005." *Journal of Climate* 21(3): 495–516.

Marengo, Jose, Laura Borma, Daniel Rodriguez, Patricia Pinho, Wagner Soares, and Lincoln Alves. 2013. "Recent Extremes of Drought and Flooding in Amazonia: Vulnerabilities and Human Adaptation." *American Journal of Climate Change* 2(2): 87–96.

Maslow, Abraham H. 1970. *Motivation and Personality*. New York: Addison-Wesley.

McClain, Michael, Reynaldo Victoria, and Jeffrey Richey, eds. 2001. *The Biogeochemistry of the Amazon Basin*. New York: Oxford University Press.

McCrae, Robert R. and Paul T. Costa. 1989. "Reinterpreting the Myers-Briggs Type Indicator from the Perspective of the Five-Factor Model of Personality." *Journal of Personality* 57(1): 17–40.

McDaniel, Reuben R. and Dean J. Driebe, eds. 2005. *Uncertainty and Surprise in Complex Systems: Questions on Working with the Unexpected*. Berlin: Springer-Verlag.

McMichael, Philip. 2012. *Development and Social Change: A Global Perspective, 5th ed*. Thousand Oaks: Sage.

Mendoza, Elsa, Stephen Perz, Cesar Aguilar, Gabriel Alarcón, I. Foster Brown, Jacob Carballo, Andrea Chavez, Jaime Chavez, Georgina Cullman, Monica de los Rios, Amy Duchelle, Hugo Dueñas, Christiane Ehringhaus, Jeffrey Luzar, Matthew Marsik, Julio Mayna, Armando Muñante, Vera Reis, and Cora van Oosten. 2007. "The Knowledge Exchange Train: A Model for Capacity Building for Participatory Governance." *Development in Practice* 17: 791–799.

Mendoza, Elsa, Stephen Perz, Marianne Schmink, and Daniel Nepstad. 2008. "Participatory Stakeholder Workshops to Mitigate Impacts of Road Paving in the Southwestern Amazon." *Conservation and Society* 5: 382–407.

Mendoza, Elsa, Stephen Perz, Sonaira Souza da Silva, I. Foster Brown, and Paula Soares Pinheiro. 2014. "Revisiting the Knowledge Exchange Train: Scaling Up Dialogue and Partnering for Participatory Regional Planning." *Journal of Environmental Planning and Management* 57(3): 384–402.

Meyer, C.A. 1999. *The Economics and Politics of NGOs in Latin America.* London: Praeger.

Michaelsen, Scott and David E. Johnson, eds. 1997. *Border Theory: The Limits of Cultural Politics.* Minneapolis: University of Minnesota Press.

Miles, R.E., and C.C. Snow. 1986. "Network Organizations: New Concepts for New Forms." *California Management Review* 28: 62–73.

Millard, Candice. 2005. *The River of Doubt: Theodore Roosevelt's Darkest Journey.* New York: Broadway Books.

Monteiro, Marko, and Elizabeth Keating. 2009. "Managing Misunderstandings: The Role of Language in Interdisciplinary Scientific Collaboration." *Science Communication* 31: 6–28.

Moon, Katie and Deborah Blackman. 2014. "A Guide to Understanding Social Science Research for Natural Scientists." *Conservation Biology* 28(5): 1167–1177.

Moore, Barrington. 1966. *Social Origins of Dictatorship and Democracy: Lord and Peasant in the Making of the Modern World.* Boston: Beacon Press.

Moore, Jerry D. 1997. *Visions of Culture: An Introduction to Anthropological Theories and Theorists.* Walnut Creek: Altamira.

Morris, Michael, Katherine Williams, Kwok Leung, Richard Larrick, M. Theresa Mendoza, Deepti Bhatnagar, Jianfeng Li, Mari Kondo, Jin-Lian Luo, and Jun-Chen Hu. 1998. "Conflict Management Style: Accounting for Cross-National Differences." *Journal of International Business* 29(4): 729–747.

Murphy, Priscilla. 1996. "Chaos Theory as a Model for Managing Issues and Crises." *Public Relations Review* 22(2): 95–113.

Myers, Norman, Russel Mittermeier, Cristina Mittermeier, Gustavo A.B. da Fonseca, and Jennifer Kent. 2000. "Biodiversity Hotspots for Conservation Priorities." *Nature* 403: 853–858.

Newig, Jens and Oliver Fritsch. 2009. "Environmental Governance: Participatory, Multi-level—and Effective?" *Environmental Policy and Governance* 19(3): 197–214.

Newman, David. 2003. "On Borders and Power: A Theoretical Framework." *Journal of Borderland Studies* 18(1): 13–25.

Newman, M.E.J. 2001. "The Structure of Scientific Collaboration Networks." *Proceedings of the National Academy of Sciences* 98(2): 404–409.

Nordhaus, Ted and Michael Shellenberger. 2007. *Break Through: From the Death of Environmentalism to the Politics of Possibility.* Boston: Houghton Mifflin.

Obuchi, Ken-Ichi, Osamu Fukushima, and James Tedeschi. 1999. "Cultural Values in Conflict Management: Goal Orientation, Goal Attainment, and Tactical Decision." *Journal of Cross-cultural Psychology* 30(1): 51–71.

O'Donnell, Guillermo, Philippe Schmitter, and Laurence Whitehead. 1986. *Transitions from Authoritarian Rule: Tentative Conclusions about Uncertain Democracies.* Baltimore: Johns Hopkins University Press.

Olson, Mancur. 1965. *The Logic of Collective Action.* Cambridge, MA: Harvard University Press.

Ostrom, Elinor. 2000. "Collective Action and the Evolution of Social Norms." *Journal of Economic Perspectives* 14(3): 137–158.

———. 2005. *Understanding Institutional Diversity.* Princeton: Princeton University Press.

Ott, J. Steven, Sandra J. Parkes, and Richard B. Simpson, eds. 2003. *Classic Readings in Organizational Behavior, 3rd ed.* Belmont, CA: Thomson/Wadsworth.

Oughton, Elizabeth and Louise Bracken. 2009. "Interdisciplinary Research: Framing and Reframing." *Area* 41: 385–394.

Overdevest, Christine and Lisa Christiansen. 2013. "Using 'Cultural Cognition' to Predict Environmental Risk Perceptions in a Florida Water-Supply Planning Process." *Society and Natural Resources* 26(9): 987–1007.

Pagdee, Adcharaporn, Yeon-su Kim, and P.J. Daugherty. 2006. "What Makes Community Forest Management Successful: A Meta-Study from Community Forests throughout the World." *Society and Natural Resources* 19: 33–52.

Página 12. 2008. "Proceso para los Acusados de la Cacería en Pando." Retrieved September 16 (www.pagina12.com.ar/diario/elmundo/subnotas/111674-35300-2008-09-16.html).

Pantoja, Nara, and I. Foster Brown. 2009. "Estimativas de Areas Afetadas pelo Fogo no Leste do Acre Associadas à Seca de 2005." *Anais do XIV Simpósio Brasileiro de Sensoriamento Remoto.* Natal: INPE. Pp. 6029–6036.

Paredes Pando, Oscar. 2015. *Sociedad, Cultura y Fronteras: Espacios Sentidos y Deseados en el Alto Acre-Madre de Dios, Brasil-Peru.* Cusco: Alpha Servicios Gráficos SRL.

Parker, Andrew and Eve Kosofsky Sedgwick, eds. 1995. *Performativity and Performance.* New York: Routledge.

Parker, Ian, ed. 1998. *Social Constructionism, Discourse, and Realism.* Thousand Oaks, CA: Sage.

Paula, Elder Andrade de. 2005. *(Des)envolvimento Insustentavel da Amazonia Occidental: Dos Missionarios do Progresso aos Mercadores da Natureza.* Rio Branco: EDUFAC.

Paulson, Neil, and Tor Hernes, eds. 2003. *Managing Boundaries in Organizations: Multiple Perspectives.* Basingstoke: Palgrave MacMillan.

Peacock, James L. 2001. *The Anthropological Lens: Harsh Light, Soft Focus, 2nd ed.* Cambridge: Cambridge University Press.

Pennington, Deana. 2008. "Cross-Disciplinary Collaboration and Learning." *Ecology and Society* 13: article 8.

Perz, Stephen. 2010. Personnel Databases. Socioeconomic, Botanical and Remote Sensing Components. NSF Roads Project, NSF HSD #0527511. Gainesville: University of Florida.

Perz, Stephen, Christine Overdevest, Eugenio Arima, Marcellus Caldas, and Robert Walker. 2007. "Unofficial Road Building in the Brazilian Amazon: Dilemmas and Models of Road Governance." *Environmental Conservation* 34(2): 112–121.

Perz, Stephen, Silvia Brilhante, I. Foster Brown, Andrea Chavez Michaelsen, Elsa Mendoza, Veronica Passos, Raul Pinedo, Juan Fernando Reyes, Daniel Rojas, and Galia Selaya. 2010. "Crossing Boundaries for Environmental Science and Management: Combining Interdisciplinary, Inter-organizational and International Collaboration." *Environmental Conservation* 47(4): 419–431.

Perz, Stephen, Liliana Cabrera, Luca Araujo Carvalho, Jorge Castillo, Rosmery Chacacanta, Rosa Cossio, Yeni Franco Solano, Jeffrey Hoelle, Mercedes Perales, Israel Puerta, Daniel Rojas Céspedes, Ioav Rojas Camacho, and Adão Costa Silva. 2012. "Regional Integration and Local Change: Road Paving, Community Connectivity and Social-Ecological Resilience in a Tri-national Frontier, Southwestern Amazonia." *Regional Environmental Change* 12: 35–53.

Perz, Stephen, Youliang Qiu, Yibin Xia, Jing Sun, Jane Southworth, Karla Rocha, and Grenville Barnes. 2013a. "Political Ecology and Land Change Science in the Study of Infrastructure Impacts." Pp. 150–168 in *Land Change Science, Political Ecology, and Sustainability: Synergies and Divergences,* edited by C. Branstromm and J.M. Vadjunec. New York: Routledge.

Perz, Stephen, Youliang Qiu, Yibin Xia, Jane Southworth, Jing Sun, Matt Marsik, Karla Rocha, Veronica Passos, Daniel Rojas, Gabriel Alarcon, Grenville Barnes, and Christopher Baraloto. 2013b. "Trans-Boundary Infrastructure and Land Cover Change: Highway Paving and Community-Level Deforestation in a Tri-National Frontier in the Amazon." *Land Use Policy* 34: 27–41.

Perz, Stephen, Rafael Muñoz-Carpena, Gregory Kiker, and Robert D. Holt. 2013c. "Evaluating Ecological Resilience with Global Sensitivity and Uncertainty Analysis." *Ecological Modelling* 263: 174–186.

Perz, Stephen, Grenville Barnes, Alexander Shenkin, Daniel Rojas, and Carlos Vaca. 2014. "Private and Communal Lands? The Ramifications of Tenure Ambiguity and Regional Integration for Tenure Formalization and Its Consequences in Northern Bolivia." *International Journal of the Commons* 8(1): 179–206.

Peters, B. Guy. 2001. *The Politics of Bureaucracy, 5th ed.* New York: Routledge.

Peterson, Gary, Graeme Cumming, and Stephen Carpenter. 2003. "Scenario Planning: A Tool for Conservation in an Uncertain World." *Conservation Biology* 17(2): 358–366.

Phillips, Nelson, Thomas B. Lawrence, and Cynthia Hardy. 2000. "Inter-organizational Collaboration and the Dynamics of Institutional Fields." *Journal of Management Studies* 37(1): 23–43.

Phillips, O.L., P. Hall, A.H. Gentry, S.A. Sawyer, and R. Vasquez. 1994. "Dynamics and Species Richness of Tropical Rain Forests." *Proceedings of the National Academy of Sciences* 91: 2805–2909.

Pieterse, J.N. 2001. *Development Theory: Deconstructions/Reconstructions.* London: Sage.

Pitman, Nigel C.A., Hugo Mogolion, Nallarett Davila, Marcos Rios, Roosevelt García-Villacorta, Juan Guevara, Timothy Baker, Abel Montegudo, Oliver Phillips, Rodolfo Vásquez-Martinez, Manuel Ahuite, Milton Aulestia, Dairon Cardenas, Carlos Ceron, Pierre-André Loizeau, David Neill, Percy Nunez, Walter Palacios, Rodolphe Spichiger, and Elvis Valderrama. 2008. "Tree Community Change across 700 km of Lowland Amazonian Forests from the Andean Foothills to Brazil." *Biotropica* 40: 525–535.

Pohl, Christian. 2005. "Transdisciplinary Collaboration in Environmental Research." *Futures* 37: 1159–1178.

Poncelet, Eric C. 2004. *Partnering for the Environment: Multi-stakeholder Collaboration in a Changing World.* Lanham, MD: Rowman & Littlefield.

Portes, Alejandro. 1995. *By-Passing and Trespassing: Explorations in Boundaries and Change.* University of Maryland: Urban Studies and Planning Program.

Prance, Ghillean T. 1994. "A Comparison of the Efficacy of Higher Taxa and Species Numbers in the Assessment of Biodiversity in the Neotropics." *Philosophical Transactions of the Royal Society B* 345: 89–99.

Prance, Ghillean T. and Thomas E. Lovejoy, eds. 1985. *Key Environments: Amazonia.* Oxford: Pergamon Press.

Rapoport, Anatol and Albert M. Chammah. 1965. *Prisoner's Dilemma.* Ann Arbor: University of Michigan.

La Razón. 2013. "Evo Expulsa a USAID de Bolivia Tras Acusarla de Injerencia Politica." Retrieved May 2 (www.la-razon.com).

Reed, M.I. 1992. *The Sociology of Organizations: Themes, Perspectives and Prospects.* New York: Harvester Wheatsheaf.

Reuer, J.J., ed. 2004. *Strategic Alliances: Theory and Evidence.* Oxford: Oxford University Press.

Reyes, Juan Fernando. 2007. "Presentación." *MAPiense* 1(1): 5.

Rioja, Guillermo. 2005. "Antropología de Frontera: Investigación Acción en la Región Trinacional MAP." *Revista de Antropología Iberoamericana* 43: 1–15.

Ritzer, George, ed. 1990. *Frontiers of Social Theory: New Syntheses.* New York: Columbia University Press.

Robertson, David P. and R. Bruce Hull. 2003. "Public Ecology: An Environmental Science and Policy for Global Society." *Environmental Science & Policy* 6: 399–410.

Robinson, Dorcas, Tom Hewitt, and John Harriss, eds. 2000. *Managing Development: Understanding Inter-organizational Relationships.* London: Sage Publications.

Robinson, William I. 2002. "Remapping Development in Light of Globalization: From a Territorial to a Social Cartography." *Third World Quarterly* 23(6): 1047–1071.

Rocha, Leonel. 2007. "Governo Barra Projeto Amazônico." *Correio Brasiliense* May 27. Retrieved June 30, 2008 (militarlegal.blogspot.com/2007/05/correio-braziliense-27052007-governo.html).

Rojas, Rafael Oswaldo. 2012. *Agricultural Expansion alongside Productive Forests in the Southern Peruvian Amazon.* PhD Dissertation, University of Florida, Gainesville.

Román-Dañobeytia, Francisco, Mijail Huayllani, Anggela Michi, Flor Ibarra, Raúl Loayza-Muro, Telésforo Vásquez, Liset Rodriguez, and Mishari Garcia. 2015. "Reforestation with Four Native Species after Abandoned Gold Mining in the Peruvian Amazon." *Ecological Engineering* 85: 39–46.

Roosevelt, A.C., M. Lima da Costa, C. Lopes Machado, M. Michab, N. Mercier, M. Valladas, J. Feathers, W. Barnett, M. Imazio da Silveira, A. Henderson, J. Silva, B. Chernoff, D.S.

Reese, J.A. Holman, N. Toth, and K. Schick. 1996. "Paleoindian Cave Dwellers in the Amazon: The Peopling of the Americas." *Science* 272: 373–384.

Rosenau, P.M. 1992. *Post-Modernism and the Social Sciences: Insights, Inroads and Intrusions*. Princeton: Princeton University Press.

Saatchi, Sassan, Salvi Asefi-Najafabady, Yadvinder Malhi, Luis E.O. Aragão, Liana Anderson, Ranga Mynenif, and Ramakrishna Nemanig. 2013. "Persistent Effects of a Severe Drought on Amazonian Forest Canopy." *Proceedings of the National Academy of Sciences* 110(2): 566–570.

Sachs, Jeffrey D. 2005. *The End of Poverty: Economic Possibilities for Our Time*. New York: Penguin.

Sahlins, Marshall. 2000. *Culture in Practice: Selected Essays*. New York: Zone Books.

Salovey, Peter and John Mayer. 1989. "Emotional Intelligence." *Imagination, Cognition, and Personality* 9(3): 185–211.

Saltelli, A., M. Ratto, T. Andres, F. Campolongo, J. Cariboni, D. Gatelli, M. Saisana, and S. Tarantola. 2008. *Global Sensitivity Analysis: The Primer*. Chichester: Wiley Interscience.

Santos, Filipe M., and Kathleen M. Eisenhardt. 2005. "Organizational Boundaries and Theories of Organization." *Organizational Science* 16(5): 491–508.

Santos, Roberto. 1980. *História Econômica da Amazônia (1800–1920)*. São Paulo: T.A. Queiroz.

Sartre, Xavier Arnauld de, and Romain Taravella. 2009. "National Sovereignty vs. Sustainable Development Lessons from the Narrative on the Internationalization of the Brazilian Amazon." *Political Geography* 28(7): 406–415.

Saterson, K., R. Margolis, and N. Salafsky. 1996. *Measuring Conservation Impact: An Interdisciplinary Approach to Project Monitoring and Evaluation*. Washington, DC: World Wildlife Fund.

Scheffer, Marten. 2009. *Critical Transitions in Nature and Society*. Princeton: Princeton University Press.

Scheffer, Marten, Jordi Bascompte, William Brock, Victor Brovkin, Stephen Carpenter, Vasilis Dakos, Hermann Held, Egbert van Nes, Max Rietkerk, and George Sugihara. 2009. "Early-Warning Signals for Critical Transitions." *Nature* 461: 53–59.

Scheffer, Marten, Stephen R. Carpenter, Timothy M. Lenton, Jordi Bascompte. William Brock, Vasilis Dakos, Johan van de Koppel, Ingrid A. van de Leemput, Simon A. Levin, Egbert H. van Nes, Mercedes Pascual, and John Vandermeer. 2012. "Anticipating Crtitical Transitions." *Science* 338: 344–348.

Schmink, Marianne, and Mâncio L. Cordeiro. 2008. *Rio Branco: A Cidade da Florestania*. Belém: Editora UFPa.

Schmink, Marianne and Charles Wood. 1992. *Contested Frontiers in Amazonia*. New York: Columbia University Press.

Schoemaker, Paul J.H. 1991. "When and How to Use Scenario Planning: A Heuristic Approach with Illustration." *Journal of Forecasting* 10(6): 549–564.

Schoenberger, Erica. 2001. "Interdisciplinarity and Social Power." *Progress in Human Geography* 25(3): 365–382.

Schonewald-Cox, Christine M. 1988. "Boundaries in the Protection of Nature Reserves: Translating Multidisciplinary Knowledge into Practical Conservation." *BioScience* 38(7): 480–486.

Scott, James C. 1998. *Seeing Like a State: How Certain Schemes to Improve the Human Condition Have Failed*. New Haven: Yale University Press.

Scott, John. 1988. "Social Network Analysis." *Sociology* 22: 109–127.

———. 2000. *Social Network Analysis: A Handbook, 2nd ed.* Thousand Oaks: Sage.

Searle, John R. 1995. *The Construction of Social Reality*. New York: Free Press.

Seeger, Matthew. 2002. "Chaos and Crisis: Propositions for a General Theory of Crisis Communication." *Public Relations Review* 28: 329–337.

Sen, Amartya. 1977. "Rational Fools: A Critique of the Behavioral Foundations of Economic Theory." *Philosophy and Public Affairs* 6(4): 317–344.

Sendjaya, Sen and James C. Sarros. 2002. "Servant Leadership: Its Origin, Development and Application in Organizations." *Journal of Leadership and Organization Studies* 9(2): 57–64.

Servindi. 2008. "Peru: Destacan Caceria de Lideres Indigenas y Sociales en Madre de Dios." Retrieved July 10 (www.servindi.org/archivo/2008/4303#more-4303).

Sewell, William H. 1992. "A Theory of Structure: Duality, Agency, and Transformation." *American Journal of Sociology* 98(1): 1–29.

Shanley, Patricia, and Citlalli López. 2009. "Out of the Loop: Why Research Rarely Reaches Policy Makers and the Public—And What Can Be Done." *Biotropica* 41: 535–544.

Shaules, Joseph. 2007. *Deep Culture: The Hidden Challenges of Global Living.* New York: Multilingual Matters.

Sheil, Douglas and Anna Lawrence. 2004. "Tropical Biologists, Local People and Conservation: New Opportunities for Collaboration." *Trends in Ecology and Evolution* 19(12): 634–638.

Shortall, Sally. 2004. "Social or Economic Goals, Civic Inclusion or Exclusion? An Analysis of Rural Development Theory and Practice." *Sociologia Ruralis* 44(1): 109–123.

———. 2008. "Are Rural Development Programmes Socially Inclusive? Social Inclusion, Civic Engagement, Participation and Social Capital: Exploring the Differences." *Journal of Rural Studies* 24(4): 450–457.

Shriar, A.J. 2009. "Roads to Poverty: Exploring the Impacts of Economic Integration on Socioeconomic Conditions and Land Use in Northern Guatemala." *Journal of Planning Education and Research* 28: 456–469.

Skocpol, Theda, ed. 1984. *Vision and Method in Historical Sociology.* Cambridge: Cambridge University Press.

Smith, Anthony. 1990. *Explorers of the Amazon.* Chicago: University of Chicago Press.

Soares-Filho, Britaldo, Ane Alencar, Daniel Nepstad, Gustavo Cerqueira, Maria del Carmen Vera Diaz, Sérgio Rivero, Luis Solórzano, and Eliane Voll. 2004. "Simulating the Response of Land Cover Changes to Road Paving and Governance along a Major Amazon Highway: The Santarém-Cuiabá Corridor." *Global Change Biology* 10: 745–764.

Sol de Pando. 2013. "El Gobierno de Evo Morales ya no Expulsará a USAID de Pando, Permite Financiamiento Norteamericano para Universidad Amazónica." Retrieved March 3 (www.soldepando.com).

Southworth, Jane, Matthew Marsik, Youliang Qiu, Stephen Perz, Graeme Cumming, Forrest Stevens, Karla Rocha, Amy Duchelle, and Grenville Barnes. 2011. "Road Paving as a Driver of Land Cover Change: Spatial and Temporal Dynamics of Forests in the Tri-National MAP Frontier, Southwestern Amazonia, 1986–2005." *Remote Sensing* 3: 1047–1066.

Spillane, James P. 2006. *Distributed Leadership.* San Francisco: Jossey-Bass.

Spracklen, D.V., S.R. Arnold, and C.M. Taylor. 2012. "Observations of Increased Tropical Rainfall Preceded by Aid Passage over Forests." *Nature* 489: 282–285.

Stem, Caroline, Richard Margolis, Nick Salafsky, and Marcia Brown. 2005. "Monitoring and Evaluation in Conservation: A Review of Trends and Approaches." *Conservation Biology* 19(2): 295–309.

Stewart, Amy and Tim Gray. 2006. "The Authenticity of 'Type Two' Multi-Stakeholder Partnerships for Water and Sanitation in Africa: When Is a Stakeholder a Partner?" *Environmental Politics* 15: 362–378.

Stoian, Dietmar, and Ariénne Henkemanns. 2000. "Between Extractivism and Peasant Agriculture: Differentiation of Rural Settlements in the Bolivian Amazon." *International Tree Crops Journal* 10: 299–319.

Straub, Stéphane. 2008. "Infrastructure and Growth in Developing Countries: Recent Advances and Research Challenges." Washington, DC: World Bank Policy Research Working Paper No. 4460.

Strunz, Sebastian. 2012. "Is Conceptual Vagueness an Asset? Arguments from Philosophy of Science Applied to the Concept of Resilience." *Ecological Economics* 76: 112–118.

Sun, Jian-Quo, and Albert C.J. Luo, eds. 2006. *Bifurcation and Chaos in Complex Systems.* Amsterdam: Elsevier.

Swenson, Jennifer J., Catherine E. Carter, Jean-Christophe Domec, and Cesar I. Delgado. 2011. "Gold Mining in the Peruvian Amazon: Global Prices, Deforestation, and Mercury Imports." *PLOS One* 6(4): e18875.

Teece, David J. 1992. "Competition, Cooperation, and Innovation: Organizational Regimes for Rapid Technological Progress." *Journal of Economic Behavior and Organization* 18(1): 1–25.

Tendler, J. 1997. *Good Government in the Tropics.* Cambridge: MIT Press.

Terra Noticias. 2008. "Abin Investigara 25 ONGS por Biopirataria." *Terra Noticias* January 30. Retrieved June 30, 2008 (noticias.terra.com.br).

Tocantins, Leandro. 1979. *Formação Histórica do Acre* . Rio de Janeiro: Civilização Brasileira.

Tomasello, Michael, Malinda Carpenter, Josep Call, Tanya Behne, and Henrike Moll. 2005. "Understanding and Sharing Intentions: The Origins of Cultural Cognition." *Behavioral and Brain Sciences* 28: 675–735.

Torgerson, D. 1990. "Limits of the Administrative Mind." In *Managing Leviathan: Environmental Politics and the Administrative State*, eds. R. Paehlke & D. Torgerson. Peterborough, UK: Broadview Press.

Trombulak, Stephen C., and Christopher A. Frissell. 2000. "Review of Ecological Effects of Roads on Terrestrial and Aquatic Communities." *Conservation Biology* 14(1): 18–30.

UNFCCC. 2010. United Nations Framework Convention on Climate Change. Retrieved June 30, 2013 (unfccc.int/2860.php).

United Nations. 2015 [1948]. The Universal Declaration of Human Rights. Retrieved November 3, 2015 (www.un.org/en/documents/udhr/).

———. 2014. *Sustainable Development Goals.* Retrieved November 3, 2015 (sustainabledevelopment.un.org).

United Nations International Strategy for Disaster Reduction (UN/ISDR). 2005. *Hyogo Framework for Action 2005–2015: Building the Resilience of Nations and Communities to Disasters.* New York: United Nations.

US Agency for International Development. 2006. The USAID Amazon Basin Conservation Initiative (ABCI). Request for Applications (RFA) Number: LAC/RSD-2006–2–LMA. February 24.

Van Houtum, Henk. 2005. "The Geopolitics of Borders and Boundaries." *Geopolitics* 10(4): 672–679.

van Oosten, Cora. 2004. *Fading Frontiers? Local Development and Cross-Border Partnerships in Southwest Amazonia.* University of Utrecht, Utrecht, The Netherlands.

Walker, Brian and David Salt. 2006. *Resilience Thinking: Sustaining Ecosystems and People in a Changing World.* Washington, DC: Island Press.

Walker, Robert, John Browder, Eugenio Arima, Cynthia Simmons, Ritaumaria Pereira, Marcellus Caldas, Ricardo Shirota, and Sergio de Zen. 2009. "Ranching and the New Global Range: Amazonia in the 21st Century." *Geoforum* 40: 732–745.

Wallace, Scott. 2011. *The Unconquered: In Search of the Amazon's Last Uncontacted Tribes.* New York: Broadway Paperbacks.

Ward, Margaret. 1995. "Butterflies and Bifurcations: Can Chaos Theory Contribute to Our Understanding of Family Systems?" *Journal of Marriage and Family* 57(3): 629–638.

Warner, Jeroen, ed. 2007. *Multi-stakeholder Platforms for Integrated Water Management.* Aldershot: Ashgate.

Weber, Edward P. and Anne M. Khademian. 2008. "Wicked Problems, Knowledge Challenges, and Collaborative Capacity Builders in Network Settings." *Public Administration Review* 68(2): 334–349.

Weingart, Peter and Nico Stehr, eds. 2000. *Practising Interdisciplinarity.* Toronto: University of Toronto Press.

Weinstein, Barbara. 1983. *The Amazon Rubber Boom, 1850–1920.* Palo Alto: Stanford University Press.

Werbach, Adam. 2005. "Is Environmentalism Dead?" Retrieved December 1, 2006 (3nov.com).

Western, D. & Wright, M.R., eds. 1994. *Natural Connections: Perspectives in Community-based Conservation.* Washington, DC: Island Press.

Wiggins, Jerry S. and Aaron L. Pincus. 1992. "Personality: Structure and Assessment." *Annual Review of Psychology* 43: 473–504.

Wilber, K. 2001. *A Theory of Everything: An Integral Vision for Business, Politics, Science and Spirituality.* Boston: Shambhala.

Willis, Katie. 2011. *Theories and Practices of Development, 2nd ed.* New York: Routledge.

Wilson, James Q. 1989. *Bureaucracy: What Government Agencies Do and Why They Do It.* New York: Basic Books.

Wollenberg, Eva, Ramses Iwan, Godwin Limberg, Moira Moeliono, Steve Rhee, and Made Sudana. 2007. "Facilitating Cooperation during Times of Chaos: Spontaneous Orders and Muddling Through in Malinau." *Ecology and Society* 12(1): article 3.

Wondolleck, Julia and Steven Yaffee. 2000. *Making Collaboration Work: Lessons from Innovation in Natural Resource Management.* Washington, DC: Island Press.

Wood, Charles and Roberto Porro, eds. 2002. *Deforestation and Land Use in the Amazon.* Gainesville: University Presses of Florida.

Wood, Charles and Bryan Roberts, eds. 2005. *Rethinking Development in Latin America.* University Park: Pennsylvania State University Press.

Wood, Charles. 2008. "Time, Cycles and Tempos in Social-ecological Research and Environmental Policy." *Time and Society* 17(2/3): 261–282.

Xavier, Miguel Gustavo. 2012. "Mudanças Climáticas e Saneamento Básico." November 30. Retrieved June 30, 2014 (agazetadoacre.com/noticias/mudancas-climaticas-e-saneamento-basico/).

Young, Oran R. 1989. *International Cooperation: Building Regimes for Natural Resources and the Environment.* Ithaca, NY: Cornell University Press.

———, ed. 1997. *Global Governance: Drawing Insights from the Environmental Experience.* Cambridge: MIT Press.

Young, T.R. 1991. "Chaos Theory and Symbolic Interaction Theory: Poetics for the Postmodern Sociologist." *Symbolic Interaction* 14(3): 321–334.

Index

About the Author

Stephen G. Perz is professor in the Department of Sociology and Criminology & Law at the University of Florida, where he is also affiliated with the Center for Latin American Studies and the School of Natural Resources and Environment. He received his PhD in sociology from the University of Texas at Austin in 1997. Since 2002, he has worked in the tri-national "MAP" frontier of the southwestern Amazon where Bolivia, Brazil, and Peru meet. There, rapid and large-scale change processes motivated the emergence of the MAP Initiative, a polycentric grassroots movement which in turn gave birth to a conservation and development model that combines research with grassroots capacity building for participatory environmental planning and management. In response, Perz has led several collaborative projects with numerous partners from other disciplines, types of organizations, and countries. Some of those projects focus on his primary research interest, the social and ecological impacts of infrastructure, but many center on the priorities of his partners, including climate change, gold mining, watershed management, and capacity building.